# SCIENTIFIC DISCOVERY: CASE STUDIES

# BOSTON STUDIES IN THE PHILOSOPHY OF SCIENCE

EDITED BY ROBERT S. COHEN AND MARX W. WARTOFSKY

VOLUME 60

# SCIENTIFIC DISCOVERY: CASE STUDIES

*Edited by*

THOMAS NICKLES

*Department of Philosophy, University of Nevada, Reno*

D. REIDEL PUBLISHING COMPANY

DORDRECHT : HOLLAND / BOSTON : U.S.A.

LONDON : ENGLAND

Library of Congress Cataloging in Publication Data

Guy L. Leonard Memorial Conference in Philosophy,
   1st, University of Nevada, Reno, 1978.
   Scientific discovery, case studies.

   (Boston studies in the philosophy of science ; v. 60)
   Companion volume to the Conference's Scientific discovery, logic,
and rationality, which contains a separate selection of papers from the
Conference.
   Includes bibliographies and indexes.
     1.  Science—Philosophy—Congresses.  2.   Science—History
—Congresses.  3.  Inventions—Philosophy—Congresses.  4.  Inventions
—History—Congresses.   I.   Nickles, Thomas, 1943–     II.   Title.
III.   Series.
Q174.B67     vol. 60     501s [507.2]     80–16945
ISBN 90–277–1092–9
ISBN 90–277–1093–7 (pbk.)

---

Published by D. Reidel Publishing Company,
P.O. Box 17, 3300 AA Dordrecht, Holland

Sold and distributed in the U.S.A. and Canada
by Kluwer Boston Inc., Lincoln Building,
160 Old Derby Street, Hingham, MA 02043, U.S.A.

In all other countries, sold and distributed
by Kluwer Academic Publishers Group,
P.O. Box 322, 3300 AH Dordrecht, Holland

D. Reidel Publishing Company is a member of the Kluwer Group

Printed in The Netherlands

# EDITORIAL PREFACE

The history of science is articulated by moments of discovery. Yet, these 'moments' are not simple or isolated events in science. Just as a scientific discovery illuminates our understanding of nature or of society, and reveals new connections among phenomena, so too does the history of scientific activity and the analysis of scientific reasoning illuminate the processes which give rise to moments of discovery and the complex network of consequences which follow upon such moments. Understanding discovery has not been, until recently, a major concern of modern philosophy of science. Whether the act of discovery was regarded as mysterious and inexplicable, or obvious and in no need of explanation, modern philosophy of science in effect bracketed the question. It concentrated instead on the logic of scientific explanation or on the issues of validation or justification of scientific theories or laws. The recent revival of interest in the context of discovery, indeed in the acts of discovery, on the part of philosophers and historians of science, represents no one particular methodological or philosophical orientation. It proceeds as much from an empiricist and analytical approach as from a sociological or historical one; from considerations of the logic of science as much as from the alogical or extralogical contexts of scientific thought and practice. But, in general, this new interest focuses sharply on the actual historical and contemporary cases of scientific discovery, and on an examination of the act or moment of discovery *in situ*.

It is appropriate, therefore, that in this second volume of the Proceedings of the Leonard Memorial Conference in Philosophy, held in Reno at the University of Nevada, October 29–31, 1978, attention is on case studies from the history of science, and on issues which relate philosophy of science to the history of science. This volume of papers from the Leonard Conference, together with the first (Thomas Nickles (ed.), *Scientific Discovery, Logic, and Rationality* (*Boston Studies in the Philosophy of Science*, Vol. 56), devoted to analytical and conceptual issues concerning scientific discovery), constitute the first major collection on these themes since *Criticism and the Growth of Knowledge* (Cambridge University Press, 1970), edited by Imre Lakatos and Alan Musgrave and based on the London Colloquium in the Philosophy of Science of 1965. They mark a new and significant direction of research in the

v

philosophy, history (and, perhaps gingerly, the sociology) of science. We are pleased that our *Boston Studies in the Philosophy of Science* can bring such a rich and varied collection to its readers. We wish, once again, to express our appreciation to Professor Nickles, both as organizer of the Conference, and as indefatigable editor of these volumes. Also, we again thank the Leonard family for their generous sensitivity to the volumes and the healthy curiosity of Guy L. Leonard.

*Center for the Philosophy and History of Science*          ROBERT S. COHEN
*Boston University*                                    MARX W. WARTOFSKY

*May 1980*

# TABLE OF CONTENTS

vii

# TABLE OF CONTENTS OF COMPANION VOLUME

## (SCIENTIFIC DISCOVERY, LOGIC, AND RATIONALITY)

# PREFACE

Scientific Discovery was the topic of the first Guy L. Leonard Memorial Conference in Philosophy, held at the University of Nevada, Reno, October 29—31, 1978. This book is a selection of papers from that conference. A companion volume, *Scientific Discovery, Logic, and Rationality*, contains a separate selection of conference papers. While both books address the same general set of issues, the contributions to the present *Case Studies* volume are more historical in nature. Not all of the conference papers could be included in the two books.

The Leonard Conferences are a memorial to Guy Leslie Leonard, a student who made the problems of philosophy and physics part of his personal quest for understanding. Our greatest debt is to the parents, Mr. and Mrs. Paul A. Leonard, to Jackie Leonard, and to the many contributors to the memorial fund. Their generosity made the conference possible.

For financial support both early and late, I also wish to thank Vice President Robert Gorrell; John Nellor, Dean of the Graduate School; Rebecca Stafford, Dean of the College of Arts and Science; and Gene Kosso and the members of the Corporate Support Committee. Former President Max Milam originally suggested a conference. For their encouragement and guidance, it is a pleasure to thank the general editors of the *Boston Studies* series, Bob Cohen and Marx Wartofsky.

Organizing a conference and then editing it creates many more debts of gratitude to persons on and off campus than I can possibly discharge in a short prefatory note, but for their substantial help in planning and carrying through the project I cannot fail to mention Jack Kelly, Chairman of the Philosophy Department; Elisa Lazzari, Department Secretary; Bruce Moran, History; Bill Scott, Physics; and the staff of the Extended Programs Division. My wife and family helped in many ways. University President Joseph Crowley added just the right touch to the welcoming ceremonies preceding Sam Goudsmit's wonderful and witty public lecture, 'Physics in the Twenties'. We were all saddened by Sam's passing into history scarcely a month after the conference — his last major public appearance.

Credit for the intellectual success of the conference belongs, of course, to the forty-odd participants who agreed, on short notice and for little or no remuneration, to travel to Reno from six countries in order to discuss with the greatest enthusiasm a topic that, until recently, was not considered to belong to philosophy at all.

*University of Nevada, Reno*                                    THOMAS NICKLES

THOMAS NICKLES

# INTRODUCTION:
# RATIONALITY AND SOCIAL CONTEXT

The authors of these essays include two cognitive psychologists, a physicist, two 'full time' historians of science, and a number of philosophers who do serious history and who employ here the method of historical case studies. The volume thus strongly evidences the 'historical turn' in recent philosophy of science and in psychology of science as well. Philosophers are not only turning to history; they are also becoming interested in areas of science which they previously had neglected. The rich symposium on the biological sciences and the mini-symposium on plate tectonic theory, for example, would have been almost impossible to schedule only a decade ago. Several of the contributors, but not all, count themselves among the 'friends of discovery' (Gary Gutting's term) – that informal group of philosophers and methodologists who hold, contrary to the logical positivists and classical Popperians, that scientific discovery is epistemologically interesting and important.

I have grouped the papers into symposia according to scientific area, as they were at the Leonard Conference itself in most cases; but there are other ways of reading them. For example, Kenneth Schaffner's paper on computer programs for medical diagnosis and for the discovery of molecular structures, and Nancy Maull's comment, are fruitfully combined with the contributions of Marx Wartofsky, William T. Scott, and David Bantz. Schaffner's essay transports the logic of discovery question to the field of artificial intelligence. He belongs to the Herbert Simon school of heuristic programming, while the other, aforementioned participants all, in different ways, argue or imply that Schaffner-Simon logics (or heuristics) of discovery face serious limitations. These discussions address several of the issues raised by Hubert Dreyfus in *What Computers Can't Do*.[1] Since logic and rationality of discovery was the main topic of the conference, I shall direct most of my introductory remarks to that question.

I

In the lead paper to the volume, Marx Wartofsky argues for a more historical and casuistic view of rationality in science. The mathematical-deductive-calculative model of rationality and the corresponding rule-guided or algorithmic

xiii

*T. Nickles (ed.), Scientific Discovery: Case Studies*, xiii–xxv.
*Copyright © 1980 by D. Reidel Publishing Company.*

model of inquiry project only one side of what should be (at least) a stereo-scopic image of science, he says. This received view construes rationality and exact inquiry in general as a matter of logical deducibility of conclusions from a more or less fixed system of propositions. Euclidean geometry is, of course, the classical model of such a system, and the calculus and symbolic logic are more recent paradigms. On the traditional view, a rational belief is a logically provable belief (or else one certified by intellectual or by sensory intuition); a scientific explanation is a logical proof from hypotheses of why a phenomenon occurred or why a law statement holds; methodological rules are nothing less than algorithms. Realization of this ideal in its Leibnizean form (for example) would enable savants to resolve all substantive scientific issues by saying "*Calculemus*" — "Come, let us calculate!"

If I understand Wartofsky, he would have us say, rather, "Come, let us deliberate! — in the hope of reaching a reasoned judgment about the case in question." Scientific inquiry, he argues, cannot be reduced to a set of proce-dural rules for calculation, such as might be programmed into a computer. The missing perspective of the stereoscopic image is that of scientific judg-ment about particular cases, informed by precedents and guided by maxims rather than rigorously rule-determined. It is this view of science, modelled on judgment in the law, in medical diagnosis, and in aesthetic appraisal, that allows for creative innovation in science, Wartofsky contends, since the calcu-lative/subsumptive image — most carefully articulated by the logical positivists in our century — cannot explain even the *possibility* of scientific behavior which is both creative and rational. Any positivist attempt to account meth-odologically for innovation runs up against a 'dilemma of explanation'. In Wartofsky's own words:

If a theory is successful in giving an explanation of discovery and invention, then it reduces creative thought in science to an algorithmizable procedure, or to a deductive consequence of the explaining theory, and thus dissolves the notion of creativity alto-gether. If the theory fails to give an explanation of this sort, then of course it is a failed theory. The dilemma is then: either the theory succeeds, and the concept of discovery is explained away, or reductively eliminated *— or the theory fails, and discovery remains unexplained.

Wartofsky models accounts of creativity not on deductive explanation but upon the down-to-earth 'tinkering' found in the crafts. The scientific theorist tinkers with ideas.

Wartofsky also observes that the calculative model of rational inquiry, a rationality of efficient means rather than of ends, owes its dominance in modern culture not only to the success of modern science and technology but also to the rise of capitalism and business society. Insofar as this is true, science itself, and perhaps especially those social scientific branches which aspire toward mathematical precision and calculational rules as the essence of scientific inquiry, are shaped by factors which have a social basis and not *simply* an 'internal' scientific foundation.

The panel discussion and several essays (especially those of Wrightsman and Moran) also raise the questions of whether and in what ways social and psychological factors have a place in the development of scientific ideas, including methodological principles and standards. While this volume demonstrates the rapid maturation of the historical approach to philosophy of science, it may also represent historical philosophy of science at a crossroads. Hitherto, appeals to history have been appeals to internal history. Today, while remaining deeply divided on the issue, more philosophers of science are prepared to give serious consideration to 'external' factors and to sociology of knowledge than at any time in recent memory. Indeed, panelists Mary Hesse and Robert Westman proclaim the need for more sociology of science. It is not yet clear, however, how such an interest is to figure in an epistemological account of science.

The panelists also address the problem, suggested by the title of their session, of how we (especially historians) can explain scientific discoveries in rational terms. The fact that several historians have made substantial discoveries intelligible to reason strongly suggests that discovery can be a more logical or at least rational — or at the very least cognitive — process than positivists and Popperians have allowed. In her retrospective comment (page 46), however, Noretta Koertge warns that this sort of rational, historical understanding of discovery does not entail the existence of a logic or methodology of discovery. On the other side, even if a discovery is achieved by a rational process, the problem of historically explaining it is not trivialized. Moreover, the problem of 'critical explanation' — explaining errors and oversights, of scientists in their research — remains especially serious.[2]

Carl G. Hempel's influential account of rational explanation (which even Laudan appears to adopt in *Progress and Its Problems*) requires the existence and knowledge of *natural laws* of rational behavior (covering laws).[3] Note that this conception of rational behavior and of rational explanation almost entails the existence of a lawful, empirical *theory* of rationality. Similarly, the covering-law explanation of discovery would require discovery laws and perhaps an empirical theory of discovery. In turn, the rational explanation of discovery episodes would require empirical, rational-discovery laws. Such laws could, of course, concern the application of a nonempirical logic or algorithm for discovery, if such were known, but Hempel and most writers deny the existence of such an algorithm. Rather than reject the algorithmic paradigm of rationality at this point, however, two generations of philosophers have concluded that discovery is not rational; hence, it cannot be rationally explained. Yet as Harold Brown observes,

The attempt by logical empiricists to identify rationality with algorithmic computability is somewhat strange, since it deems rational only those human acts which could in principle be carried out without the presence of a human being. . . . Rather, those decisions that can be made by the application of algorithms are paradigm cases of situations in which rationality is *not* required; it is exactly in those cases which require a decision or a new idea which cannot be dictated by mechanical rules that we require reason.[4]

The panel discussion centers on the topic of rationality. Ian Hacking provocatively contends that the questions philosophers raise about rationality in science are pseudo-questions, that science has nothing special to do with rationality, since the meaning of 'rational' is already fixed in our language. By implication he appears to reject the widespread view (expressed most clearly by Ernan McMullin in the panel discussion) that scientific behavior is, in the large, paradigmatic of rational behavior, that we can somehow model our general theories of rationality on scientific practice. Hacking may also be objecting to recent analyses of rationality in terms of change (Stephen Toulmin[5]) and cognitive progress (Larry Laudan[6]) as wrongheaded. Whether or not that is Hacking's concern, we should at least raise the question whether rationality in general (and not just 'scientific rationality') *necessarily* involves shifts to positions which are better adapted to changed circumstances or which are cognitively more progressive. Do not such analyses very much reflect the political-social-intellectual progressivism which characterizes our historical era to the point of being scientistic and of questionable value in understanding the behavior of more tradition-bound societies and institutions? Is tradition-bound conservatism *automatically* irrational when *cognitively* superior options are available?[7] These questions are legitimate, for the recent innovations in theory of rationality, though modelled on science, purport to be *general* accounts of rationality, applicable in outline to all peoples at all times, whatever their aims or goals.

It is fair to say that most conference participants do not share Hacking's views on rationality in science. They hold the issues concerning competing models of rationality to be genuine, although in serious need of clarification. It *is* clear that a good many philosophers reject the traditional identification of rationality with logicality (in Toulmin's useful phrase) and that many do hold scientific inquiry to be somehow paradigmatic of rational decision-making. There is no reason why scientific paradigms need bias the discussion, as long as other areas of inquiry (law, medicine, *etc*.) and everyday life are not neglected.

What then is to replace the Platonic-Cartesian-Kantian conception of rationality as derivability from a fixed logical system or at least the logical coherence of a propositional system? Almost by definition, a serious alternative to the received view must abandon the idea of a fixed system of propositions. Given this abandonment, general theories of rationality, for all times and places, are still possible (e.g., Laudan's theory), although they must now be procedural theories rather than substantive theories which convey the *content* of rational belief. However, one might take the still more radical step of rejecting altogether the idea that there can be a general *theory* of rationality, according to which any rational move must be the direct instantiation of a general principle or rule, substantive *or* procedural, which in turn belongs to a consistent, hierarchical system of principles. Naturally, any such alternative conception is bound to look disappointingly unsystematic and fuzzy when contrasted with the old conception; and its articulation must depend heavily on concrete examples (on showing rather than telling) instead of on the setting forth of universal principles. In fact, such a 'de-theorized' conception of rationality is not really an *alternative* to more traditional theories of rationality at all, since it abandons the theory game. Let me briefly explore some of the features that such an alternative might have.

First, it would reject extreme 'rule-rationality' theories for a more 'act-oriented' conception allowing greater scope to particular, informed, skilled judgments of the Polanyi-Grene-Toulmin-Wartofsky variety. Such a view need not deny that rational judgments are in a sense universalizable. It simply rejects the simplistic argument that universalizability implies the existence and availability of a simple system of rules which can be used in a calculative manner.[8] Such a view has the advantage that it is not automatically bedeviled by the twin problems of cultural diversity and conceptual change, for unlike traditional theories, this view at least allows that rational persons may differ. That is, rationality no longer is a slave to logical consistency with other judgments, including one's own previous judgments.[9] This view can easily avoid a scientistic bias; and rationality becomes a function partly of individual goals and circumstances, including noncognitive goals and environments.

Second, while both procedural and substantive principles would continue to play important roles in rational deliberation, these principles would not form a tight, linear, hierarchical system. It is well known that our rational intuitions are inconsistent,[10] but rather than take this to mean that we must completely abandon some of them and that we need formal, axiomatic theories of rationality (as in set theory), or a rigid ranking of principles as to supremacy (as in ethical theories with more than one leading principle), may it not be preferable to leave open the possibility that our deepest intuitions and guiding principles will simply conflict in particular cases, as our legal and moral principles do in fact? In cases of conflict (as in many other cases), we simply must evaluate the particular cases on their merits, but with appropriate consideration of precedents, appealing to one principle more strongly here and another there, depending on the circumstances. The principle which takes precedence in one case may not in another. Nor is it a matter of one principle excluding the other. In many cases, rational deliberation must take both into account and determine their relative weights for that situation.

Third, a serious alternative both to traditional theories of rationality and to recent innovations like Laudan's would allow noncognitive goals their proper places in deliberation. There has been a notable tendency on the parts of philosophers of science in general, and of a good many historians as well, to *compartmentalize* rationality — to treat cognitive, scientific decisions and their rationality as totally insulatable from other spheres of life, which may themselves come in for separate treatment. While it certainly is useful, even necessary, for normative and analytical purposes, to draw various 'internal/external' distinctions (and there is not just one but several distinctions that may be drawn, depending on one's purpose and context) and to focus on the purely cognitive elements of decision-making, such a procedure can lead to trouble in our attempts to understand and explain the actions of real historical agents. For an agent's situation frequently is quite complex, with different constraints on the decision pulling in different directions, without the possibility of the vectorial summing that calculational theories insinuate. And yet the agent must make a single decision having ramifications in all of these dimensions. The choice of a problem and a program of research, for instance, rarely is insulated from such matters as the availability of proper research tools and funding (*e.g.*, a well-stocked library or laboratory, complete with research assistants), professional prestige, relation to the work of one's colleagues, family security, political, ethical, and religious implications of the research. Can it be a *reasoned* choice which fails to take into account all relevant factors? How could a reasoned decision simply ignore so-called 'external' factors? (If you choose a project which you believe scientifically important but for which no funding is available, your work can hardly proceed .at all!) In application, the compartmentalized view of rationality leads to the absurdity that one and the same historical decision or action may be rational *qua* scientific goals, irrational as to family life, religiously non-rational, rational again as to professional prestige, but politically problematic.

To repeat, this sort of compartmentalization of rationality into different spheres of life is useful and proper when one attempts to develop a normative methodology of individual spheres, but philosophers and historians too frequently forget that this compartmentalization is an analytical device. There is many a slip 'twixt normative analysis of one area of human interest and historical, interpretative understanding of actions which have projections along many, incommensurable axes. The attempt to understand and explain the actions of agents as whole persons in complex historical situations calls for a more synthetic approach in which nonscientific and noncognitive goals receive proper emphasis in rational deliberation.[11] Otherwise, our 'theories' of rationality are bound to be cognitively biased, even scientistic.


## II

Let us now turn to the historical cases studied by the contributors to this volume. As much has been written about Copernicus as about any historical, scientific figure, yet there remains a lively debate over the rationality of Copernicus's opposition to the received, highly successful, Aristotelian-Ptolemaic world-view. In the panel discussion, Robert Westman and Maurice Finocchiaro touch on the problem of explaining why Georg Rheticus, Copernicus's first disciple, became a Copernican, in the context of a discussion of the role of psychoanalysis in the explanation of scientific behavior. The first paper of Symposium I, on theoretical and methodological innovation in the Copernican era, find Bruce Wrightsman re-examining the case of Copernicus himself. Wrightsman goes beyond the standard emphasis on the idea of God-the-great-artist as underlying Copernicus's demands for aesthetic coherence, systematicity, and consistency in an adequate theory of the heavens. Wrightsman argues, impressively, that Copernicus's theory of space — in particular his conception of 'the place of the universe' and hence much of his innovative break from the Aristotelian tradition — was determined in detail by subtle reasoning from his theological beliefs.

In the following essay, on the activities at the German Renaissance court of Wilhelm IV of Hesse-Kassel, historian Bruce Moran raises a fundamental question: What are the sources of those values and goals — of measurability, precision, completeness of data and its classification, and cooperative enterprise based on free communication of ideas — which became the values of the new science, as advocated by men like Francis Bacon? (Readers of *Scientific*

*Discovery, Logic, and Rationality* will recall Dudley Shapere's main point that *everything* has been learned or discovered, including methodological principles and values themselves.[12]) Historians have asked and supplied various answers to this question, but Moran produces a wealth of new data in support of his thesis that one major source of these new methodological values and goals was the courtly network of German prince-practitioners such as Wilhelm IV. Moran insists that the emergence of these values can be understood only by giving due attention to the social context of technological innovation, competition, and communication; and that if one stops with a familiar sort of 'internalist' history, one inevitably is led to see 'modern' scientific methods, goals, and values as springing full-blown from men like Bacon and Descartes.

Cognitive psychologist Paul McReynolds concludes Symposium I by pointing out the influence of the technology of clocks, including automata driven by clockwork, on conceptions of human behavior and thought — and the stimulus of the clock metaphor to methodological thinking. While the metaphor occasionally was used in a negative way, to show what human thought and action were not, it is remarkable how psychological thinkers of widely differing schools found a positive use for the clock metaphor. The metaphor played a major role in the creation and development of motivation theory.

Howard Gruber, also a cognitive psychologist, opens Symposium II, on the biological sciences, with a paper on Darwin's early development. His essay offers valuable lessons for the cognitive psychologist of science on how to work with historical materials. (In the panel discussion, Gruber warns against the now popular 'hit-and-run', short-term case studies, which can be superficial and misleading.) Gruber's 'evolving systems approach' treats Darwin's development as a series of systematic thought-stages, which evolve under the fairly steady pressure of new insights — the larger changes being transformations to a new thought-stage which preserves invariant certain structures of the old.

Next Michael Ruse, a philosopher and historian of biology who also has devoted many years to Darwin, argues (partly against his former self) that philosophers who ignore discovery issues can only entertain a distorted conception of science and can only fail to gain a full understanding of actual scientific cases. Ruse contends that one cannot fully understand (epistemically) Darwin's theory and the debate it engendered without knowing Darwin's route to discovery.

Lindley Darden's paper on theory construction in early genetics (Darwin, De Vries, Bateson, Morgan, *et al*.) is one of the few papers in the philosophicohistorical literature to address explicitly the problem of theory formation. The logical positivists and (especially) their behaviorist and operationist allies were interested in concept formation, but, ironically, philosophers' interest in the formative aspect of science died precisely at the time in the early 1950's when Carl G. Hempel, in his masterful critical review, announced the failure of the positivist search for a criterion of cognitive significance and concluded that "Concept formation and theory formation go hand in hand; neither can be carried on successfully in isolation from the other." [13] One might have hoped that the interest in concept formation would have matured into an interest in theory formation, but that did not happen, and the topic is only now becoming respectable. Darden, building on her earlier published work, demonstrates that analogical thinking and especially the development of 'interfield connections' can be a fruitful source of new theoretical ideas.

Kenneth Schaffner's major paper on discovery in the biomedical sciences and Nancy Maull's trenchant comment address several fundamental issues concerning discovery, pursuit, logic, and rationality. [14] Schaffner argues that both the generation and the preliminary evaluation phases of discovery are philosophically interesting and important. Unlike some other 'friends of discovery', who are content to speak of 'rationality' of discovery in a broad sense, Schaffner defends the possibility of a *logic* of discovery powerful enough to address deep problems. He is not committed to logics of a strictly algorithmic variety, but he nevertheless makes a strong case for substantive discovery procedures — in the form of heuristic computer programs such as those already developed for medical diagnosis and for the determination of molecular structures. On the surface these programs appear to be discovery routines of a theoretically profound sort, but Maull suggests that the conceptual depth is built into the program by the human programmers in such a way as to reduce the machine's innovative behavior to data classification. I already have mentioned the larger issues raised by Dreyfus on the prospects for artificial intelligence. Are the impressive examples which Schaffner describes just another instance of early success providing a springboard to overly optimistic claims about what machines can and will accomplish in the foreseeable future? One point that emerges from the Schaffner-Maull discussion and (indirectly) from Bantz's paper is that in most cases it is more difficult to create a machine or program which can suitably discover and formulate a new problem than to produce a machine or program which can solve the problem, once formulated. This demonstrates with special clarity that the discovery and formulation of a new problem is itself an important scientific achievement.

Concluding the symposium on the biological sciences is William Wimsatt's exciting discussion of heuristic strategies and their biases in population biology. He is chiefly concerned to criticize the reductionistic biases built (sometimes implicitly) into leading heuristic research strategies in that area — specifically, their role in the units of selection controversy. Wimsatt's paper, which may be his most important work to date, is that rare example of philosophical writing in such close contact with the ongoing scientific research that it can provide methodological guidance to the scientists involved. Such papers provide concrete support for the claim that, on some occasions at least, philosophers of science can assume an advisory role as well as their usual descriptive, rational reconstructionist, and critical roles *vis a vis* scientific research. Not that the advice of philosophers will be infallible on any occasion, but then we have learned that in science infallible advice is simply not to be had — from anyone.

Edward MacKinnon opens the symposium on the physical sciences by reviewing highlights of his (published and forthcoming) technical studies of the routes to discovery of de Broglie, Heisenberg, and Schrödinger. MacKinnon underscores the role of physical models even in the work of Heisenberg, who, because of the style of his published papers, ironically is thought to have rejected model-thinking in physics in favor of a strictly positivistic concern with observables. So much for the 'finished research report' as an accurate reflection of scientific activity! MacKinnon points out that the sophisticated reasoning involved in de Broglie's, Heisenberg's, and Schrödinger's paths to discovery rules out irrationalist models of discovery. Yet the reasoning falls into no simple, inductive or deductive patterns. MacKinnon therefore advocates a 'quasi-phenomenological' historical reconstruction which will disclose the roles of two types of underlying 'reasoning structures': the 'mathematical formalism' and the 'physical framework'.

Physicist William T. Scott interprets his current work on cloud physics in the light of Michael Polanyi's conception of scientific activity. Scott also wants to demonstrate the fruitfulness to his area of research of Rom Harré's claim that the primary aim of science is the elucidation of mechanisms and structures rather than the generation of deductive systems of propositions. Research on cloud formation, observes Scott, provides excellent examples of Polanyi's 'tacit knowing' and 'tacit integration of clues and cues' — which constitute Polanyi's analysis of the elusive 'intuition' involved in scientific discovery. Scott attempts to correct the widespread misinterpretation of Polanyi as an irrationalist. To leave room for knowledge which cannot be fully articulated in Polanyi's sense, Scott contends, is not in itself an irrationalist position.

David Bantz's substantial essay on the Heitler-London 'discovery' of quantum chemistry is one of the few treatments of chemistry by a philosopher of science. His problem is why the rather qualitative theory of Heitler and London should have been considered a major discovery at all, and he presents and criticizes standard positivist and Kuhnian answers to this question. His discussion carries him to earlier forms of chemical representation employed by such men as Walther Nernst and G. N. Lewis. Bantz's paper nicely brings out the conceptual depth of important problem situations and the difficulty of knowing in advance at what level to formulate and attack a problem.

The volume concludes with two papers on plate tectonic theory, itself the product of recent, revolutionary changes in geophysics. Rachel Laudan opens the continential drift symposium by arguing the case for a rational, theoretical pluralism, based on the fruitfulness of T. C. Chamberlin's 'method of multiple working hypotheses' for developing alternative hypotheses in geology. This research, of course, produced the revolution in that discipline. Laudan focuses on the work of J. Tuzo Wilson, who formulated the transform fault hypothesis. More generally, she maintains that only a theoretical pluralism can explain fundamental scientific change, by offering a rationale for the generation of alternative, 'maxi-theories'. The theories of Kuhn, Lakatos, and Feyerabend fail to solve adequately this problem of scientific discovery.

Hank Frankel applies his inexhaustible knowledge of modern geology to the developmental stages of Harry Hess's construction of the seafloor spreading hypothesis, the major turning point in the development of plate tectonic theory. Frankel argues that Larry Laudan's account, in *Progress and Its Problems*, as a problem-solving activity and Laudan's account of the evaluation of competing research traditions in terms of their problem-solving effectiveness nicely fits Hess's research methods. It was the seafloor spreading hypothesis — the solution to problems which Hesse had long pursued — which converted him to continental drift from a life-long opposition to it.[15] The hypothesis also solved the main problem facing drift theory — how can continents plow through the seafloor? — by showing that they do not have to!

## NOTES AND REFERENCES

[1] Hubert L. Dreyfus, *What Computers Can't Do*, Harper & Row, New York, 1972. The second edition (1979) contains a new preface surveying recent work in artificial intelligence.

[2] On critical explanation, and the problems of historical explanation in general, see

Maurice Finocchiaro, *History of Science as Explanation*, Wayne State Univ. Press, Detroit, 1973, Chap. 6; and his ' "Rational Explanations" in the History of Science' (forthcoming). Finocchiaro addresses the topics of rationality and judgment in his *Galileo and the Art of Reasoning*, D. Reidel, Dordrecht, 1980, and in a forthcoming paper on Huygens's theory of gravity. Stanley Cavell has much to say about reason, rules, and judgment in his important book, *The Claim of Reason*, Clarendon Press, Oxford, 1979.

[3] Hempel, *Aspects of Scientific Explanation*, Free Press, New York, 1965, pp. 463ff. Laudan, *Progress and Its Problems*, Univ. of California Press, Berkeley, 1977, pp. 184ff, 217.

[4] *Perception, Theory, and Commitment*, Univ. of Chicago Press, Chicago, 1979, pp. 132, 147–148.

[5] Stephen Toulmin, *Human Understanding*, Princeton Univ. Press, Princeton, 1972.

[6] Larry Laudan, *Progress and Its Problems*, Univ. of California Press, Berkeley, 1977.

[7] For this point see Harold I. Brown, 'On Being Rational', *American Philosophical Quarterly* 15 (1978), 246. See also I. C. Jarvie, 'Toulmin and the Rationality of Science', in R. S. Cohen *et al.* (eds.), *Essays in Memory of Imre Lakatos*, D. Reidel, Dordrecht, 1976, pp. 311–334.

[8] That event *a* caused event *b* implies the existence of laws relating *a* and *b* under some description (*e.g.*, a chemical or microphysical description) but does *not* imply that the laws are simple or available at any particular level of description, or indeed, that they are even known. Similarly, a reasoned decision may be taken to imply the existence of general normative principles, but not simple ones accessible at a particular level of description. For the point about causal statements, see Donald Davidson, 'Causal Relations', *Journal of Philosophy* 64 (1967), 691–703, and my article, 'On the Independence of Singular Causal Explanation in Social Science: Archaeology', in *Philosophy of the Social Sciences* 7 (1977), 163–187.

[9] Neither are the procedural theories of Laudan and Toulmin, although the charge of scientism suggests that there is a substantive bias involved in their selection of procedures governing rational choice. As for inconsistency, Dudley Shapere has pointed out that the internal consistency constraint on theories has been violated fruitfully in many instances. See his 'The Character of Scientific Change', Section VI, in T. Nickles (ed.), *Scientific Discovery, Logic, and Rationality*, D. Reidel, Dordrecht, 1980, and his 'Notes Toward a Post-Positivistic Interpretation of Science', in P. Achinstein and S. Barker (eds.), *The Legacy of Logical Positivism*, Johns Hopkins Univ. Press, Baltimore, 1969, pp. 115–160. See also Toulmin's *Human Understanding, op. cit.*

[10] See, *e.g.*., R. D. Luce and H. Raiffa, *Games and Decisions*, John Wiley, New York, 1957, Chap. 13.

[11] As Brown points out in his perceptive article (*op. cit.*), to which I am indebted, the view of rational deliberation which we both defend has parallels to Aristotle's discussion of deliberation in the *Ethics*. Indeed, the discussion at *Nic. Ethics 112$^b$* links deliberation in choice to the method of analysis (as opposed to deduction or synthesis) in mathematical discovery. See also the excellent discussion of judgments of equity as superior to legal justice at *1137$^b$*. We see that the 'no theory' view of rationality is in fact quite old, although I perhaps place more emphasis on the choice of goals or ends (instead of means only), and on noncognitive goals, than the Stagyrite did. Westman forcefully points out the neglect of noncognitive goals in philosophical discussions in his critique

of Laudan, in P. Asquith and I. Hacking (eds.), *PSA 1978*, Vol. II, Philosophy of Science Assn., East Lansing, Michigan, 1980.

[12] Shapere, *op. cit.*

[13] See Hempel's 'Empiricist Criteria of Cognitive Significance' (a conflation of two earlier papers), in his *Aspects of Scientific Explanation*, Free Press, New York, 1965, p. 113. In reply it will be said that the positivists were interested in the 'justification' of concepts, not in their 'discovery'. But I would argue that the criteria of admissibility of concepts could, and did, serve as a heuristic guide to behavioral scientists in the formation of new concepts. At the level of full-scale theories, however, the general, purely formal criteria of admissibility offered little methodological guidance for theory construction.

[14] Their contributions are discussed at several points of my 'Introductory Essay: Scientific Discovery and the Future of Philosophy of Science', *Scientific Discovery, Logic, and Rationality, op. cit.*

[15] I am indebted to my colleagues, Jack Kelly and Bruce Moran, and to Maurice Finocchiaro for helpful discussion.

MARX W. WARTOFSKY

# SCIENTIFIC JUDGMENT:
# CREATIVITY AND DISCOVERY IN SCIENTIFIC THOUGHT

## I. INTRODUCTION

Although creativity in science, mathematics, and technology is crucial to the fundamental processes of discovery and invention, it has largely been ignored by the philosophy of science, or it has been regarded as a question which lies outside the domain of philosophy of science proper. This has been the scandal of contemporary philosophy of science. But it has not been a hidden scandal, tacitly acknowledged and whispered about behind closed minds. Rather, it has been an open scandal, indeed, a theoretically justified one, so that its justification has made it appear non-scandalous, and even reasonable. Two questions present themselves here: *first*, how did the scandal arise? How is it that such an admittedly important feature of science as creativity, in its distinctive scientific modes as discovery and invention, could be excluded from systematic treatment by the very discipline whose task it is to understand science? And what rationales have been given to justify this exclusion? *Second*, if discovery and invention are to be proper subjects for the philosophy of science, how are they to be treated? How shall they be systematically *included*? What frameworks are necessary for understanding this feature of science?

What is needed, in my view, is a fundamental category-shift. Because discovery in science is an epistemological question, *i.e.*, one concerning scientific knowledge, and indeed, the *genesis* of such knowledge, the shift that is needed is one in the epistemological categories in terms of which philosophy of science studies or understands scientific thought. Thus, in this paper, I propose to deal with the question of creativity in science in terms of a neglected epistemological category: *scientific judgment*. I will contrast this with what has been the dominant epistemological category in contemporary philosophy of science, namely, *scientific explanation*. I will begin with a consideration of the problem which is posed by the question of creativity in science. I will deal, *first*, with how and why the question has been ignored or avoided by both empiricist and rationalist philosophies of science; *second*, with the question of how creativity is to be characterized, and with how it has been approached theoretically; and *third*, I will suggest how the question

1

*T. Nickles (ed.), Scientific Discovery: Case Studies*, 1–20.
*Copyright* © 1980 *by D. Reidel Publishing Company.*

of creativity can be confronted in terms of the epistemological category of scientific judgment.

In order to set the context of my discussion, let me suggest at the outset that the category of scientific judgment is analogous to that of aesthetic judgment in the arts, clinical judgment in medicine, judicial judgment in law, technological judgment in engineering and applied science, and practical judgment in moral, social, and personal contexts. I would also add that I am rejecting the traditional dichotomy in the philosophy of science between the context of discovery and the context of justification. I will say more about this later in the paper.

## II. HOW TRADITIONAL PHILOSOPHIES OF SCIENCE HAVE EVADED OR IGNORED THE QUESTION OF CREATIVITY IN SCIENCE

The creative aspects of scientific thought have always been problematic for both empiricist and rationalist philosophies of science. Therefore the question has been ignored, or systematically placed outside the domain of philosophy of science, and creativity has been regarded as a mysterious, inscrutable, and even irrational phenomenon. The logical positivist and logical empiricist philosophy of science sharply demarcated the domain of creative thought – the 'context of discovery' – from the domain of scientific explanation proper – the 'context of justification'. Thus, introducing this demarcation many years ago, Reichenbach (1938) wrote:

The philosopher of science is not much interested in the thought processes which lead to scientific discoveries .... That is, he is interested not in the context of discovery but in the context of justification.

Despite his critique of such 'justificationism', Popper (1959, p. 31) nevertheless maintained this same distinction in another way. He distinguished between the creative or imaginative activity of discovery, on the one hand, and on the other what one may call, in Popper's case, the 'context of falsification', thus maintaining the dichotomy, except now as one between discovery and falsification. It is true that for Popper, by contrast to the logical empiricists, the activity of discovery, based on what Popper (1962, 1972) called *conjecture*, was systematically very important, even fundamental to his account of science. But as a process – psychological or otherwise – it remained inscrutable, and therefore outside the analysis of science proper. The *sources* of conjecture were unimportant to Popper; what was important was the subsequent *test* of the conjecture, the attempt to refute or falsify it.

Thus, both major movements in contemporary philosophy of science — logical empiricism and so-called critical rationalism — left the context of discovery at the door of the house of philosophy of science proper.

By contrast, Polanyi (1958, 1966) emphasized the element of 'understanding' and the heuristic processes of scientific thinking. But he viewed this as largely dependent upon a non-rational or sub-rational kind of thought, what he characterized as 'tacit knowledge'. Such knowledge was, on Polanyi's view, inexplicable and not subject to specification by a rule or an algorithm. True, for Polanyi, such tacit knowledge could become explicit, or 'focal' knowledge, but then it in turn depended upon what served in this new context as tacit knowledge. In short, the very possibility of explicit, 'focal' knowledge required as its background, so to speak, a domain of inexplicit knowledge, in terms of which we are able to grasp, or understand what is explicit. His account — for example, in his major work, *Personal Knowledge* — had much in common with the older concept of *Verstehen*, which had been introduced in the *Geisteswissenschaften*, as distinct from *Erklaren* and *Kennen*.

Until recently, the dominant philosophies of science had focussed on the notion of scientific explanation, *i.e.*, they remained essentially within the context of justification. The task of philosophy of science was specified as logical reconstruction, *after the fact*. The process of discovery, being *before the fact*, was therefore systematically excluded from consideration. With few exceptions — *e.g.*, Polya, Hadamard — no attempt was made to analyze or to understand the features of discovery and invention in an epistemological way. It was only when the question was raised concerning the nature of scientific change, or of revolutions in science, that this reconstructionist framework with its emphasis on scientific explanation, came to be questioned in a serious way. The first question I want to deal with, therefore, is *how* and *why* did this neglect of the question of scientific discovery, of creativity in scientific thought, continue for so long?

To answer this question, we must examine the premises of the traditional philosophies of science, and the nature of the project which they defined. Insofar as philosophy of science was taken to be a project in the rational reconstruction of science, it took its task to be the reconstruction, in *logical form*, of the nature of scientific explanation. The model for such rational or logical reconstruction was taken from the rationalist tradition — from the model of mathematical physics developed first in the fourteenth century and then realized in the work of Galileo, Newton, *etc*. The so-called nomological model of explanation defined explanation in Kantian terms as deducibility under a concept. That is, science was conceived as a system of true statements.

The empirical or factual statements — those which gave putatively true descriptions of particular states of affairs in the form of singular statements, *i.e.*, the so-called observation statements, or protocol-statements or basic statements, were to be ordered in such a way that they could be deduced from Universal statements — the so-called law-like statements or theoretical statements. Theory thus explained the facts, when it could be shown that singular statements describing the facts could be deduced from universal statements asserted in the theory.

Apart from the sharp internal criticism which led to the weakening of the nomological model and its criterion of cognitive and empirical significance, the historical problem of change of theories, the rejection of old theories, and the acceptance of new ones in their place, led to a reconsideration of the traditional view, since the older view gave no account of such change. Kuhn attempted to give an account of such scientific change, of what he called 'revolutions in science'. In his work, *The Structure of Scientific Revolutions*, and in subsequent discussion, Kuhn seriously attempted to take the element of scientific discovery into account, and he dealt with the question of the acceptance and rejection of theories in the context of psychological and sociological *responses* to creative thought. But his account finally asserted only *that* such changes take place, and described the boundary conditions under which scientific revolutions took place. He failed to explain *why* such changes took place; and like the older philosophers of science, Kuhn also left the nature of creativity in scientific thought untouched, though he fully took into account its subsequent role in science. Thus, the logical empiricists, the Popperians, and Kuhn all left the question of scientific discovery to one side. They *assumed* it, but none of them attempted to analyze it nor to explain it. In fact, like Polanyi, Kuhn left the process of scientific change to the domain of the non-rational, if not indeed the irrational. And at least in his earlier versions, Kuhn could only speak of 'paradigm-shifts' as 'leaps of faith', borrowing Kierkegaard's irrationalist concept here from the philosophy of religion. Such an account, therefore, does not deal with how discovery or innovation in scientific thought comes into being, but only with the process of how such discoveries come to be accepted, how they come to believed, *after* they have been introduced.

Now, belief is indeed an interesting epistemological issue. How we come to believe what we do, how we change our beliefs, what justifies or warrants our beliefs, how we choose among alternative and conflicting beliefs — these are all important questions for epistemology, as well as for psychology and sociology. But the question concerning belief *becomes* a question, in the first place, only *after* a discovery is made, or a new theory is proposed which challenges older beliefs. From the point of view of the question of creativity in science, the question of belief is a subsidiary question, a secondary question, concerning the 'context of justification'.

All the recent literature on scientific discovery, therefore, deals very little with discovery itself, but rather with what happens subsequently. Nevertheless, there are theories of discovery which deal in some ways with this question. In this brief paper, I can only give the broad outlines of some of the traditional alternative approaches, as a background to my own discussion.

### III. ALTERNATIVE THEORIES OF DISCOVERY AND INVENTION IN SCIENCE

Here, I want to sketch three major alternative theories — *empiricist* or *inductivist*; *rationalist* or *hypothetico-deductive*; and *intuitionist*. These theories have much in common and are not simply mutually exclusive. But I will focus on their differences.

One should add another 'theory' — which, strangely enough, will be seen to be part of each of these theories, and which I will call the 'no-theory' theory. It is a metatheory which asserts that *there can be no theory of discovery*, on systematic grounds, and in principle.

A. *Empiricist theories*: There are at least two alternative empiricist theories of discovery, at least insofar as we speak of the discovery of scientific *laws*: we may call these the *inductivist* theory and the *descriptivist* theory. Stated simply, the inductivist theory alleges that scientific laws or hypotheses are not so much discovered as they are simply generated by inductive generalization from a set of observed instances. Thus, after observing, say, *n* similar instances of a given phenomenon, the inductive 'leap' is made to a generalization or a universal statement concerning all (past and future) instances of the phenomenon. Francis Bacon already saw the flaws in such 'induction by simple enumeration', and the logical fallacy, in deductive logic, of inferring a universal statement from singular statements is well known. From Hume to Goodman, the problem of induction has been sharply enough stated not to require repetition here. Yet, the inductivist approach still has many contemporary varieties, and the more sophisticated ones are of two sorts: the first is to take such generalizations as having no more status than that of inductively supported 'hunches', and then to test them by eliminative procedures — as Bacon and Mill already suggested, (*e.g.*, by Mill's 'joint method of agreement and difference'). The second version, somewhat similarly, takes the method of inductive generalization as a *heuristic technique* for suggesting hypotheses to be tested. It therefore assigns no particular inductive force at all to the generalizations, but only regards them as methodologically fruitful. Thus, the *number* of instances on which the inductive generalization is based plays no role at all. (One story about Norbert Wiener has it that when Wiener was asked: "On how many instances would you be willing to base a generalization?", he answered: "Two instances would be nice, but one is enough!") Plainly, this method can hardly be called inductive in the usual sense, in that it lays no claim at all to inductive force.

The second empiricist theory of discovery eschews any inductive claims at all, and in fact reduces the notion of discovery to that of systematic description. We may therefore call it descriptivist. This view sees laws in science as no more than descriptions of functional correlations between variables. A 'law', properly speaking, is no more than the summary record of previously observed instances. The law-like statement therefore has the logical status of a universal statement only on borrowed time, so to speak, and only in a manner of speaking. In fact, in the account of stochastic processes, or of easily repeatable phenomena, the 'law' is not taken as a universal statement but only as a report of ongoing correlations. Only the mathematical fiction of convergence upon a limit, as, *e.g.*, a statistical claim about warranted expectations concerning future instances, gives such a statement of observed correlations the status of anything more than a record or a report. On such an account, there can hardly be said to be any act of discovery, but only patience, fortitude, and care. Of course, the very choice of phenomena to be observed, or the experimental insight concerning how to arrange things so that one elicits interesting or fruitful correlations are both inventive and creative procedures in science. But the descriptivist theory gives no account of such inventiveness or creativity, and therefore can give no account of its own procedures as, in any sense, procedures of discovery.

B. *Rationalist theories.* By contrast to the inductivist or descriptivist approaches of empiricism, rationalist theories of science have proposed a 'method' of discovery, already familiar to ancient Greek mathematical theory, and made popular by Kant and by Peirce, *i.e.* the so-called method of 'hypothetical inference' or what Kant called 'transcendental deduction' and Peirce called 'abductive inference'. (This has recently been 'rediscovered' as a novelty, in recent discussion about "inference to the best explanation: see, *e.g.*, Harman, 1965.) The use of the term "inference" is misleading here, if one takes it in its stricter logical connotations, as a procedure of deduction according to a rule of inference, or as an algorithm or a computational procedure. For "inference" to an explanatory hypothesis is precisely what cannot be done according to a rule, although there is a rule for determining whether such an "inferred" hypothesis fulfills its function: namely, whether what it is supposed to explain follows deductively from it.

One of the traditional objections to the notion of inductive generalization is that it has no rule of inference and therefore is not a form of deduction at all. And this holds equally for hypothetical or abductive inference. But that is to say that insofar as there are empiricist or rationalist theories of discovery, they do not propose that discovery in science is reconstructible logically, nor as a rule-governed or algorithmic procedure. And of course it is on these grounds that discovery has been judged to lie beyond the domain of the philosophy of science proper, since philosophy of science has been regarded as a project in logical reconstruction by both the logical empiricist and the rationalist schools.

C. *Intuitionist theories*: An alternative to the empiricist and rationalist theories of discovery, but one which is in essential agreement with them in practice, and shares in the 'no theory' theory, is the view that discovery in science, and creativity in general, is a totally non-rational (or pre-rational or irrational) process, that it involves a mysterious and unanalyzable act of intuition or insight, and therefore that the very attempt to analyze it or to understand it is mistaken. In one version, we may call it an *inspirationist* theory. Here discovery, like the act of poetic creation described in Plato's *Ion*, is an 'act' which is not an act at all, but rather a situation in which forces or ideas greater than oneself simply express themselves through the individual, without his or her will or judgment. For Hegel, the 'cunning of Reason' simply utilized individuals to realize the Idea of History. In contemporary terms, the discovery or invention, the act of creation in scientific thought, simply reveals itself to the properly prepared mind, to the appropriate 'receptacle' (See Mach, 1896.) Discovery therefore remains, on this view, an inscrutable act of grace (like being chosen by God for salvation, or by the World Spirit for an historical role.) In this approach, the scientist is not the creator but rather the passive instrument of a force which lies beyond his or her consciousness.

These are what I see as the major empiricist, rationalist and intuitionist theories of discovery (of course, presented in oversimplified terms). I think none of them is adequate. But I believe their inadequacy lies fundamentally in the way in which the problem of creativity is conceived, and I turn next to a consideration of how this problem has been formulated.

## IV. THE PROBLEM OF CREATIVITY

The conceptual problem which confronts all these theories of discovery in science arises from a sort of dilemma, which I will call the *dilemma of explanation*: If a theory is successful in giving an explanation of discovery and invention, then it reduces creative thought in science to an algorithmizable procedure, or to a deductive consequence of the explaining theory, and thus dissolves the notion of creativity altogether. If the theory fails to give an explanation of this sort, then of course it is a failed theory. The dilemma is then: either the theory succeeds, and the concept of discovery is explained away, or reductively eliminated — or the theory fails, and discovery remains unexplained.

I believe that the problem arises from the conception of what a theoretical explanation should do — namely, from the view that a theory should yield an account of the phenomenon under consideration as a deductive consequence of the theory. If we weaken this requirement on theories, or rather if we broaden the conception of what a theoretical explanation is, then we may perhaps avoid the dilemma. Suppose instead that we require of a theory which 'explains' discovery that it give us an account, *not* of how a particular discovery follows as a deductive consequence of the theory and from some set of initial conditions, but rather that it leads us to understand the strategies or rules of art that were involved in the creative thought processes in science. This kind of understanding is something like a dramatic reenactment in the imagination of the process of discovery itself. But it is not a reenactment of a step by step sequence of thoughts which follow each other in accordance with a rule of inference, *i.e.*, deductively. Rather it is a reenactment of the *sense* of a process of thought, or of an argument; it is, in effect, a matter of understanding of the sort involved in making judgments. To put this another way, the requirement on a theory which 'explains' a phenomenon — whether discovery or something else of the same sort — is not an extensional formulation of the 'logic' of discovery, but rather an intensional account of the meaning-relations within the process. We cannot speak here of deducibility in the strict logical sense, and certainly not of an algorithm of discovery. We can speak instead of a *heuristic* account, *i.e.*, one which guides us in understanding the creative process by reconstructing the strategies, the methodological rules, the 'rules of art', the modes of judgment, that were involved in a

given process of discovery or invention in science. It is this kind of understanding, this process of comprehending how a scientist may have proceeded
in thought, that falls within the category of *scientific judgment*.

How does the category of scientific judgment offer an alternative approach
to the problem of explaining or understanding creativity in science? In effect,
it introduces an alternative to the nomological or deductive model of scientific
explanation, and broadens the conception of what it means to 'explain'. But
in a deeper sense, it is an alternative account of rationality, or of the criterion
of rationality. And this is where the crucial epistemological question lies.

For the purposes of this brief paper, I will state my thesis only in outline,
to be expanded in a longer version of this presentation. The reason that
creativity has remained so problematic is that the definition of rationality
which has come to dominate science, and which has indeed played a powerful
role in the development of modern science, is that which defines rationality
in terms of logical or mathematical systematicity. That is to say, a rational
construction (whether in science or mathematics or philosophy) is one in
which each step of the argument is seen as a necessary consequence of the
previous steps. This 'necessity' has been variously defined. In classical seven-
teenth-century rationalism, the test of necessity was simply *what could not
be conceived to be otherwise*. This is hardly a systematic appeal, but rather an
appeal to some rational intuition. The more explicit notion of necessity,
which weakens it by relativizing it to a rule, is simply: what follows according
to a rule of inference or a computational procedure. Here, of course, the
modal sense of necessity is replaced by the notion of deductive consequence,
defined extensionally in terms of preserving the truth values of variables, and
the general rule of inference that false conclusions cannot follow from true
premises.

This issue is, of course, not merely logical, nor merely epistemological, but
ontological as well. The criterion of rationality which defines it in terms of
necessary or logical connection among statements has its ontological correlate
in the notion of causal connection in the whole ontology of mechanism which
proposed a world construction on the basis of the model of mathematical
structure. Thus, when Galileo proclaimed, in Pythagorean-Platonic fashion,
that "The book of nature is written in the language of mathematics", he was
making not merely an epistemological claim about how we come to explain
natural phenomena, but an ontological claim as to what the structure of the
world is such that it *can* be explained or understood in terms of mathematics.

By this criterion of rationality, *every* phenomenon, *including* the phenomenon of discovery or invention, is finally understood or explained only if

the explanation is in these terms, *i.e.*, in terms of a reconstruction in explicit deductive logical or mathematical form. But it is this criterion of rationality, precisely, that leads to the dilemma of explanation, as we have seen.

In proposing that the category of scientific judgment should replace that of scientific explanation, in the above deductive sense, in order to be able to give a rational account of creativity in science, I am proposing that we critically revise the very criterion of rationality which lies at the basis of the current view of explanation. But how does one proceed to critically revise such a criterion? It seems to me that the first step is to see the presently dominant criterion of rationality not simply as *given* or as finally *achieved*. Rather, it needs to be seen in its historical perspective. That is to say, rationality itself, as a criterion of human thought or understanding, has a history; and what I am proposing is that an appropriate critical understanding of the presently dominant criterion of rationality requires an epistemological analysis which is not only conceptual or analytical, but historical as well. In short, my critical discussion and the thesis I want to propose here are based on an approach which I call *historical epistemology*.

The thesis, briefly stated, is this: that the present criterion of rationality, defined in terms of logical reconstructibility or in terms of the deductive model of explanation, developed out of the success of mathematization *both* in science and in economic life. Mathematization in science, as a tendency, goes back to the Pythagoreans of course. But its utter dominance and its brilliant successes came in the sixteenth and seventeenth centuries, side by side with the major development and dominance of exchange and money economies, and with the development of commerce and trade in Europe. This notion of abstraction and reduction to logico-mathematical terms was already well developed in philosophical and mathematical thought, it is true. But the full realization of such formal abstraction — its interpretation in terms of physical and economic magnitudes and in terms of the arithmetization of geometry — developed side by side with the growing and revolutionary dominance of the exchange economies of the great mercantile and early capitalist states of Europe (the Adriatic cities, the Hansa cities, England, *etc.*). At the same time, the engineering arts — shipbuilding, cathedral construction, fortifications, ballistics — had all developed over several centuries, contributing their share to the general mathematization of practical activities, which marked the early stages of European capitalism. The whole conception of the world in terms of the visual representation of it, was radically altered also by the introduction of linear perspective into painting in fifteenth century Florence, in which the mathematical model of (Euclidean) geometrical optics became the rule of visual representation of the three-dimensional visual world projected on a two-dimensional plane surface. Here, the practical activities of the painters and architects (*e.g.* Brunelleschi), and that of the navigators and cartographers played a role as well. (See Edgerton, 1975).

I am *not* claiming that the criterion of mathematical rationality simply derives from this mathematization of *praxis* in social, economic, technological, and artistic life: but rather that the historical adoption of this criterion, as the dominant one, is related in complex and important ways, to these developments in social practice. As the proverb says, "One hand washes the other". Theoretical developments in mathematics and physics played a large role in the mathematization of *praxis*; but also, conversely, this mathematized praxis played a large role in the encouragement and legitimation of the mathematical criterion of rationality. The relation between the two is (if one may use the expression without abusing it) *dialectical*.

What I am suggesting therefore is that the historical development of the contemporary criterion of rationality needs to be recognized as historical, and as a choice among alternatives. The reasons for the choice are important subjects for serious historical as well as epistemological research.

What is all the more interesting, historically, is that side by side with this criterion of rationality — of clear and distinct ideas, of the principle of sufficient reason, of the symmetry principle in physics and mathematics — there exists a complementary tradition which is quite different: namely, that of divine inspiration or of non-rational intuition. Descartes's insight into the mathematizability of physics came to him, he tells us, in a dream, and from an angel. Newton's theology, his Cambridge Platonism, was as much a theory of divine inspiration, of a form-giving God who creates both the world and true thoughts about it, as his mathematics and physics were expressions of the 'Mathematickal Philosophy'. In short, the traditions of mathematization and of inspiration, of explicit formulability and ineffable intuition, of rationalism and irrationalism, exist side by side, historically; and their apparent contradiction remains a problem throughout the rationalist tradition in philosophy. This same contradiction — or at least, this same incoherence — is preserved in the contemporary dichotomy between the context of discovery and the context of justification, between creativity in science and mere rational reconstruction of science.

There is an alternative tradition, which one may call the heuristic tradition — that of *Ars Inveniendi* — which Francis Bacon and others promulgated at the same time, in recognition of the creative inventiveness of the artisan, the mechanic, the craftsman. But this tradition has been largely subordinated in

the development of modern science, and of modern philosophy of science. It
has been regarded as an activity of a lower sort, a kind of craftsmanlike skill
in judgment, or in discovering solutions to problems by 'tinkering'. As a
practical art, it has been regarded as of lower status than the 'theoretical',
since by definition, the 'theoretical' is defined by the criterion of rationality
we have been describing. It has therefore not been given theoretical status,
nor regarded as theorizable *in principle*, as we have seen. Yet its role in
scientific thought — where the 'tinkering' has itself to do with concepts and
theories — has been fundamental to discovery. Einstein at one point writes:
"All our thinking is of this nature of a free play with concepts" (1949).
But if indeed, this kind of creativity is a truly 'free play', then it is not
the subject of a rule-governed procedure, or of an algorithm. But is it therefore subject to no 'rational' comprehension? Is it therefore to remain forever
mysterious and unapproachable?

## V.  HOW THE QUESTION OF CREATIVITY CAN BE CONFRONTED IN TERMS OF THE EPISTEMOLOGICAL CATEGORY OF SCIENTIFIC JUDGMENT

Thus far in my argument, I have tried to show how and why the problem
of creativity has been neglected or systematically evaded in contemporary
philosophy of science. I have stated that the 'problem' becomes a problem
because it is posed in the problem-context of the contemporary criterion of
rationality. And I have further suggested that this criterion itself needs to be
critically understood as an historically achieved and historically adopted
criterion, testifying to the great theoretical and practical successes of mathematization in science and socio-economic life. That the criterion is historical
is not, in itself, a normative critique. The only normative criticism I have
offered thus far is that the theory of explanation based on this criterion can
give no account of discovery and invention in science. But this in itself is not
yet a critique, for it may be true that there can in principle be no account
of creative thought which does not reduce it to an algorithm (and which
eliminates the concept for all practical purposes). What needs to be shown,
if a criticism of the current dominant view is to be made, is that the 'no
theory' theory of discovery is false.

   And this can be shown only if a feasible account can be given of what a
theory of discovery would be like, which retained our notions of discovery *as*
creative, *i.e.*, as introducing by an act of thought something not yet present,
and not merely to be explicated from what is already presently available.

The category of judgment, in general, is one which suggests the synthesis, the bringing together of things — ideas, concepts, the subject and predicate of a proposition — in such a way that a relation among them not previously realized is brought to light. The *discovery* of such a relation suggests that the relation already exists, and that the creative act consists in *recognizing* it, literally, in 'uncovering' it, revealing it. Thus scientific discovery in general has been regarded as the bringing to light of relations among things in nature, where these relations are themselves already objectively present. The activity of *dis*-covering these relations, however, requires the creation of new concepts, of new instruments of analysis, of new techniques of experimental intervention — in short, the introduction of new possibilities of interaction with the natural world, both theoretically and practically. Thus, we may say that the act of discovery does not create or invent a new world, but rather a new way of construing the world. In general, such a point of view is realist in its epistemology, or, we may say, objectivist.

By contrast, one may argue in conventionalist-pragmatist fashion (as, *e.g.*, Nelson Goodman, 1978, does) that the world is not one way or another, but rather that for our various purposes and interests we may construe it in an infinite number of ways. There are, in effect, as many 'worlds' as there are viable constructions of 'it' (where 'it' or '*the* world' has no univocal or preferred reference); and where the viability of a construction depends on the rules adopted for the construction, or the system within which the construction is made and interpreted. The more radical interpretation of such a conventionalism, relativizing constructions to choices among 'paradigms', claims in effect (as Kuhn, 1962, does) that with the choice of a new paradigm, the scientist's 'world' literally changes.

Changing the world is a revolutionary task, as we all know. Changing the natural world has been the task of human beings in their everyday practice, in their science and technology, and in their social activity as well (since the social world, the world of human interaction and praxis is a part of *the* world, but the world of nature transformed by human purpose and action). Such changes are not introduced by theoretical fiat, though theory in science plays a role in effecting a change in our conception of the world and therefore in orienting and changing our practice in it, and opening up possibilities of practice which did not before exist.

It is this reorientation of our conception, this recognition of possibilities of action, this guide to new and different practice that is essentially involved in the category of judgment. Thus, I take judgment to be not merely an epistemological category in the sense of what concerns our thinking, but also in the sense of what concerns our action; and this, in terms of what we come to recognize *in thought* as what it may be possible to do in practice. The category of judgment is therefore essentially the category of thought as a guide to, or a suggestion of possible practice. Judgment is therefore essentially a category of what I would call the *practical imagination* — the capacity to invent alternative modes of action in terms of present possibilities and conditions; or the capacity to invent new conditions beyond present limits.

In scientific judgment, then, the act of creation consists in imagining new ways of relating present facts, or of imagining *new facts* in such a way as to realize them by some mode of action — *i.e.*, by experimental practice, by technological innovation.

A theory of discovery in these terms is one which helps us to understand how such a project in the scientific imagination can be approached; and helps to explain historically how, in the history of science, such projects of discovery and invention were approached. It is therefore not simply a reconstruction of science *done*, but of the *doing* of science. And therefore, it focuses on the ongoing *activity* of scientific thought, and not simply on its products. In principle, however, such a theory cannot itself yield discoveries — that is, it is not a discovery-generating theory, in the sense that discoveries follow from it, by the application of a rule. Rather, the relation between such a theory of discovery and discovery itself is a heuristic relation: if it is a 'logic' at all, it is a *suggestive* logic and not a *deductive* logic. Now a logic of suggestion is in effect a logic of maxims — what it would be wise, or better, or more fruitful, or more helpful to do. Such rules of art, of judgment, of practical wisdom, such suggestions of methodological preference, may be *expressed* in maxims; but as statements, maxims are of the form that what follows from them *deductively alone* has very little force, and is too unspecified in its reference. Therefore, the attempt to construct a deductive logic of such heuristic expressions is bound to end in vacuity. But scientific judgment (like judgment in law, or aesthetic judgment, or clinical judgment) is practical precisely in the sense that its emphasis is casuistical — that is, its quality lies in the ability to interpret general maxims in their application to specific instances, where the interpretation is not given by a rule, or a definition. The crucial element in discovery makes such an algorithmic rule-following procedure impossible in principle. For in the case of the creative act of discovery or invention, it is precisely a *new* situation that has to be dealt with, a *new* problem that has to be resolved. The repetition of a procedure based on success in past instances is therefore inadequate.

What then is learned in practice, if not simply what has succeeded in the past? What is learned, I believe, is a certain sensibility with respect to structures, a certain grasp of nuance, an ability to see whole through a maze of complexities, and most important, an openness of mind to new possibilities — the element of play that Einstein refers to. The paradigmatic mode of giving an account of scientific judgment is the same as the mode of acquiring it: *by example*. Since maxims without concrete interpretations are vacuous, and rules of art without instances of application are likewise empty, an account of scientific judgment as it is involved in creativity in science cannot, in principle, be given in a paper like this one which sets out to talk *about* discovery. Rather, the account would be an account of examples, of actual exercises in judgment given specific boundary conditions, or problem situations, or a reconstruction of historical cases. In this paper, I can only be broadly heuristic myself in suggesting such an approach. But is such an approach, properly speaking, a *theory* of discovery? Does it 'explain' how discoveries come to be made? No. Certainly not in the sense of explanation that would yield the discovery as the 'conclusion' of a deductive inference. Does it help us to understand the creative process in scientific thought? In a qualified sense, yes. It is a guide to how one may *approach* the understanding of creation in science. It is itself a prolegomenon to the practice of understanding discovery, namely to the direct confrontation with those. problems and questions which have provided the contexts for discovery in the history of science; and to the formulation of problems and questions which provide the contexts for new discoveries in science. It is true therefore that an approach to creativity in science from the point of view of the category of scientific judgment does *not* give an explanation of discovery in the sense demanded by the dominant theory of rationality. Rather, it shifts ground radically in claiming instead that 'explanation' now requires the reenactment of the process itself, the reconstruction of the *practice* of creation, and not merely the logical reconstruction of the results of that practice.

*Department of Philosophy and*
*Center for Philosophy and History of Science*
*Boston University*

## BIBLIOGRAPHY

Edgerton, Samuel: 1975, *The Renaissance Rediscovery of Linear Perspective*, Basic Books, New York.
Einstein, Albert: 1949, 'Autobiographical Notes', in P. A. Schilpp (ed.), *Albert Einstein: Philosopher-Scientist*, Tudor Publishing Co., New York.

Goodman, Nelson: 1955, *Fact, Fiction, and Forecast*, Harvard Univ. Press, Cambridge.

Goodman, Nelson: 1978, *Ways of Worldmaking*, Hackett Publishing Co., Indianapolis.

Hadamard, Jacques, 1945, *The Psychology of Invention in the Mathematical Field*, Dover, New York.

Harman, Gilbert: 1965, 'The Inference to the Best Explanation', *Philosophical Review* 64, 88–95.

Kuhn, Thomas: 1962, *The Structure of Scientific Revolutions* (2nd ed., 1970), Univ. of Chicago Press, Chicago.

Mach, Ernst: 1896, 'On the part played by accident in invention and discovery', *The Monist* 6, 161–175.

Polanyi, Michael: 1958, *Personal Knowledge*; *Towards a Post-Critical Philosophy*, Univ. of Chicago Press, Chicago.

Polanyi, Michael: 1966, *The Tacit Dimension*, Doubleday, Garden City, New York.

Pólya, G.: 1945, *How To Solve It*, Princeton Univ. Press, Princeton.

Pólya, G.: 1962, *Mathematical Discovery*, John Wiley, New York.

Popper, Karl: 1959, *The Logic of Scientific Discovery*, Hutchinson, London.

Popper, Karl: 1962, *Conjectures and Refutations*, Basic Books, New York.

Popper, Karl: 1972, *Objective Knowledge*, Oxford Univ. Press, Oxford.

Reichenbach, Hans: 1938, *Experience and Prediction*, Univ. of Chicago Press, Chicago.

# DISCUSSION OF WARTOFSKY'S PAPER

PHILIP QUINN: I wonder just what conclusion we are supposed to draw from your historicizing rationality. Suppose that I accept the story you told about how the rationalist mathematical tradition arose. I do not see that anything yet follows from that about the merits of the standards of rationality that are carried by that tradition. I at least take a theory of rationality, good reasons and so on, to be an explicitly *normative* theory, and if we have one that we accept, we can't help but judge the past in terms of it. So I wonder whether you want to infer from this historicizing that there is something wrong with the theory of rationality; or is your thought that we need another, distinct theory to deal with creativity just an independent thing?

WARTOFSKY: No, on the contrary. I think we have good reasons for having adopted that criterion historically in place of something else. It has been extremely powerful, and its successes have been great, as I said in the paper. I am not proposing that we give it up. What I am proposing that we give up is the dogmatism which asserts that this *covers* the ground of human rationality and that anything that is not included as deducible in terms of this conception is therefore regarded as outside the domain of rationality. Now rationality, as I think Ernan McMullin pointed out during the panel discussion, is a much wider practice than that of deduction. What we are to include in it is not predefined by some analytical characterization of what rationality ultimately and *apriori* must be. Nor is it to be defined by what is biologically built into our genes, as the inevitable framework and limits of our capacities. When I say the theory or conception of rationality has a history, I mean that it also has a future. The failure to recognize one's ancestry, one's historical ancestry, is in part an effect of what has happened during the last one hundred years in the standard philosophy of science. The alternative tradition of what you might call 'practical judgment' or 'practical imagination' has remained a subsidiary and external domain to the philosophy of science. It was and is ruled out of court on principle. What I was trying to do was suggest the reasons why it was ruled out. I can give some more; but the Aristotelian distinction between the theoretical and the practical is not simply an analytical distinction. It is a distinction that grows out of the nature of Greek life.

17

WILLIAM T. SCOTT: I think your emphasis on judgment is just right. However, you made a remark about Polanyi that I think I can correct. It isn't that he thinks tacit knowledge is subrational. The tacit part, the unspecifiable part, is the clues you use when you are focusing on something; but what you focus on, of course, is often in the rational category. It is how you make the judgment of what is rational that is tacit. We know that we can do it yet cannot say it. I do not think this denies rationality. I entitled one of the papers I wrote about Polanyi's thought 'A New Affirmation of Rationality' [*Archives de Philosophie* **35** (1972), 7–31 and 245–262]. I think that the concept of judgment would be a central concept in his thought — very similar to what you were saying — even the judgment as to whether something is logically deducible from something else. It takes a judgment to know that the deduction has been done right.

WARTOFSKY: Yes, I am a little troubled by how to use the term '*tacit* knowledge' here. In that particular paragraph I do not say that its referent is *ir*rational; I say that it is either sub-rational or non-rational. But here I may be using the term 'rational' so as to imply 'explicit' or 'explicable'. It may be that the tacit components of our knowledge are explicable, although not explicit, since they can become focal. In that sense, they may be incipiently rational in the sense of explicable. On the other hand, I would not want to constrain the domain of rationality to only that which is explicable and neither would a great many people in philosophy of science who would take background knowledge itself as a constitutive element in our scientific understanding — Popper and others.

LYNN LINDHOLM: I'm not sure how much it helps for us to say that the idea of discovery which you put forward as practical judgment is not an attempt to articulate a logic of discovery but rather to outline a sensitivity toward a direction, because either this sensitivity to a future direction is not going to give you any practical guidelines or it is going to forbid some activity and therefore to be imposed on you to that extent.

WARTOFSKY: I would say that the practical guidelines are quite clear, although maybe I was not explicit enough. There is one negative practical guideline: do not think that you are going to learn anything about how discovery takes place by reading a rule-book on discovery. You discover how discovery takes place by studying how discovery *took* place and by participating in that activity. You have a range of alternatives. It depends on whom you choose to study with, what masters you choose, whom you take as examples. A good deal of the whole enterprise of research programs is to try to pick out the good masters and to separate them from the bad ones.

LINDHOLM: But aren't you going to have to spell out a theory of discovery when you start making these choices?

WARTOFSKY: On the contrary, I think making this choice *is* spelling out a theory of discovery. I do not think that there is another thing which you then have to do.

THEODORE KISIEL: This question depends on the rule-book approach, particularly Pólya, to balance off Polanyi — the two Hungarians. I believe that they had this argument back around 1945. I think that there should be some attempts made to spell out at least some sort of rules, along the lines of Pólya. He makes the point that he is never going to have a definitive set of rules, because they are always going to be fallible; it is going to be an evolving set of rules. But I think of something like a typology of problem situations or a typology of similarity relations and analogies. For instance, along with Pólya, you have an attempt by W. J. J. Gordon at Harvard to establish the basic types of analogies which are used, let us say, in industrial research. I think that things like this — some sort of attempt to specify at least some of the possible heuristic strategies is something important to keep in mind. I think you have to balance off Polanyi and all this talk about tacit knowledge with an attempt to specify those heuristic strategies.

WARTOFSKY: It depends in part on what one means by a 'rule'. 'Rule' is a very tricky term, and I do not want to get into etymology or meaning analysis. The way we use 'rule' ordinarily is that it is something one follows. In order to follow a rule, there are two possibilities. Either there is an algorithm for *following* the rule or we have to be able to understand the rule in such a way as to be able to follow it without an algorithm. I am not against the 'rules of art' approach, but it is precisely rules of art that are involved here and not, for example, rules of inference. Rules of inference have to be such that they are algorithmizable or they are not rules of inference at all. A rule of art, however, is a different kind of rule, and the kind of suggestive help that a rule-book of that sort could be is exemplified by a rule-book on manners. A rule-book of good scientific manners might be helpful, but in fact it is not going to tell you what to do in a given instance. Therefore, the rule-book by itself can at most put you in the ballpark.

JOSEPH AGASSI: But Marx, you have two standards. By one standard, the rule-book is all right but not good enough; by the other standard it is no good at all. Make up your mind which it is!

WARTOFSKY: I just tried to indicate that if you take the rule-book as a book of rules to be followed, then it is no good at all.

AGASSI: But otherwise it is some good?

WARTOFSKY: The only good it is is heuristically good. If one takes it heuristically, it points to the kinds of things that one should pay attention to without specifying what one is going to then do. It points you in the direction of learning how to proceed, so to speak, in the situation itself.

AGASSI: Is this not true of all rule-books — of cookbooks, of manners, of diplomacy, of politics?

WARTOFSKY: Yes.

ROBERT WESTMAN: I thought that you were going to take the ballpark metaphor to its logical conclusion. Is baseball an instance of your point? Do you have to *play* baseball to know the rules?

WARTOFSKY: No, that is different. You have to play baseball *according* to the rules. Baseball is not a discovery procedure. Baseball is a kind of a game which has to follow the rules or it is not baseball.

WESTMAN: But learning how to play third base involves discovering . . .

WARTOFSKY: How to play third base is not in the rules of baseball. The rules that can be violated by a third baseman are. How to play third base as against playing left field is not in the rule-book and is not part of the rules of baseball.

PANEL DISCUSSION

# THE RATIONAL EXPLANATION OF HISTORICAL DISCOVERIES

[The panelists, in speaking order, were: Noretta Koertge, *Indiana University* (moderator); Mary Hesse, *University of Cambridge*; Ian Hacking, *Stanford University*; Maurice Finocchiaro, *University of Nevada, Las Vegas*; Ernan McMullin, *University of Notre Dame*; and Robert Westman, *University of California, Los Angeles*. Their remarks have been revised and edited somewhat for publication but not so much as to destroy the spontaneity and the conversational tone of the proceedings.]

NORETTA KOERTGE: The nature of history in general and the history of science in particular has long been the subject of philosophical controversy. Exactly how does a historian reveal the past to us? Some say that history is simply narrative — one tells the story of what happened. Others add that in historical explanation one puts the reader into the skin of the historical character, getting him to feel the historical situations vicariously. Some people say the historian should provide an objective analysis of the historical actor's predicament — by using situational logic we can show that anyone would have responded similarly. Others claim that there are covering laws which guide the unfolding of history.

All of this is very problematic. One might hope, however, that simpler patterns would emerge if we were to concentrate on the nature of explanation within the history of science. On the face of it, scientists appear to be engaged in a more orderly and rational activity than those involved in politics, waging wars, or molding social institutions. However, there is one crucial aspect of science which seems particularly ·recalcitrant to analysis, namely, the process of discovery.

One major task of the historian of science is to make the reader understand something about the development of new scientific ideas, new scientific theories, new ways of looking at the world. The question which I am going to pose to the panel is: How does a historian lead us to understand scientific discovery? Is there anything systematic which can be said about how historians should recount the development of science?

Each of our panel members has made contributions to the history of ideas. I hope that as they address this topic they will draw on their experience as

21

*T. Nickles (ed.), Scientific Discovery: Case Studies*, 21–49.
*Copyright* © 1980 *by D. Reidel Publishing Company.*

working historians as well as their philosophical expertise. Our first speaker
will be Mary Hesse.

MARY HESSE: Since I do not have time for a long speech, I'll have to start
things out dogmatically, and, I hope, also provocatively.

Let me begin by asking, What is the aim of scientific discovery? I believe
that it is the discovery of true causes. That will have to be described in greater
detail, but there is not time to do that now. How do we know true causes?
We know true causes by empirical predictions, successful tests, and feedback
from unsuccessful tests. That defines for me what we, in our society, mean
by science, and therefore it defines what we mean by rationality in scientific
procedures. As long as this process was accepted in the history of science – I
mean in the actors' procedures that we study in history of science – and were
believed to lead to a unique or at least a convergent theory of true causes,
then a discovery could be judged as rational insofar as the process was adopted
and the goal was reached. In other words, it seems to me, the logic of discovery
was identical with the logic of induction. By induction I mean something as
sophisticated as you like – not just the Baconian rules but all the sophisticated
methods of which perhaps some recent examples are the applications of artificial intelligence to the diagnosis of disease and to the study of molecular
models in biochemistry. Now the point is that induction is not mechanical
and does not lead to unique results and theories. It may lead to a multiplicity,
a plurality of theories, all of which satisfy the empirical tests and lead to
successful prediction.

I think this much would be held in common by two philosophers who
talked about discovery and justification, namely Reichenbach and Popper.
When Reichenbach rejects the notion of discovery as not being identical with
either induction or justification, I think he is rejecting it because he takes
discovery to be a purely individual, subjective, psychological question;
whereas Popper rejects discovery because it smacks of induction, and he
rejects induction. But the upshot of the more recent discussion, post Popper,
is that we now have two sorts of questions, which are the respective concern
of philosophers of science and historians of science.

The first is this. Imagine that we are looking at an historical case of, say,
eighteenth century science. One question to ask is, What was the empirical
test procedure actually adopted by the actors – whatever they might have
said about it? There is an ambiguity, to be sorted out by historians, about
what they *thought* they were doing, and what they can be *seen* to be doing
if we research into their editors' criteria of acceptance for publication, the

interaction of the scientists at the time, and so on. Was the actual test procedure accepted by the actors the empirical one I have described? If so, they come into the ballgame of science as I've defined it. They probably thought that they were after true causes, or at least that's the way they would have expressed it, but their test procedure would still have been empirical. That's the first question. That sort of question seems to me one on which philosophers can throw great light by being clear about present models for scientific rationality and giving historians some criteria by which they can make evaluations from their own investigative standpoint concerning what we should regard as rationality in judging the procedures of the past.

Then we have another type of question. How did the actors themselves see their scientific rationality, and how were their conceptual frameworks, theoretical models, and so on, chosen? Was it by reference to their test procedure, or by other means — by conformity with metaphysical principles, by constraints that were more widely around in the history of ideas of the period, and so on?

That seems to me primarily an historical question (we have a little division of labor here) because the historian has to know not only the intellectual history, but also the social, political, religious and other history of the period. The historian has to know all this in order to give an explanatory account of scientific activity, partially in terms of rational causes, that is, the actor's intentional choices as seen in that period, but partly also in regard to those conceptions, those world-views, those taken-for-granted common sense notions — that the planets must revolve in circles, for instance — which were not chosen but were intrinsic to the socialization into the thought of the time. That's why I would make a strong plea for a serious consideration of the sociology of knowledge as an essential part of the historian of idea's toolkit. I am a bit unhappy about a separation or a division of labor between those historians who deal with the history of ideas rationally and those who deal with the social history of science. I wish one could see these coming closer together.

In my final two minutes, I want to tell a little myth about Sir Karl Popper, a bold knight, who was trying to rescue the maiden Science from a number of dragons. There are at least three of these dragons which I think he went out to kill and which have been taken by his later henchmen to be thoroughly dead and buried, but which I would like to see tended to and revived. The first dragon is inductivism. It seems to me that questions about the logic of induction, which would include how we get theories by devising new classifications, developing new concepts and applying probabilistic arguments, would be part of what the logicians of discovery are looking for. I would like to see that dragon revived, and I don't think maiden Science will suffer from its revival.

Popper also intended to kill something he called 'psychologism'. This seems to me to be just one part of what I am asking the historians to take on board again, namely the nonrational, nonindividual, nonintentional constraints on any intellectual community, particularly the scientific community, at any given time.

Thirdly, Popper thought he had slain the bogey of holism, by which I mean whatever is not consistent with an individualist model of rational man — a Robinson Crusoe model of the scientist. In Popperian theory the scientist is not understood as part of the scientific community interacting with a particular language which is essential to its theories. Rather it seems that the scientist could, in principle, if he had enough time, enough instrumentation, and enough data, develop his conjectures and refutations himself. I think that there is nothing in Popper which would be inconsistent with a story of a Robinson Crusoe scientist of that kind. Here we need a bit more sociology.

I hope I have been provocative in at least two ways, namely by saying, "Let us revive inductivism!" and "Let us revive sociology of knowledge!"

IAN HACKING: I suspect that our title — 'The Rational Explanation of Historical Discoveries' — is supposed to steer us to a discussion of what is rational in science, and that is my chief difficulty with this panel: I have a negative attitude towards questions about rationality in science.

I do know a number of people who have made discoveries of the first magnitude. If I were asked to explain *why* they did it, I would have to say that they are very clever people, they work hard, and they combine a high seriousness with a certain light-hearted attitude toward their expertise. The word 'rational' would never cross my mind in describing their work or their character, unless in a sentence like this: (Now I'm going to choose a sentence which I might utter with respect to someone whom I don't know at all well): "I thought Sam Goudsmit was irrational to stay in Reno, but now that I've been here and know the man, I can understand the charms of the place for him." *That's* where the word 'rational' occurs in English, as Gilbert Ryle says in his lecture, 'A Rational Animal'.[1] Ionesco, or some other playwright of the absurd, might put into the mouth of one of his characters the sentence, "How terribly rational of Goudsmit to invent the electron spin!" But Ionesco would be teasing us, and I think that philosophers from Aristotle on have been teasing us with talk of the rational or rationality.

Nevertheless, in my opinion there have been two important contributions to the study of rationality in this century, both negative. One is by Jerzy Neyman, the other by Imre Lakatos. Both are parts of the final step of taking Hume seriously. Neyman wrote about low-level science. Ronald Fisher, using the fact that petty officer is a non-commissioned rank in the Royal Navy, wrote sneeringly in an early edition of one of his books (he changed it later) about decisions made by the petty officers at His Majesty's Dockyards. He meant that Neyman's theory applied to small decisions. Neyman taught that in small matters there is no inductive inference, only inductive behavior — behavior which has good long-run operating characteristics. When we have sufficiently firm and detailed beliefs and can at least devise a model of the world in which one can assess the relative frequency of making a right decision, then one can use Neyman's or, indeed, Pearson's techniques. If I believe that my horoscope in the *San Francisco Chronicle* gives good advice 95% of the time, I'll follow it today. Well, here it is, as a matter of fact. For today, the advice under *Aquarius* is, "Avoid betting!"

So if you have firm beliefs about the future, then you can have detailed beliefs about the future, and you can engage in inductive behavior such as avoiding betting. That is what Neyman teaches, and of course I endorse it. There is no such thing as inductive inference, nor a logic of statistical inference or whatever. There is simply believing that such and such is often a rewarding course of action, and then acting in the hope that the present action will be one of the rewarding ones. What about the grounds for belief in the general propositions themselves, which would provide the grounds for all the particular ones? My view is that there *are* no grounds. Neyman provided for us the theory of petty officers, who get their command from above.

Lakatos gave us the view from the bridge. There is no way of assessing theories now. Rationality for theories is retroactive. One can say only, "That turned out to be a marvelous theory. It created our canons of truth, made our picture of the world; it determined how we have come to be." But there is no forward-looking theory of theory appraisal, except the statistical one, illustrated by: "People who are clever at eighteen are often clever at twenty-five." and "People who were wise at thirty are often wiser at fifty." One wants the latter to assess the former, and that is *all* that a theory of rationality, problem solving, *etc.*, could possibly amount to.

MAURICE FINOCCHIARO: I want to begin with some remarks about an issue raised this morning in the discussion of Larry Laudan's paper. He argued that, historically, logics of discovery have been conceived as providing logics of justification and concluded his paper by saying:

Some of those recent writers who would revive an interest in the logic of discovery see it as something very different from the logic of justification. In this sense, they are radically at odds with the traditional aims of the logic of discovery. The older program for a logic of discovery at least had a clear philosophical rationale: it was addressed to the unquestionably important philosophical problem of providing an epistemic warrant for accepting scientific theories. The newer program for the logic of discovery, by contrast, has yet to make clear what philosophical problems about science it is addressing. Unless we are prepared to countenance the view that there can be a philosophy and a logic of virtually anything, then we should be wary about efforts to revive the philosophy and logic of discovery. Unlike the traditional logics of discovery, latter-day approaches seem to be a program in search of a motivating problem.

During the discussion following Laudan's paper, Thomas Nickles tried to suggest that the problem that recent writers on the logic of discovery are trying to articulate could be taken to be the problem of the historical explanation or understanding of discoveries, that is, of discoveries that have actually taken place in the historical development of science. That is really the direction to move, in my view, as I will explain. I will try to accept Laudan's challenge and answer by saying that there *is* a motivating problem and that the problem has many philosophical aspects. The problem is one that at least some of the recent writers on discovery are concerned with.

The general problem is that of trying to articulate the logic and methodology of the rational explanation of historical discoveries, as Nickles called it, or, as I might call it, 'the logic and methodology of the historical explanation *and understanding* of scientific discoveries'.

One question we might ask in this connection is, What things should be taken as *explananda*? What things should we try to explain? One possibility would be discoveries themselves. Another possibility is to just try to explain *beliefs*, which is the suggestion made by Laudan himself in his book, *Progress and Its Problems*. Another possibility is to take reasoning as our *explanandum*. Different things will happen to our explanations, depending on what we take as the things to be explained. Even if we do take discovery as the *explanandum*, there is the further question of just what about discovery we want to explain. Should we try to explain the *occurrence* of the discovery? If we do, then we probably take a step in the direction of the covering law theory of explanation. We could take our *explanandum* to be the previous *non*occurrence of a discovery; that is, we might try to explain why it did not occur before. I have made such a suggestion in my book, *History of Science* *as Explanation*[2], by adopting certain ideas of Agassi. Then again, if we take discoveries as our *explananda*, we could try to explain not their occurrence or previous nonoccurrence but their intellectual content. If we take reasoning as our *explanandum* then we are mainly concerned with reconstruction of the arguments.

Other questions arise over what our *explanans* should be like. Now the phrase 'rational explanation' is often used, but it needs clarification. When most people, including probably Nickles and Laudan, speak of a rational explanation, I think they mean the explanation of a scientific belief in terms of a theory of rationality. They do not mean what we should perhaps call a 'reason-explanation', an explanation of an agent's belief in terms of the reasons which led him to the belief. To explain a given belief in terms of a theory of rationality in the way that Laudan suggests is really to give a methodological justification of that belief. In his book, Laudan uses the example of why Newton rejected Descartes's vortex theory of planetary motion. Laudan's answer, which he uses to illustrate what a rational explanation would be like, is that Newton correctly judged that the Cartesian vortex theory was grossly incompatible with the facts. I think that should properly be called either a 'methodological explanation' or a 'method-explanation'. It's really the explanation of a belief in terms of a methodological principle rather than the explanation of the acquisition of a belief in terms of reasons, *i.e.*, a belief involving a proposition in terms of which Newton might have *arrived* at that later belief. All I'm doing here is distinguishing what I would call method-explanations from reason-explanations. Reason-explanations are simply explanations of beliefs in terms of the reasons which *led* the agents to those beliefs. Most so-called rational explanations are really method-explanations.

Now having distinguished these two, and having focused on reason-explanations, we notice that not all reason-explanations are rational, because someone may very well acquire a belief on the basis of an invalid reason. Here we would have an 'irrational' reason-explanation, if you want to call it that. It is part of the philosophical task to clarify these concepts.

Then there are questions about how reason-explanations relate to method-explanations. Some people would suggest that a method-explanation is just a more general description of a reason-explanation. We also want to distinguish both reason-explanations and method-explanations from psychological explanations. To use an example of Robert Westman's, we might say that Rheticus, who was a follower of Copernicus in the sixteenth century, accepted Copernicanism because he regarded Copernicus as a father figure. That is an example of a psychological explanation. It is obviously not a reason-explanation, nor is it a method-explanation, but at the same time there seems nothing 'irrational' about it. Indeed, it might not be improper to call it a 'rational' explanation; so here we have a rational psychological explanation.

Since I have no more time, let me reassert that I think there *is* a problem which represents one direction to move in connection with the logic of discovery, and the problem is that of articulating, clarifying, and formulating the logic and the methodology of the explanation of discoveries that have actually taken place.

ERNAN McMULLIN: The controversy over what to do with discovery in science originated in the sharp distinction drawn by most philosophers of science of the positivist period between appraisal (the phase of science which lent itself to the logical analysis taken to be the goal of philosophy) and discovery (the phase of science relegated to the psychologist and the historian). The implication quite often seemed to be that discovery lay outside the boundaries of the rational. This implication has troubled philosophers more and more in recent years, for a variety of reasons. Historians of science, for example, seem to be able to make quite good sense of the discoveries they write about. How do they do this, if indeed discovery is as dark a topic as it appeared to the theorists of falsification and justification? I take this to be the main topic of our symposium.

One preliminary point may be worth making. The term 'discovery', as Gary Gutting has reminded us at this conference, conveys the suggestion that a positive appraisal has been made. One does not ordinarily speak of a new theory as a 'discovery' until it has proved itself in some way. A conjecture, no matter how ingenious, does not qualify as a 'discovery' as that term is most often used by scientists. As Professor Goudsmit put it, we give credit not to the first person who hits on an idea but to the first who can make it stick, that is, show it to be plausible. Clearly, such showing involves criteria of appraisal. So that if discovery be taken in *this* sense, it is not distinct from appraisal, and it is a topic for the philosopher of science who is concerned with justification, theory-acceptance, and the rest. In this case, the controversy collapses; there would be no disagreement about the appropriateness of a 'logic of scientific discovery'.

But obviously, this is not where the disagreement lies. When Popper and Reichenbach set 'discovery' outside the purview of the philosopher of science, what they meant by 'discovery' was the initial creative formulation, the shaping of a conjecture, the apt conceptual modification, prior to the question of explicit assessment. Peirce had earlier called this 'abduction', contrasting it with both deduction and induction. Popper was, of course, right in holding that the formulation of plausible conjecture is not a matter of logic (as he understood the term, 'logic'). What Hanson, Blackwell, and other writers argued in response to this was that discovery in this sense nonetheless displays characteristic patterns, that it works within constraints (though admittedly nót logical constraints). Not only does it belong to the realm of rationality; in some sense it can be said to be central to it. For the creative ability to shape a 'likely story' makes far more of a demand on rational powers than does the mere ability to follow the rule-bound steps of deduction.

My concern in this brief comment is with the contribution of history of science to this debate. Let me begin by noting two significant redundancies in the title of of our symposium. Making a discovery is a human act bounded by conditions of space, time, and cause as other human acts are. It is thus necessarily historical. To add the term, 'historical' to 'discovery' is presumably meant to underline that we are investigating not a timeless structure like deduction but rather an event subject to the historian's skills, one about which the first and most appropriate questions to be asked are historian's questions.

The term, 'rational', also appears redundant. The historian necessarily seeks for a 'rational' (that is, a reasonable) explanation. No other sort would be acceptable. There is no need, it would seem, to speak specifically of 'rational' explanation. 'Rational' is, however, a protean term. In this context, it can have two rather different meanings. A 'rational explanation' may be just a 'reasonable' one, one in accordance with the norms of good history-writing. Or it might mean: an explanation in terms of reasons, or more broadly an explanation which would bring out the rationality of the event being explained. It is this latter sense which is presumably intended here. But first, a word about the distinction itself.

Historical explanation is not limited to the giving of reasons and the detailing of motives. It can also invoke circumstances that affect the way in which reasons come into play, as well as causal relationships of a nonintentional sort, ones which are not mediated cognitively in any way. Thus in 'explaining' the discovery of the double-spiral structure of DNA, the historian might cite the combination of disciplines represented in the Cavendish research group by contrast with the groups in Pasadena and at King's College, the presence in Cambridge at just the opportune moment of Pauling's son, the tensions between Wilkins and Franklin at King's, the ambition to 'get there first' which drove Watson and Crick . . . . . In principle, an historian could even construct a 'rational' explanation of a totally 'irrational' act by citing physiological or psychological causes. The (historical) rationality of his explanation must not, therefore, be taken as a sufficient testimony of the (intentional) rationality of the actions he is explaining.

One cannot, then, cite the fact that historians of science can produce rationally adequate explanations of scientific discoveries to prove that these discoveries are themselves rational. Something more is needed. This is where the other sense of 'rational explanation' comes in. Historians of science not only can construct rationally adequate explanations of specific discoveries, they can construct these explanations in terms of broadly cognitive factors. Why should it have occurred to Copernicus to take the sun as center of the planetary system? Well, there was his dissatisfaction with equants, and his consequent attempt to replace them with 'proper' circular motion. There was his perception of the relationship between the stations and retrogressions of the planets and the position of the mean sun, as well as of the coincidence of the one-year period in one of the epicyclic motions of each of the planets, and, of course, there was his Pythagorean belief in the cosmic primacy of the sun as source of life. All of these together do not give anything like a deductive account of how he reached his conclusion (his 'conclusion', after all remained an hypothesis). Nor do they serve necessarily as warrant in our eyes for the conclusion. They are intended to help us understand only how *Copernicus* arrived at his results, what factors influenced him, what might have suggested what.

A discovery, let it be said again, is a unique and complex human event, not just an abstract statement. Though we may speak of quantum theory or penicillin as 'discoveries', this is a derivative usage. Reasons operate in discovery, then, not as abstract premises but as causes. It was not the inadequacy of equant models that mattered, it was Copernicus's perception of that inadequacy. The cognitive factors that historians are wont to cite are quite various: not only the perception of explicit reasons for and against, but the noticing of analogies, the skilled recognition of patterns, the adopting of particular methodological principles, the making of unnoticed assumptions. . . . . To make a discovery intelligible after the fact is not to justify it or estimate its plausibility, but to show why it happened the way it did, to retrace the line of inquiry and see why the agents involved made the choices they did and how they came to formulate their ideas as they did. There has been an intense debate in recent years, for example, about how Galileo discovered his law of falling bodies. Was it a matter of inductive inference, as Drake steadfastly maintains? Or did Galileo come to a gradual realization that the time-honored Merton odd-number rule could provide him with the first principle he needed? We may never know for certain just how it came about. Indeed, Galileo himself, had we been able to ask him, might not have been able to reconstruct the exact steps of thought he had followed. But the important thing is that the historian does have resources to meet the challenge of explaining, in a tentative way, the course followed by even the most creative of inquiries.

This leads me finally to formulate the two theses I would like to propose for discussion. The first is that it *is* possible to situate scientific discoveries within a particular realm of rationality, namely that of the historical reconstruction. The reconstruction is of a special kind because it involves reasons, suggestions, criteria; though noncognitive causes may be invoked by the historian, it is ordinarily possible to construct a fairly satisfactory explanatory account on the basis of cognitive factors only. And these can be connected in a network which may make it possible to prescind from the historical singularities of the case and focus on the idea-relations only. History of ideas often makes people uncomfortable, with its apparent reduction of the creative to the expected. But (*pace* Michel Foucault), ideas *do* have histories, and the understanding arrived at by tracing, for instance, the lines from Neoplatonic active principle to Newtonian gravitation, is a genuine one.

My second thesis is that the 'rationality' thus elicited is very different from the 'logicality' demanded of the good theory, for example. The order of discovery is not that of rule-guided inference but rather that of suggestion, analogy, tentative trials. Sometimes discoveries *do* follow the neat paths of rule, and the business proceeds "as though by machinery", in Bacon's evocative phrase. But far more often, the historian has the more difficult task of ferreting out what might have suggested a particular explanatory model, or what might have led to a reformulation of the problem in slightly different terms and the like. The connections are loose and inevitably speculative on the historian's part.

Inductive inference offers no particular challenge to the historian. He does not usually feel called on to 'explain' how the experimentalist goes from the set of data to the generalization. The assumptions about curve-fitting and the like are relatively simple and widely shared. The historian can assume that inductive discovery needs no special attention from him unless it exhibits some unusual features. It is enough to chronicle how the observations were made; the 'leap' from that to the experimental law is very likely one the reader can then make for himself.

But theoretical discovery (abduction or retroduction in Peirce's terms) is quite another matter. Here it is not just a matter of generalizing from some to all of the same kind. There is a conceptual innovation, and we have to be helped to understand how the scientist might have made it. The historian can lead us to a knowledge of what the scientist's resources were, of what he had to draw on in making the conceptual leap he did. Koestler defined the act of creation as: "the juxtaposition of hitherto unrelated matrices of thought".[3] There is a fitting together in a new way of elements already in some sense available. This is, perhaps, an overly limiting conception. But it does bring out one essential aspect of creativity: that it requires a well-stocked mind, that it realigns and reshapes structures already to some extent present. This is where the historian can help.

The 'rationality' here is of the shaping, groping reason. It is the rationality of the metaphor, not of the theorem. It is important to insist that this *is* rationality, and not just sleep-walking, to allude once more to a Koestler theme.[4] To the extent that the historian can reconstruct the 'meaning' of a sleep-walking sequence, it is not as a rational episode. But when an imaginative and competent historian reconstructs the discovery-sequence of a Kepler or a Copernicus or a Galileo, it does not read in the least as sleep-walking, Koestler to the contrary notwithstanding. This is *not* to say, let me once again emphasize, that the story should unroll in predictable fashion. What the historian aims to do is only to make sense *post factum* of what has happened, not to write the scenario in advance. There may be leaps of guesswork and intuition that can only be motivated, not 'explained' in any stronger sense. But what is crucial is that it *does* in the end 'make sense'. The rationality of creative hypothesis-construction is far broader, evidently, than is the logicality of prediction, generalization, justification, falsification. Why call it 'rationality'? Because it epitomizes the approach of reason to the problem of understanding what always lies a little beyond the safe but limited reach of logic.

To what extent does the discovery-process exhibit patterns that can be catalogued by the alert philosopher? The question recalls the Dray-Hempel debate of twenty years ago, about the degree to which the historian should have available to him a set of law-like generalizations to help him explain events of the past. Suppose the event to be explained is Lavoisier's discovery of oxygen or Bohr's creation of the planetary model of the hydrogen atom. Are there 'laws' or patterns (sociological? logical? psychological?) *that* are being exemplified? Does adequate explanation depend on the availability-in-principle to the historian of such laws? Or is one to say that in the case of creative inquiry, the 'normal' law-likeness of the past breaks down? Though it would seem that history of science ought to lean to Hempel rather than to Dray, it does not in fact do so, for the reasons we have just seen. Though there may be some high-level generalizations about the construction of metaphor or the pursuit of problem-solutions that are applicable to such episodes as the ones we have been scrutinizing, it is fairly obviously not the case that there will be a set of generalizations of which Bohr's discovery is a simple instantiation. The historian's explanation of an episode such as this cannot, even in principle, be deductive or statistical. But it is nonetheless a genuine explanation for all that, one that illuminates by showing how *this* might have suggested *that* or why *this* might have seemed a plausible next move to *that* person. Creative human activity is by definition not rule-guided. And that is the simple reason why we cannot confine the rational to the logical.

ROBERT WESTMAN: Thanks to affirmative action, there is a token 'straight' historian on the panel, who has received in the last forty minutes a great deal of advice. I suppose my situation is in some ways paradigmatic of the situation in which historians in general could find themselves, if they knew what was being said about them. The first piece of advice that I received today was from Ian Hacking's horoscope, which says, 'Be clear and direct in explaining your views. Maintain moderate pace." I don't know whether I can follow all of this advice, but I would like to comment on the advice that has been given to me as some sort of 'representative' of the history of science.

I sit here in a state of amazement at some of the things that have been said, perhaps because I hold a terribly naive view of what many philosophers seem to be saying to historians. Many philosophers seem to be saying that the kind of history of science you should be doing is the kind which ought to demonstrate the rationality of science. And then I have to worry: "Now what in God's name is rationality?" But when I start reading the philosophical literature, I discover that there is much disagreement. There is a good reason why one discovers such diversity, and that is because philosophy of science, not unlike other disciplines, is not monolithic. It is filled with controversy and with divergent opinions. This is also true of the history of science, and it is true of the sociology of science. So if one is crazy enough to want to work at the interface of these three disciplines, as I think I would like to do, I am faced with the problem of making, if I may use the word, *rational* choices among the pool of possible kinds of explanation which have been generated in these three disciplines. If a philosopher seeks to read something in the history of science, he is faced by the same kind of diversity. Last night Professor Goudsmit spoke of a certain kind of historiography of science which was available in the 1920's. Since then considerable evolution has occurred, and our discipline is now quite heterogeneous.

Now I do not know whether Mary Hesse is taking a new position, but at least I do not recall reading anything you have written, Mary, where you openly advocate the legitimacy of seeking sociological and psychological explanations in historical work. But if that is what you are saying now, I must express profound joy!

I do think there is one lesson that I have learned from the philosophical literature, and that is to look for explanations and not to be content with mere exegesis, to articulate merely what $X$, $Y$, and $Z$ actors were saying.

Ian sweeps away an accumulation of cobwebs by saying that, after all, strictly speaking, Hume was right — there is no inductive inference. There is a rationality of small decisions and that is fairly straightforward, although it is so simple that it is hard to see at times. And Lakatos's contribution — that really one can only assess the rationality of some scientific achievements in retrospect — seems to agree with what we see amongst the other leading methodologists of science. After all, to begin with, as Lakatos himself says, a research program which looks degenerate could become regenerate sometime later; so one needs to wait a *long* time before one can say with certainty that one research program has actually superseded another. Until Larry Laudan clears up some of the fine distinctions needed between the rationality of pursuit versus the rationality of adoption, he too will be left with Lakatosian retrospective rationality.

When we get to Maurice Finocchiaro's contribution, we appear to be faced with a welter of explanations. He paraphrases one which I offered some years ago,[5] and I have to say that my own reference group — historians of science who have read my articles — have tended to *ignore* what Finocchiaro and I consider to be the interesting explanations. Why, for instance, was Rheticus the only Copernican for thirty years after the publication of *De Revolutionibus*? I think that is an important problem. I shall say a little more later on about how I went about trying to explain Rheticus and why I think I had good reasons for following the route that I chose.

Finally, I hope that I can avoid characterizing everyone's position too simplistically, but when Ernan McMullin says that you philosophers should respect the diversity of history of science, I cannot help but agree with him. The general tone of the whole panel is very encouraging.

Let me now say something about some of the principles which guide me when I look for explanations as an historian. One of the precepts that guides me is my conception of man. It seems to me that this is implicit in everyone's search for explanations, whether they are philosophers or historians. To begin with, it seems to me that, roughly speaking, one can speak of two different kinds of conception of man – *plastic man*, whose character is ever molded by causes from his social environment, from his background (he never has a thought in his life, he is constantly conditioned; he is overdetermined); and *autonomous man*, who is never influenced by a cause and who is a totally rational agent. Now I propose these as ideal types, because it seems to me that these are the extreme positions.

I take it as my working problem right now to seek out some pool of explanations which will allow me to explain why actors behave as they do in terms of psychological and social causes, at the same time allowing for the fact that actors can think and can think rationally. Now when I get to the last part – thinking rationally – I start to get a little worried, because, if I inspect all the models of rationality on the scene, I have a problem of how to make a rational choice among them. So perhaps the thing I ought to do, in fact one of the things that I have done, is to work with several of them. This is something that most of my colleagues in the historical discipline do not do, because they do not see the relevance of such normative models to their own work. I think that this is because there is some misperception of what philosophers do. I probably share in some of those misperceptions. There is a communication problem, which is not unusual between disciplines. A short run strategy for historians to pursue is to familiarize themselves with various accounts of rationality which have been proposed by philosophers and to try them out. That might be one sensible way of proceeding, just seeing what kind of work we can get the model to do for us. It seems to me that, as historians of science, we ultimately do presuppose *some* account of science, and we do presuppose some account of rationality. So at the very least, we ought to be aware that we have a problem, even if we do not know how to solve it.

[*At this point the session was opened to audience participation.*]

MICHAEL GARDNER: I have a question for Dr. Westman. A friend of mine who is an intellectual psychohistorian faced the problem of trying to decide, as an historian, which psychological theory she should use as the basis for her psychohistory. She is not a professional psychologist. I wonder how you decided to use what I take to be a Freudian framework? In the discussion between you and Laudan yesterday in San Francisco, at the Philosophy of Science Association symposium on Laudan's book[6], he claimed that

Freudianism wasn't sufficiently progressive, and you seemed to be disputing that. I wonder what evidence you really have that it is a good framework for explanation? Certainly the studies of the effectiveness of its therapeutic techniques are not very encouraging, to say the least. That is one piece of evidence.

WESTMAN: I would need all the rest of the time to answer that. First of all, there is no simple answer to your friend about how she should go about making a good choice among psychological explanations, assuming that she thinks that to be a reasonable thing to do. Evidently, there were reasons she had in the first place for wishing to seek that kind of explanation. A few years ago in Pittsburgh, when I presented an earlier draft of my paper in which I advanced a Freudian-type explanation of how Rheticus became a Copernican, somebody said to me (I think it was Adolf Grünbaum): "Don't you need a complete justification of your explanatory account before you start doing the history?" And I replied that if I wanted to be a Kant then I would have to budget at least twenty more years to writing *The Metaphysical Foundations of Psychoanalysis* before I could write history using these as explanations. My own strategy is that you really need to make a commitment to some form of explanation and see what kind of work it will do for you. In my own case the reason for my commitment was that, at the time I wrote the article, I was undergoing psychoanalysis. I have since completed it, and it did me a lot of good. There is no substitute, I think, for first-hand experience, regardless of attempts to do comparative statistical studies appraising 'outcomes' of psychotherapy versus spontaneous remission. I think I know a lot about how analysis works, and I am not sure that what philosophers write about it is always accurate. Grünbaum has a book coming out, which promises to open up a fruitful debate about how 'scientific' psychoanalysis is. His recent article contains a foretaste of themes to be treated more thoroughly in his book.[7] I think the statement that Laudan made in his book and the statement that you made just now are really arbitrary, about how progressive psychoanalysis is as a discipline. There are questions about just what those tests are testing, whether they are well run tests, and what counts as a successful outcome. I think a rational strategy at this time for an historian who wants a psychological explanation, is to try one and see what it will do for him and *then to be open to criticism*. My account is open to criticism, but I have received little as yet.

FINOCCHIARO: In connection with this question, I think it is perhaps misleading to ask by what criteria you choose a general psychological theory to explain historical events. What I am thinking of here is that, as Westman suggested, you can try various theories and then if one of them works, your explanation can be regarded as support or further confirmation of that particular psychological theory.

KOERTGE: I think there are some real dangers in that kind of explanatory eclecticism. First of all, I would propose the following: an explanation is only as good as the theory which it presupposes. That doesn't mean that you have to justify a theory before using it, but it does mean that Westman's psycho-analytic explanation of Rheticus's behavior stands or falls with psychoanalytic theory. This, of course, is why we now reject Herodotus's explanations of historical events in terms of the intervention of the gods!

I also want to introduce a note of caution about the idea of using its success in history as a bit of evidence for psychoanalytic theory. If you are proceeding as an across-the-board psychoanalytic historian, then your success in coming up with psychoanalytic explanations might be impressive.

But if you are being terribly eclectic — if you introduce a certain explanatory theory only when it appears to fit and cheerfully discard it when it doesn't work — such selective successes would *not* count as evidence for psychoanalytic theory any more than does success in certain isolated runs at Pratt and Rhine's laboratory count as evidence for parapsychology. We have to be very careful here, because not all instantiations of a theory confirm it.

WESTMAN: Right. I don't want to get into a debate with you about what you take to be a good confirmation, but I do want to say that if I had not done that, if I had not at least tried to look for a candidate explanation of the type that I was looking for, I think that I would know less than I know now. I actually think that I explained something. I do think that one place the account can be criticized is from the point of view of the evidence. Nobody has tried to find counterevidence to the claim.

HOWARD GRUBER: I am a psychologist, and I would like to turn the problem around the other way. When I began to read the history of science, it was from the point of view of cognitive psychology. I never really concerned myself at all with the underlying motives for someone's work that might be explicable by Freudian theory or some other psychoanalytic theory. I asked a very different sort of question: How does a scientist actually go about doing his work, how does a scientist think? My reason for getting into the literature of the history of science was that psychologists were addressing the problems of the psychology of thinking with a terribly limited data base — basically what happens in the experimental laboratory in experiments that take about twenty minutes and involve problems that an average college sophomore will be able to solve. Even if traditional psychologists were finding the answers they were looking for, we are interested in another range of thinking — hard problems which only a few people in a century can solve. There is no *apriori* reason for thinking that the same processes are at work on this other scale.

In the twenty years or so that I have been working, what I have been doing is constructing a cognitive psychology suitable for the purposes I mentioned. And meanwhile, of course, cognitive psychology itself has been evolving, and I believe that the history of science has evolved. There has been a convergence, and there is now something like the makings of a cognitive psychology adequate to deal with the questions we've been talking about. I wouldn't want to push that too far, but there is quite a difference between the way things are now and the way they were twenty years ago. I would suggest to your friend that she steep herself in what is now happening in all of these disciplines, choose some work to do and try to do it. I don't know whether this is a counterproposal, because I am talking about a different task. Maybe if the task were that of explaining motives — although I doubt it — there is one theory well enough developed that you could apply it to the case.

Finally, I want to ask Westman a question. Do you consider in your work that you are testing psychoanalytic theory by what you do?

ROBERT WESTMAN: I may have given the impression that a lot of my work invokes explanation of a psychohistorical type. The fact of the matter is that I have only used it in a very small segment of my work in which the task was to explain the theory choice of a particular individual. I felt that I had exhausted the more traditional kinds of explanation — for example, the kind that E. A. Burtt would put forth: Rheticus was a Neoplatonist or Kepler was a sun worshipper.[8] These types of explanations do not do the work. They do not because there were many Neoplatonists and Neopythagoreans at the same time, and they were not Copernicans. Being a Neoplatonist was clearly insufficient. We do know one fact, and that is that Rheticus had a very close personal relationship with Copernicus. I tried to dig out some facts about it and construct an hypothesis about it. That was the problem that I was trying to solve. I proposed a candidate explanation for it. I do not think that that is incompatible with trying to solve other problems with explanations from cognitive psychology, which is clearly generating a lot of rich theoretical proposals. But I do not think that I want to go so far as to say that it would count as a confirmation of psychohistorical explanations *in general*, because there are many kinds of them. The type of explanations I am seeking are not sought by many people. There is a growing discipline of psychohistory and a journal for it now, *The Journal of Psychohistory*. But as applied to scientists, there are only myself and Frank Manuel. I think Frank Manuel did not get as much mileage out of his explanations as he might have. So I do not think our accounts are rival accounts.

FINOCCHIARO: I'd like to ask Professor Gruber a question. He has been working on the psychology of reasoning, as one might call it, and he is inclined to say that the psychology of reasoning must adopt an historical approach as well as a merely experimental approach. Is that a fair way of formulating your point?

GRUBER: Significant intellectual events take place over a pretty long time span. It takes a few years to do any interesting work, and you must study the way the work evolves over that period of time. Since it is in fact evolving in a sociohistorical context as well as a changing personal context, there is no way of really seeing what the person is doing otherwise. I don't use the word 'explain' here. Even a good, sophisticated description requires a knowledge of the history of what you are looking at.

EUGENE LASHCHYK: I think there are limitations to attempts like Professor Westman's to explain discoveries using a psychoanalytic approach. If you attempt to explain why so-and-so *believed* such-and-such a theory, at the time in which he lived, then psychoanalytic theory might be helpful in explaining why he believed it. But I do not think that you could explain psychoanalytically why Einstein came up with the special theory of relativity. In other words, I do not think that you can explain conceptual innovations using psychoanalytic theories.

WESTMAN: I take it that you are acknowledging that I can get some mileage out of a psychohistorical explanation of choice or decision at a certain time by an actor. But you are wondering whether I can give a good psychohistorical account of discovery. Well, first of all, I have not tried to do it. Secondly, I do not know of anyone else who has tried to do it with respect to the history of science. I know of some bad attempts in some other areas of history, and I would not recommend them. But I would not dissuade anyone, on the basis of that, from pursuing the possibility of developing a good account. I offer, by analogy, the fact that Joseph Ben-David wrote a book on the scientist's role in society — a very provocative book, a very dense book, hard to read.[9] But it pushes what I think is a very simplistic line about the scientist's role, and it has a very unsophisticated, 'whig' view of history. But I think one can give a better account of what a 'role' is, and I think one can thereby vindicate a certain kind of sociology of knowledge as a result. So I give the same answer to what I think was really a challenge, for somebody to try to do a psycho-historical account of scientific discovery. Try to do it; try to do it well.

EDWARD MacKINNON: One question that was raised concerning the title of this meeting — 'The Rational Explanation of Historical Discoveries' — seems to me a bit misleading. I think that if this issue had been raised, say, ten years ago, the question might have been: What have philosophers of science to offer to historians of science? If we put the question in that context, I think it looks a bit different from what is now intended. As Professor Westman pointed out, one could take a scientific discovery and give a historical account, a reconstruction of what happened, but one wants to do something more than that. What is the something more, in terms of explanation?

In the past people have tried to have some standard forms of rationality which they imposed — the deductive model, statistical models, *etc.* Attempts to impose these straightjackets tended to be counterproductive, and historians reacted against them. Now what is happening on this panel and in this whole conference is that people are looking at rationality in science in the way Ernan McMullin and Larry Laudan have suggested: at the outset we take scientific breakthroughs as key examples of human rationality in its developed form, and proceed to ask whether we can uncover certain patterns of ration-ality, different types of patterns which we can relate to different sorts of problems and then relate those in turn to some of the norms of rationality that we have — so that ultimately something better might emerge as a theory of rationality. The something better that might emerge, an overall theory of rationality and specifications of the type of reasoning geared to certain types of problems, could be of some help to historians of science in providing them what Westman called a pool of available explanations and norms.

McMULLIN: I'd like to second that. That's exactly the sort of thing I had in mind. Let me just amplify it briefly in a couple of ways. First of all, let me say once again that when we look at scientific discoveries as historical, we are looking at them as human actions that have to be explained as other human actions have to be explained. That is such a simple point, and yet it is a point that sometimes was overlooked in the past. There are many kinds of human actions, and here is a rather intricate kind of one, one that is conditioned by a complicated community, and mediated in all sorts of ways. Nonetheless, when we are trying to understand how Newton hit on gravitation or how he made some of the discoveries he did in mathematics, for example, we are looking at an individual human action within a highly individual context, and therefore, we are guided in our explanation by the same kind of norms that the historian is guided by. We had better be careful about that. That is one thing that philosophers surely have learned in the last twenty years − at least, I *hope* they have!

Though there are presumably many characteristic forms of rationality, this ought not be taken to support an anarchistic sort of pluralism, as though the diversity were so great that a reflective historian or a reflective philosopher could say *nothing* of an analytic sort about discovery. That's the direction in which the Dray discussion was sometimes carried against Hempel.

Here I would be inclined to disagree very strongly with what Ian had to say. Much of experimental science proceeds by curve-fitting, by extrapolating from a few data-points to a smooth curve. That is the paradigm of inductive inference, and it is surely the sort of rationality that one can most easily discern at the discovery level in the history or present practice of science. I must confess myself to be at a loss, then, when Ian tells us that there is no such thing as inductive inference, that there is only inductive behavior. The *warrant* for an experimental law (of which there are myriad examples in such sciences as chemistry) is in the first instance the set of observations which led to its formulation. The inference from these observations to a law-like statement, often presented in the form of an equation, is an inductive inference. It takes for granted the language in which the observations are expressed. It does not conclude with certainty. Nonetheless, it is an *inference*: that is, the experimental evidence is accepted as warrant for the tentative law-statement.

The theoretician who wants to explain why hydrogen and oxygen combine in the proportions of 2:1 to produce water is not interested in the inductive behavior of the experimentalist who works with these gases. He starts from a proposition expressing a correlation in the observed combining properties of gases, not a proposition about the inductive behavior of scientists. The proposition is provisional in several ways, but nonetheless expresses genuine knowledge about the world, the kind of knowledge from which science takes its origin.

I would further disagree with Ian in his summary of Lakatos's view of theory appraisal. He has Lakatos say that the rationality of theories is retro-active and thus there is no way of assessing theories *now*. But this is surely not correct. What Lakatos says is that one has to evaluate the past performance of a research program in order to determine its (present) worth. He doesn't say: "That turned out to be a marvellous theory," as Ian suggests; rather, he expresses it in the present tense: that *turns* out to be (on the basis of its past performance) a marvellous theory. Such an assessment is forward-looking in the sense that it gives us grounds to expect that the theory is a 'good' one, that is, that it will *continue* to guide us well. So that when Ian says that there is "no forward-looking theory of theory appraisal," he is mistaken. He may not like such theories. But *all* theories of theory-appraisal (Lakatos's included) are forward-looking. That is what appraisal *means* in this context. Once again, the criteria are not those of deductive logic but that is what makes them interesting. Nor are they as weak as Ian's "People who are clever at eighteen are often clever at twenty-five." The difference between *theory* appraisal and straight inductive inference from an empirical correlation like a chemical law is that success of theories derives from the postulated entities to which they give unique access and which are their main claim to *knowledge*. A theory of scientific rationality must give pride of place to this ability of theory to reveal in a provisional but nonetheless real way the basic structures of the world around us.

HESSE: I would like to comment, because now I am really very puzzled. I feel very much in the same position as philosopher as Bob does as an historian. A lot of things have been turned around about models of rationality. My difficulty as a philosopher has recently been in trying to make sense of models of rationality in the social sciences. These do not seem to be covered by our traditional, Hempelian-Popperian models in the natural sciences and yet my difficulty is to find *even one* model which is distinctively different from what has been explicated by the Carnap-Hempel-Popper and post-Popper discussions. It seems to me that we suffer from a *paucity* of models of ration-ality, and if the historians of science have come up with models which are other than ultimately the test procedure of the method of empiricism, I would love to hear about them. I do not know what they are, and yet they keep being presupposed in this discussion!

WESTMAN: Mary Hesse's problem regarding the alleged paucity of models of rationality raises for historians the question of just why they need be concerned at all with philosophers' formulations of rationality. To begin with a sociological observation: it is conventional for historians to dissociate themselves vigorously (at times) from philosophical model building. No doubt this is to be explained by the fact that historians are concerned, by and large, with *context-bound* rationalities. The sources of evidence, the criteria available to historical actors in a given context, are what set the framework for the decisions that they will make. Finding out what the actors themselves took to be evidence, finding out what they believed and what traditions they were plugged into, reconstructing the temporal sequence of moves made in a scientific discovery — these are the activities which commonly occupy many historians. A good historian is happy if he can *recover* the context with accuracy; often there are debates within the discipline as to whether the right contextual curtains have been hung for the actor's performance.

Need for great *trans*contextual appraisals occurs in historical writing with surprising infrequency. In the aftermath of Kuhn and the many *philosophical* discussions of incommensurability, this is worth emphasizing. Historians are at their most normative when they write *textbooks*, that is, when long time spans need to be covered, epochs characterized, and scientific achievements judged. It is in the extended historical account, frequently written for pedagogical reasons, that many historians reveal their true colors, their normative imaging of science. Talk of 'scientific' *vs.* 'mystical', 'our modern scientific viewpoint' (as though there were unanimity) or the 'puerile' beliefs of Kepler's *Mysterium Cosmographicum* are all instances where unconscious and frequently unarticulated transcontextual appraisals are rendered. One way in which the rationality modeling of philosophers can be of value to historians is in breaking down an implicitly held, *monolithic*, normative image of science. If many historians today are concerned to show the social conditioning of scientific knowledge, the historian himself should not be exempt from that process. *His* image of science has also been socially conditioned through his education, his idealization and his failures in doing science. The historian who is a brilliant reconstructor of context may yet be a latent positivist when engaged in an enterprise demanding transcontextual judgments. Here the very fact that philosophers *disagree* about the nature of rationality and science can function as a healthy antidote to the tendency of an historian to invoke a normative consensus in his own time. Knowledge that philosophers have great difficulty in agreeing on a putatively objective account of *simplicity* – is it low probability, high probability, no probability, or degree of informativeness? [10] – is useful to know when considering an historical actor who makes a simplicity claim. One is already sensitized to the fact that allegations of simplicity may translate in nonobvious or problematic ways. Hence, where the historian can educate the philosopher by sensitizing him to (what we might call) the 'rich, local rationality of the battlefield', the philosopher can teach the historian (by example) that our great generals *still* cannot agree on how to fight the next war.

GARY GUTTING: My question arises first of all out of Father McMullin's comments and also from something that Professor Hesse said. Father McMullin said that we probably should not talk in any strong sense about a *logic* of discovery, not rules which generate the discoveries; rather, we should talk about the *rationality* of discovery and about models of that. Professor Hesse answered by asking what are these models of rationality. The question that occurred to me is, What exactly do we mean by a model of rationality? Let me sharpen the question in this way. If we did mean a logic, then I could understand what we meant. There would be a system specifying what rules to apply and what criteria are involved in deriving conclusions. Now it needn't be an algorithm, but at least there are the rules. That is one conception, but apparently Father McMullin and many other people do not want that. At the other end of the spectrum, you could say that a model of rationality is just a matter of a kind of skill or a knowhow, and some of Father McMullin's comments went in that direction. And there you could talk in a Kuhnian sort of way. There are some exemplars – models in the sense of good instances of scientific work – and you have students look at those and then they pick up the skill, they learn the trick. Now presumably, what we are after is something in between. It is not a logic, and it is not just a know-how. That is what the model of rationality is, but I don't really have a very firm grasp on what that sort of thing would be. The question is not only whether there are any models of rationality exemplified by actual science and how you find them, but what *would* that kind of a thing be?

McMULLIN: Let me take a shot at that, very briefly. First of all, let me reinforce the point that I tried to make earlier. One danger that I think the profession has perhaps got over is that of supposing that to explicate a rationality is the same as to explicate a logic. A second danger that is still with us, maybe, lies in the suggestion that there are certain kinds of skills or acts on the part of scientists that somehow lie outside the bounds of reason or are 'intuitive' in a sense that would suggest that they are irrational. This was suggested in one paper this morning. I think. Suppose that we have a trained scientist who exhibits a certain kind of skill, *e.g.*, in theory assessment, of a sort which he cannot fully explicate. It is a skill that he has learned over a long time and is 'tacit' in Polanyi's sense of that term.[11] I would want to argue that it is preposterous to suppose that that skill is not a rational skill. To my mind it is a paradigm of one form of rationality. But the difficult problem Gary Gutting poses is whether anything more can be made of it. Can it be laid out in any 'patterned' form at all? Koestler tries to do this in *The Act of Creation*, using analogies of all sorts. Another more promising approach, perhaps, is that of artificial intelligence. A further one might be suggested by models from cognitive psychology. One is trying to find a certain sort of pattern in discovery that is different from the pattern of discursive inference, a pattern which recurs in episodes of this sort. It is, first and foremost, an empirical question, one of great concern, however, to the philosopher. But the issue of whether or not discovery should count as 'rational' does not rest upon whether or not a pattern of this sort can be found.

GUTTING: The way you put it, it sounds like a psychological question.

KOERTGE: Let me just ask you a question about how you are using 'logic'. Would you say that reasoning by analogy, as described by Mary Hesse, falls under logic?

McMULLIN: It would fall under logic to the extent that certain kinds of prescriptive rules could be suggested for it, at least probabilistic rules.

HESSE: I am not sure what rationality is unless it is something to do with conformity to some rules or other, not necessarily deductive rules. (I think this is where we might all differ from Ian, because Ian essentially wants to restrict rationality to conformity to deductive and statistical type rules.) Any use of 'rationality' which does not refer to some generally accepted rules seems to me to be probably disreputable. For example, conformity to some local ideology which is accepted and enforced: that has been the definition of rationality in very many societies, and it is so sometimes in ours, if we are honest. Another meaning might be, that which has somehow evolved in the history of the human species as the unthinking responses which are conducive to survival. Conformity to this might be described as rational, but it would not be intentionally rational. Or else rationality might be defined so as to conform to rules about which one might want to ask, are the rules rational? You see, maybe even theologians were rational. This is what bothers me. I still come back to Popper's old problem about demarcation. According to

Popper, scientists have rules, and he gave some grounds for the rules being rational. We should talk about logic more rigorously in developing models of rationality. At present I don't see any alternative logics to the logic of empiricism and the logic of deduction. And therefore I don't see any reputable models of rationality apart from them.

HACKING: I think the word 'rationality' gets applied to elderly aunts and so forth, and I don't think that rationality has anything to do with science at all.

McMULLIN: It's a shame that we are getting to such important points only at the end. Let me respond to Mary's point and perhaps also to Ian's. I would want to argue that science as an organized activity with a certain history and a certain kind of record of success (we have to be careful not to beg questions over success) *defines* a progressive sort of rationality, one which has disclosed itself historically over centuries and is not yet fully formed. Science is one paradigm of rationality. I don't want to argue that it is the only one. But it assuredly constitutes a very large part of what we *mean* by rationality. In this connection, it may be recalled that the rational animal of Aristotle was not a person who could deduce the conclusions of a syllogism but a person who could hit on the principles of a syllogism, which is very different matter. Rationality for him too was defined in terms of the abilities required in order to carry out scientific demonstration successfully.

Two final comments on rationality. Certain kinds of people, trained in a certain kind of community, acquire a certain sort of skill, which is tested by certain sorts of outcome. A person with a skill of theory-assessment can look at the tangled problem of lunar origins, for example, and decide which of the three current theoretical models offers the best chance of being the correct one. That is a 'rational' act on his part, and a great deal has been said by philosophers of science in recent years about the structures of this form of rationality. On the other hand, the rationality of discovery, of the creative act generally, is far harder to analyze, as this conference and the disagreements we have just been having amply illustrate. But the fact that it is hard to analyze does not in the least, it seems to me, diminish its claim to be an operation of the *reason*, of that human faculty which allows us to reach out to understand our world.

KOERTGE (*Retrospective final comment*): Robert Westman has playfully suggested that this panel was convened so that philosophers could once again give advice to historians about how to do their business! I believe, however, that many philosophers of science today expect that the advice would go in the other direction.

Historians of science, so the argument goes, are remarkably successful in making intelligible to us individual cases of scientific discovery. (Recall for example, Ravetz's detailed technical account of how Copernicus arrived at the heliocentric theory, Watson's narration of his discovery of the double-helix, or Hacking's story of the emergence of probability).[12]

If individual historical discoveries can be *understood*, continues the argument, then surely there must be something we as philosophers can say about a logic of discovery or the rationality of discovery. (This line of argument is particularly attractive today since certain philosophers, *e.g.*, Lakatos and Laudan, have already advocated basing normative philosophical theories of scientific method and scientific acceptance on quasi-empirical studies of past scientific practice.)

I would agree that the success of historians (and we should look carefully at the extent to which they in fact are successful) gives us reason to hope that there are *patterns* of discovery, but I see no reason to expect that such patterns would be rational in any standard sense of the term, although they may very well be cognitive, *i.e.*, reflect some of the typical ways in which we think.

To illustrate this demarcation, let me present some examples of simple reasoning patterns (not necessarily ones connected with scientific discovery) which I do *not* consider to be rational in any direct or straightforward sense.

According to an old associationist theory of psychology (its truth need not concern us now), when presented with a stimulus such as the word *black*, one's thoughts often move spontaneously either to its opposite, *white*, or to a similar concept, *e.g.*, *gray*, or *dark*. Such an inference pattern may well be common and we can even give some rules for it, but I would hardly call it either logical or rational. Yet the universality (or high probability) of the pattern could be used to support explanations. Thus one might imagine invoking such a pattern while explaining Lavoisier's discovery of the idea that something was *gained* on combustion in terms of an inversion of the phlogiston account. (Such a proposal would be vastly over-simplified and probably completely wrong, but perhaps it illustrates my point.)

I would put a similar gloss on Koestler's theory of bisociation.[13] It may well be the case that many discoveries involve the juxtaposition of old frameworks in a novel way. (Einstein's own account of his discovery of special relativity fits nicely.) Pointing out such a pattern (and describing it precisely) may help us both to understand past creative acts and even to make new discoveries ourselves. But again I would feel uncomfortable about calling the procedure of juxtaposition a logical or rational one, at least not without further argument.

But isn't this just a quibble about words? After all, as McMullin pointed out, isn't the very use of such reasoning patterns a distinctive feature of the behavior of man, that good old rational animal? In reply, I would propose the following minimal requirement: A necessary condition for calling a procedure or pattern rational is that, by conforming to the procedure, one is on the average more apt to attain one's aim than if one doesn't conform.

Is Koestlerian bisociation rational in this sense? To answer we would need to compare its success rate with other patterns of discovery, such as inversion, concept-stretching, analogical reasoning, Gestalt closure, *etc., etc.* What we might hope to do is to define fifteen-odd common patterns of discovery and then we could claim that using each of the fifteen in turn was rational.

What we may find, however, is that a large proportion of scientific discoveries conform only to the 'pattern' of set-breaking, *i.e.*, their only distinctive feature is that they are radically different from what went before. But does not the very success of historians render this negative result unlikely? I think not. One characteristic device which a historian uses is to break what on the face of it appears to be a giant conceptual leap into small discrete innovations. For some reason we find a series of tiny quantum leaps (no matter how radical) more understandable than a single massive *fait accompli*. Even set-breaking discoveries can often be fragmented and to that extent rendered more intelligible.

To sum up. The history of scientific discoveries is a rich source for both psychologists and philosophers. We should be very careful, however, to distinguish among inference patterns which are legitimate in all possible worlds (these I would call part of logic), those which seem to be heuristically successful in our world (these I would call rational), and those which are typical of human thought but which have no obvious systematic benefit for science.

## NOTES AND REFERENCES

[1]  G. Ryle, 'A Rational Animal', Auguste Comte Memorial Lecture, University of London, Athlone Press, 1962.

[2]  M. Finocchiaro, *History of Science as Explanation*, Wayne State University Press, Detroit, 1973.

[3]  See Arthur Koestler, *The Act of Creation*, MacMillan, New York, 1964.

[4] See Koestler, *The Sleepwalkers*, MacMillan, New York, 1959.

[5] R. S. Westman, 'The Melanchthon Circle, Rheticus and the Wittenberg Interpretation of the Copernican Theory', *Isis* 66 (1975), 165–193.

[6] This symposium appears in Peter Asquith and Ian Hacking (eds.), *PSA 1978*, Vol. II, Philosophy of Science Association, East Lansing, Michigan, 1979.

[7] A recent article of his contains a foretaste of themes to be treated more thoroughly in his book: 'How Scientific Is Psychoanalysis?', in Raphael Stern, Louise S. Horowitz, and Jack Lynes (eds.), *Science and Psychotherapy*, Haven, New York, 1977, pp. 219–254.

[8] See E. A. Burtt, *The Metaphysical Foundations of Modern Science*, Doubleday, New York, 1924, pp. 58–59. See also T. S. Kuhn, *The Copernican Revolution*, Vintage, New York, 1957, pp. 129–130.

[9] J. Ben-David, *The Scientist's Role in Society*, Prentice-Hall, Englewood Cliffs, New Jersey, 1971.

[10] E. Sober, *Simplicity*, Oxford University Press, London, 1975.

[11] See Michael Polanyi, *Personal Knowledge*, University of Chicago Press, Chicago, 1958, and *The Tacit Dimension*, Doubleday, Garden City, New Jersey, 1966.

[12] See J. Ravetz, *Astronomy and Cosmology in the Achievement of Nicolaus Copernicus*, Warsaw, 1965, and 'The Origins of the Copernican Revolution', *Scientific American*, October, 1966, 88–98; J. D. Watson, *The Double Helix*, Antheneum, New York, 1968; and I. Hacking, *The Emergence of Probability*, Cambridge University Press, London, 1975.

[13] See *The Act of Creation, op. cit.*

BRUCE WRIGHTSMAN

# THE LEGITIMATION OF SCIENTIFIC BELIEF:
# THEORY JUSTIFICATION BY COPERNICUS[1]

One of the most important and enduring philosophical issues in the history
of science has been the purpose and status of scientific theories. It is an issue
with a long history; but in its modern form it can be traced back to the
publication of Copernicus's major work, *De Revolutionibus Coelestium
Orbium* (1543). The appearance of that work generated a continuing controversy, both by the nature of its central claim — terrestrial mobility — and
by the discrepancy between Copernicus's novel claim for the *truth* of his
theory and the then-prevailing view of the hypothetical nature of all such
theories. The latter point of view was expressed in the anonymous letter to
the reader prefixed to the beginning of the printed work by its editor, Andreas
Osiander.[2] Between these two points of view about the status of scientific
theories lies the root of the modern Realist-Instrumentalist debate and the
more recent discussions concerning the rationality of scientific theories.

It is not my intention here to attempt a resolution of those issues, nor
even to survey their recent history.[3] My purpose, rather, is to explore an
important issue arising from those discussions of Copernicus's work but often
neglected in appraising it, that relates to the concern of this conference:
namely, the process by which Copernicus conceived of and came to believe in
the reality of terrestrial motion and the way he justified that belief. In short,
the Copernican 'discovery'.[4]

Let me begin by stating the philosophical problem that confronted Copernicus, which has engaged modern attention, and show its bearing on the
question of discovery: Given (a) two competing scientific theories such as the
Ptolemaic and the Copernican, both of which represent observational data
equally well and yield equally reliable predictions, and given (b) the fact that
the novel Copernican theory flatly contradicted accepted physical principles
as well as conventional beliefs and sacred scripture, all of which justified its
rival; how, then, could one decide which of these two theories is correct?
More importantly for my purposes is the question: how did Copernicus
decide? His assertion of the truth of his theory could not be based upon
conventional methodological principles, since such principles only served to
falsify his claim. Nor did there exist before the nineteenth century any
conclusive factual support by which to settle the matter.[5] This leads to the

51

*T. Nickles (ed.), Scientific Discovery: Case Studies,* 51–66.
*Copyright* © 1980 *by D. Reidel Publishing Company.*

central issue confronting Copernicus: what will justify "the assertion of an unsupported conjecture in the face of fact and well-supported contrary conjectures"?

P. K. Feyerabend answered that question by insisting that acceptance of the new theory can only be based on 'metaphysical belief'. The position of Thomas Kuhn is better known but causes similar visceral discomfort to his critics. He argues that the decision is a *choice* between rival paradigms and cannot be resolved by criteria that are entirely theory neutral or value free. What motivates the choice of the novel theory therefore is a kind of 'conversion' experience. Kuhn's critics have strenuously criticized this aspect of his work, insisting that if there are no theory-neutral criteria of scientific judgement, no common methodological standards available for selection between competing theories, the decision becomes arbitrary and subjective, a kind of 'religious change' that is 'irrational' and a matter of 'mob psychology'.[6] To such critics, the admission of these elements into the appraisal of theories threatens the very rationality of the scientific enterprise. Kuhn's position thus seems to confirm what positivists have always maintained: that theories are mere 'instruments' of computation and prediction without explanatory power and that one can rationally reconstruct and appraise a theory only in terms of its testable implications. The generation of a scientific theory, therefore, is not a part of the scientific process itself since it is fundamentally irrational. Hence, there has arisen the sharp distinction between the 'context of justification' and the 'context of discovery' with the former, alone, deemed suitable for logical and rational analysis and the latter relegated to the dumping ground of the emotive.[7]

It is my contention that this distinction is false, value loaded, and inimical to a rational understanding of scientific progress. To ask for those considerations that persuaded Copernicus and led him to assert the truth of a new theory in defiance of formidable falsifying arguments is not to ask for a description of psychological states but for those *reasons* and *arguments* by which he advanced from theory inception to completion. It will be my purpose in this paper to demonstrate the logical 'pattern of discovery'[8] in Copernicus's work by showing the process by which he conceived, formulated, justified, and thus, came to believe in and assert the truth of his theory.

Let me begin this exploration of that process by asking, what motivated Copernicus's search for an alternative theory? What raised doubts in his mind about the credibility of the prevailing Ptolemaic view and what aims in science did he entertain that led those dissatisfactions to undermine his belief in its validity and motivated his life-long search for a *true* system of the universe?

Historians typically point to the problem of the calendar and the urgent need for its reform as the source of his dissatisfaction. Jerome Ravetz, for example, recently argued the case that Copernicus, like every other astronomer, knew that the calendar was hopelessly inaccurate and that dates calculated on the basis of the Alphonsine tables bore little relationship to the observed motions of sun and moon.[9] And because of the complexities of observed motion, calendaric reform was not possible without a better theoretical basis for the laws of motion and the prevailing theories either failed to account for observed motions or could not explain those motions. Thus (according to Ravetz), Copernicus was prompted to search for a theory that would accomplish both demands and would provide the needed basis for calendar reform. He then concludes: "Copernicus failed to set down clearly and concisely what made him believe the earth really rotates in orbit around the sun." And further: "To ask . . . 'whence came the marvelous insight that showed him the truth' is to invite no answer or a purely speculative one".

But the source of that insight is not at all mysterious; in Book I and in the dedicatory letter to Pope Paul, Copernicus clearly describes the source of his 'insight' and the concerns by which he was led to search for a more 'reasonable' alternative and to adopt the assumption of terrestrial mobility. But he has almost nothing to say in those places about calendar reform.[10] What he *does* say there is that the unsatisfactory state of astronomy led him to search for alternative views among the writings of the ancients and that in such writings he learned of Pythagorean teachings about the motion of the earth.[11] By the time of the Fifth Lateran Council (1512–1517), which undertook the task of reforming the calendar, Copernicus had already formulated the basic principles of his system. The result of those early investigations – the *Commentariolus* – written between 1510 and 1514, contains the postulate of terrestrial motion as his sixth assumption.[12]

It is clear then that Copernicus's search for principles by which to renovate astronomy did not derive from the need for calendar reform but, as he tells us, because of the fundamental inconsistency in Ptolemy with the principle of absolute motion. The concern was philosophical, not practical. Ptolemy's *Almagest* accounted for the observed phenomena well enough; but it did so only by contradicting the cardinal physical principle of uniform circular rotation. The Eudoxian-Aristotelian scheme of homocentric spheres, while consistent with that principle, could not satisfactorily account for the phenomena. For these reasons astronomical hypotheses throughout the middle ages were regarded as just that: merely hypothetical. Osiander's letter thus represented the generally prevalent skepticism about the status of astronomical theories, as well as his own theological convictions that *truth* could be known only by divine revelation.

It is precisely on those grounds that Copernicus departs from the traditional conception of the task of astronomy and asserts the truth of his theory: it is his belief in the intelligibility of the universe as a revelation of God that informs his conception of science as the search to discover truth. This belief, anterior to all his dissatisfactions with prevailing theories, was the primary motivator of his demand for consistency of theory and data with physical and metaphysical principles. It was the starting point for his life-long search for a *true* system of the universe.

These convictions were present very early in his investigations. As early as the *Commentariolus*, Copernicus wrote:

The planetary theories of Ptolemy and most other astronomers, although consistent with the numerical data, seemed likewise to present no small difficulty. For these theories were not adequate unless certain equants were also conceived; it then appeared that a planet moved with uniform velocity neither on its deferent nor about the center of its epicycle. Hence a system of this sort seemed neither sufficiently absolute nor sufficiently pleasing to the mind.

Having become aware of these defects, I often considered whether there could perhaps be found a more reasonable arrangement of circles, from which every apparent inequality would be derived and in which everything would move uniformly about its proper center, as the rule of absolute motion requires. After I had addressed myself to this very difficult and most insoluble problem, the suggestion at length came to me how it could be solved with fewer and much simpler constructions than were formerly used, if some assumptions (which are called axioms) were granted me.

Then, after listing the seven assumptions on which his system is based, including the all-important sixth assumption of terrestrial motion, he adds:

Accordingly, let no one suppose that I have rashly asserted, with the Pythagoreans, the motion of the earth. Strong arguments will be found in my exposition of the circles. For the arguments by which the natural philosophers attempt to establish its stability depend for the most part on appearances; it is just such arguments that fail here, since I treat the immobility of the earth as due to an appearance.

When his major work finally appeared in 1543 containing those 'strong arguments', the concern for consistency was still uppermost in his mind. In the letter of dedication, Copernicus points specifically to that lack of consistency in prevailing theories and to the subsequent necessity of constructing *ad hoc* devices like the equant to preserve the fiction of uniform motion. This expedient not only violated the rule of motion but was offensive to Copernicus's demand for a *true* theory. The resulting system, he said, was an absurdity:

It is as if in his picture, an artist were to bring together hands, feet, head and other limbs from quite different models without a common relationship to a single body. The result would be a monster, not a man.

Since the prevailing theory was therefore no system at all, lacking any common, unifying principle of motion, to assert the reality of *that* theory would be to make a mockery of reason *and* faith. For Copernicus could not believe that the God whom he affirmed in *De Revolutionibus* as the 'universal Artisan of all things' and the 'Best and Most Orderly Workman' would be so clumsy as to have created such a monstrosity.[13] It was thus on the basis of his belief in a creator God who was the 'Best and Greatest Artist' that he sought for 'purer and more convenient assumptions' that would be consistent with the principle of uniform motion.

It is in just these places where the demand for unity, simplicity, necessity, and consistency is expressed that Copernicus discloses the underlying beliefs that propelled his search and the criteria by which he judged the validity of rival theories and validated his own. On the surface, such criteria appear only as aesthetic values; but for a philosophically-minded Christian astronomer, thoroughly trained in the doctrines of Aristotle, such criteria had physical and metaphysical significance whereby they functioned as criteria to judge the validity of theories. For Copernicus, the study of the universe could never be reduced to mere technical astronomy; cosmology was also physics (philosophy) and metaphysics or, in Aristotle's word, *theology*.[14] Thus, to reduce the diverse motions of celestial bodies to a single, unifying theory in which all the component parts become so interdependent as to establish the necessity of their observed motions, becomes a necessary condition for a theory to be true. Those criteria commended themselves to Copernicus not simply on aesthetic grounds but because those qualities were intimately related to his Christian beliefs about the unity, wisdom, and power of God, whose creation reflects these very qualities of its creator.

Behind such beliefs stands a long tradition of Christian speculation derived from two sources: the classical philosophical tradition stemming from Plato and Aristotle and the Judeo-Christian tradition of the Bible. From the latter (*e.g.*, *Psalms 19, Romans 1*) came the central idea that the creation reflects its creator. From the former (especially from the Platonic-Pythagorean wing), Copernicus derived the belief that the real elements of the universe were geometrical qualities, best exemplified in the shapes and movements of celestial bodies. These were the 'Forms' that were implanted and innate in the human mind so that they could be 'recollected' and thus *recognized* in the universe. Aristotle's views, to which Copernicus everywhere adheres (with certain important alterations), were not dissimilar, except that for Aristotle, of course, the Forms are incarnate *in* things and are thus *discovered* by transaction between the human organism and the environment. For Copernicus, as a Christian, this means that the true forms, having been created by God, are *revealed* in the creation. Since the 'Best and Greatest Artist' and 'Artificer of all things' has thus designed and created the universe and has created human beings in his own image, the *true* construction of the universe is intelligible to man and can be apprehended by mathematical reasoning. Knowledge of the universe is thus the result of divine disclosure; and it was the essence of Copernicus's religious tradition that one not only can but *should* study the creation to discover the true design of its creator:

... it is the loving duty (of the philosopher) to seek truth in all things, in so far as God has granted that to human reason.

These words from his letter of dedication clearly and simply state Copernicus's entire view of the aim of science. He states this aim again in his dedicatory letter where he describes his purpose as seeking to understand " ... the movements of the world mechanism created for our sake by the Best and Most Orderly Workman of all". In this place, Copernicus directly ties this belief to his 'annoyance' that "the philosophers, who in other respects had made a careful scrutiny of the least details of the world, had discovered no sure scheme for the movements of the mechanism ... ".[15]

By themselves, such theological/aesthetic criteria will not decide the matter of scientific truth any more than will accuracy or observational agreement. But for Copernicus, no theory can be true that blatantly violates such principles. They function then for him as *necessary* but not sufficient conditions of truth. Thus, the absence of such qualities in Ptolemy falsified that theory for him and their presence in his own theory served to increase his confidence in its truth.

Copernicus now must argue for its plausibility by demonstrating that the assumption of the earth's motion not only unites the system but makes all its parts totally interdependent:

And so, having laid down the movements which I attribute to the earth ... I finally discovered by the help of long and numerous observations that if the movements of the other planets are correlated with the circular movement of the earth, and if the movements are computed in accordance with the revolution of each planet, not only do all their phenomena follow from that but also this correlation binds together so closely the order and magnitudes of all the planets and of their spheres and the heavens themselves that nothing can be shifted around in any part of them without disrupting the remaining parts and the universe as a whole.[16]

So much has Copernicus accomplished within the domain of traditional mathematical astronomy, without requiring philosophical or theological adjustment. But that would leave it a mere hypothesis, and Copernicus would have the truth. While such arguments and demonstrations establish the plausibility of his theory and undoubtedly strenghtened his conviction in its truth, his reasoning is insufficient to establish that it is true or even probable. To do that he must now argue as a philosopher and theologian; first, by refuting the traditional but powerful objections to terrestrial motion; second, by demonstrating the necessity of that motion to integrate celestial movements and to show that it is the only way to do so; third, to provide an alternative physical principle that will account for that motion, and fourth, to demonstrate the validity and necessity of that physical principle by showing its consistency with accepted metaphysical/theological axioms.

Copernicus could anticipate the powerful physical and theological objections that would be raised against his theory. Indeed, he candidly acknowledges the difficulty of his position by admitting the apparent 'absurdity' of his theory on grounds of popular belief, common sense and tradition, the most daunting of which is the unanimous geocentric testimony of sacred scripture. Hence his appeal to the Pope that his work be judged only by mathematicians and his expression of scorn for those ignorant of that art who will "shamelessly distort some passage in Holy Scripture ... to attack my work."[17] Though he disclaims fear of such criticism here, he expressed it privately to others and hesitated to publish his work for over thirty years, until persuaded to do so by his friends. It is a tribute to the strength of his conviction and to his sense of obligation to share his discovery that he did venture into the risky domain of philosophical and theological argument.

Copernicus proceeds to build probability by refuting the standard philosophical objections to a moving earth (Book I, Chaps. 7–8), after which he concludes: "From all these considerations, it is more probable that the earth moves than that it remains at rest." Probability is gained, however, not only by the refutation of objections but by the fact that, in his system, all the phenomena physically follow from his assumption of motion. This leads to the inescapable conclusion that his system is not simply another possible hypothesis, but the *only* possible one:

And so we find an amazing symmetry with this mathematical system of the universe and a certain tying together of the harmony of movement and the size of the spheres *such as can be found in no other way*. (Bk. I, Chap. 10; emphasis mine)

The problem now confronting Copernicus is this: if the earth indeed moves and no stellar parallax is observed nor other disconfirming physical consequences occur, what physical principle that is consistent with the uniformity of motion will account for that? So far, his only *physical* arguments have been falsifying ones; now it is necessary for Copernicus to provide an alternative principle of motion that will account for and necessitate terrestrial movement.

Here Copernicus turns once again to the ancients and draws from the Platonic-Pythagorean tradition a doctrine from which to derive such a principle of motion. In a major departure from Aristotle's physics, Copernicus asserts that motion is determined by geometrical form, not by substance. The entire argument in Book I of the *Revolutions* hangs on the argument from geometrical form and centers on the concept of sphericity. Chapter 1 maintains that the universe as a whole is spherical; Chapters 2 and 3 affirm that the earth is spherical; Chapter 4 contends that the motion of celestial bodies is uniform and circular. So much is traditional Aristotelian doctrine; but then comes his crucial physical argument:

We now know that the motion of the heavenly bodies is circular. *Rotation is natural to a sphere and by that very act is its form expressed.* (Emphasis mine)

This physical principle provides Copernicus with the necessary basis for explaining the motion of the earth. By arguing that the spherical shape of a heavenly body is itself a sufficient condition for its rotation and from the fact that sphericity is exhibited by the earth, he has made terrestrial rotation *necessary*. By extension, since the earth is embedded in a spherical shell which, by virtue of *its* form, *must* rotate, the earth also revolves around the sun.

In this significant, but qualified, departure from Aristotle's Physics, Copernicus has in one stroke circumvented and negated Aristotle's elaborate effort in *De Caelo* to construct a system of homocentric movers and unrolling spheres by which to explain the transmission of motion. For Copernicus, it is no longer necessary to be concerned with the effective transmission of motion from mover to moved. By making the spherical form itself the sufficient determinant of motion, the mover has, so to speak, become internalized, inherent in a heavenly body as the power which causes them to rotate and to cohere.[18] Further, the sphere of the Prime Mover, which, in Aristotle, imparts motion to the entire system can be dispensed with.

But at this point, a new problem arises. If everything in the universe is spherical in form, and therefore in circular rotation, what is now to be the 'benchmark', the fixed reference point, from which absolute or real motion can be determined? What, in short, will now determine the 'place' of celestial objects, now that the 'Unmoved Mover', which heretofore had fulfilled that function, is eliminated? For if nothing rests in the system, no distinction between real and apparent motion is observationally or theoretically possible, and Copernicus would be forced into a relativistic position, in contradiction to his repeated insistence upon consistency with 'the rule of absolute motion'.

For both philosophical and theological reasons, Copernicus cannot settle for a relativistic universe as did Cusa, for his system must have *astronomical* as well as theological significance, which Cusa's universe did not have. Copernicus will establish its theocentricity in another way, as we shall shortly see. But for astronomical and theological reasons, the question of which motion is real and which merely apparent — that of the earth or that of the stellar sphere — *must* be established if his theory is to be consistent and to command assent. Since his entire argument hangs on the assumption of the earth's motion, something else in the universe must be taken as immobile. And inasmuch as the diurnal motion formerly attributed to the sphere of the fixed stars is now accounted for by the moving earth, it must be the sphere of the fixed stars that is at rest.[19]

But what will justify this? How can the sphere of the fixed stars be immobile? Given his obsession for philosophical consistency and given his own physical doctrine that "rotation is natural to a sphere", how can the sphere of the fixed stars *not* move? The presence of such a glaring inconsistency in one whose aim was the search for such consistency poses a genuine dilemma for Copernicus. And it is impossible for him to resolve it by demonstrating on physical or optical grounds whether it is the observer or the observed that is moving. So he must do so on metaphysical and theological grounds. That is, the physical inconsistency can only be warranted metaphysically by the theological function served by an immobile stellar sphere.[20] He hints at this theological justification when, at the close of his arguments against objections to terrestrial motion (Bk. I, Chap. 8), he appeals once more to the Platonic-Pythagorean tradition and to its doctrine of the *nobility* of immobile heavenly objects:

Further, we conceive immobility to be nobler and more divine than change and instability, which latter is more appropriate to earth than to the universe. Would it not then seem rather absurd to ascribe motion to that which contains or locates and not rather to that which is contained and located, namely the earth?

What lies behind this argument is a final, undisclosed metaphysical/theological assumption which he shares with Aristotle: namely, that everything in the universe has a *place*, including the universe as a whole. He writes:

Given the above view — and there is none more reasonable — that the periodic times are proportional to the sizes of the orbits, then the order of the spheres, beginning from the most distant is as follows: Most distant of all is the Sphere of the Fixed Stars, containing itself and everything, and being therefore itself unmoveable. It is the place of the universe ... (Bk. I, Chap. 10)

This doctrine comes from Aristotle, who had defined 'Place' (*topos*) as "the innermost motionless boundary of what contains." [21] That 'place' was, of course, the Prime Mover in Aristotle's system; it was defined by him as without 'place', being uncontained by any further receptacle. Yet, if it moves, as it surely must in order to impart motion to the contained, then, according to Aristotle's definition, it changes its 'place', which means it had 'place' to begin with, which would lead Aristotle into a blatant contradiction. As is well known, Aristotle avoided this contradiction by positing the Prime Mover as an *unmoved* Mover, justifying it on metaphysical grounds and explaining it as a teleological rather than an efficient mover. Copernicus follows a similar procedure; however, he has dispensed with the Prime Mover by his sphericity principle of motion and, in so doing, has lost the very thing that determines 'place'. And without that, he has lost the possibility of determining absolute motion. Either Aristotle's definition of 'place' had to be altered or his doctrine of the motion of the outermost sphere had to be rejected. Copernicus resolved this by rejecting the motion of the outermost sphere, which, in his system, is occupied by the stellar sphere. This sphere, accordingly, becomes the 'place of the universe'. [22]

Dispensing with Aristotle's concept of the Unmoved Mover means that Copernicus has also dispensed with Aristotle's concept of God, as he must do to be consistent with his own Christian theology. [23] This means that the concept of 'place' has acquired not only a different physical identity but has also gained a different theological meaning, a meaning acquired from the biblical tradition. In that tradition, the key phrase, 'place of the universe', had come to be used as a synonym for 'God' as the result of nearly two millenia of Jewish and Christian speculation on the name of God.

Because the covenant name God gave to Israel — 'YHWH' — was ineffable and not to be misused (by injunction of the second commandment of the Decalogue), late Judaism came to avoid direct references to God and developed several substitute terms of address (*Elohim, Adonai*) and other terms of reference by which to distinguish between God, Himself, who is beyond all perception, and His visible self-manifestations. The term 'Shekhina' (Glory) was one such 'name', commonly used to refer to a visible manifestation of God, often described in appearance as a radiant cloud or light. Another 'name' was 'heaven', and still another, with similar cosmological connotations, was 'place' (Māqōm).[24]

In the literature of first-century Palestinian Judaism, this practice had led to considerable cosmological/theological speculation, in which God was endowed with both personal and spacial attributes. The connotations of 'place' to which this led, and which entered deeply into the Christian tradition, can be seen in such statements in the Mishnah as: "Why do we call the Lord, 'Māqōm'? Because the Lord is the Dwelling Place of the world, but the world is not His dwelling place".

During the middle ages, such ideas were developed much further in Jewish mystical speculation, represented by the *Caballa*, which became widely studied during the Renaissance, syncretized with the Platonic-Pythagorean tradition, and its doctrines spread by such thinkers as Mirandola, Agrippa, Reuchlin, Fludd, Campanella, and Bruno. It is possible and highly likely that Copernicus encountered such ideas during his years of study in Italy and that they gave him an important insight into the way he could make his cosmological system firmly consistent with Christian theology. By identifying the outermost stellar sphere as the 'place of the universe', he is, like Aristotle, endowing that sphere with theological as well as physical significance, making that sphere which is closest to the Abode of God (the Empyrean) serve as the ultimate determinant of all change and motion in the universe. Thus, while the immobility of that sphere is inconsistent with the physical principle of sphericity, it is *theologically* consistent with the religious axioms of Copernicus's system and with Christian doctrine itself, which then was the final arbitrator of truth and the ultimate legitimator of any system of thought. The religious rationale Copernicus gives for his scientific efforts in the letter of dedication to the Pope and in his introduction to Book I of the *Revolutions* plainly reflects these discreet, but significant, theological justifications.

That this contention is correct is supported, moreover, by an additional fact: Copernicus has one other spherical body in his universe that is likewise immobile – the sun.[25] While the sun does not, for Copernicus, fulfill any discernible astronomical or physical functions (as it is later to do for Kepler), it does perform a theological function similar to that of its cosmological counterpart, the stellar sphere: it stands at the center of Copernicus's system as a visible symbol of God's presence in, and his sovereignty over, the entire universe. As such, and because of its long and rich association with God in the Judeo-Christian tradition, it provides an appropriate symbol for a *theocentric* universe.[26] Copernicus saw that arrangement not merely as mathematically useful or physically necessary but *providential*:

In the center of all the celestial bodies rests the sun. For who in this most beautiful temple could place this lamp in another or better place than that from which it can illuminate everything at the same time? Indeed, it is not unsuitable that some have called it the light of the world; others, its mind, and still others, its ruler. Trismegistus calls it the visible God; Sophocles Electra, the all-seeing. So indeed, as if sitting on a royal throne, the Sun rules the family of the stars which surround it.

For Copernicus, as later for Kepler, there is a marvelous cosmic correlation between the physical universe and God.[27] What else but such a religious vision could have evoked that ecstatic outburst of praise contained in the last line of Book I, Chapter 10: "How exceedingly fine is the divine work of the Best and Greatest Artist!"

In such fashion did Copernicus disclose his motivations and the manner in which he formulated, reasoned, and justified his novel theory. I have attempted to demonstrate, thereby, the importance of the entire process of 'discovery' to our understanding of science and to elaborate the previously noted but oft-neglected role of Copernicus's theology in the process of that discovery.[28] In the course of his argumentation, a definite pattern of reasoning emerges which renders such epitaphs for discovery as 'irrational' and 'subjective' patently inappropriate. One may not like Copernicus's reasons for coming to believe in and justifying his system but that is not a rational ground for refusing to accept them as *reasons*. We must therefore remind ourselves that scientific investigation had much broader implications for Copernicus than it has for many today and included those purposes which we classify as religious and extra-scientific. Such considerations, however, were crucial for Copernicus and were demonstrably instrumental for his achievement. By his own statements, they were the primary motivations for his research, the ultimate source of the truth he discovered, the basis of his confidence in his conception as true and the final justification for believing in, asserting, publicizing, and commending that theory to others. In the absence of any available or then-conceivable confirming evidence, it was only on such grounds that belief and persuasion were possible. And without that belief, there would have been no *Revolutions* and, perhaps, no revolution in science, since the same considerations that led Copernicus to belief were those which made his theory *believable* to his earliest adherents, especially Kepler. Indeed, these criteria were not idiosyncratic or unique to Copernicus, and therefore 'subjective'; rather, they were, to use Herbert Feigl's illuminating term, 'inter-subjective', the 'shared basis of values' among scientists by which claims to truth have always been tested.[29] In that light, the wisdom of I. Bernard Cohen's statement becomes apparent: the 'logic of discovery' converges on the 'logic of the discovered'.[30]

Finally, we should remind ourselves of one other fact: the chief purpose of science is to discover new things, not merely to test the products of discovery. To ignore the process by which discoveries have been made is to debilitate science education and to conceal, rather than reveal, what makes science such a fascinating and a truly creative human enterprise.

*Department of Theology, HPS*
*Luther College, Iowa*

## NOTES AND REFERENCES

[1] Earlier versions of this paper were given at the Midwest Junto of the History of Science Society, at the University of Western Ontario, to a Philosophy Colloquium at Nottingham University, England, and informally to members of the Copernicus Research Institute in Warsaw, Poland, during a sabbatical fellowship in 1977.

[2] On Osiander's involvement with Copernicus, see my 'Andreas Osiander's Contribution to the Copernican Achievement' in R. S. Westman (ed.), *The Copernican Achievement*, Univ. of California Press, Los Angeles, 1975.

[3] A good survey of the issue may be found in A. Musgrave and I. Lakatos (eds.), *Criticism and the Growth of Knowledge*, Cambridge Univ. Press, Cambridge, 1970, and in F. Suppe (ed.), *The Structure of Scientific Theories*, 2nd ed., Univ. of Illinois Press, Urbana, 1977.

[4] I use the word 'discovery' in a qualified way, since Copernicus, of course, did not discover terrestrial mobility.

[5] Bessel discovered stellar parallax in 1818. In 1616 Galileo observed the phases of Venus, which are predicted by Copernicus's theory, but that has no bearing on his discovery or his own justification of it.

[6] *Criticism and the Growth of Knowledge, op. cit.*, pp. 33, 56–57, 93–118. Feyerabend's views are summarized in his 'Problems of Empiricism', in R. Colodny (ed.), *Beyond the Edge of Certainty*, Prentice-Hall, Englewood Cliffs, New Jersey, 1965.

[7] This distinction was first formulated by Hans Reichenbach in his *Experience and Prediction*, Univ. of Chicago Press, Chicago, 1938.

[8] From the title of N. R. Hanson, *Patterns of Discovery*, Cambridge Univ. Press, Cambridge, 1958.

[9] Jerome Ravetz, 'The Origin of the Copernican Revolution', in *Scientific American 236* (August, 1977), 88–98.

[10] At most, the reform initiated by the council and coordinated by Paul of Middelburg gave additional impetus to the investigations he already had in progress. His *Commentariolus* appeared before the report of Paul to Leo X was issued, in which Paul reports receiving an opinion from Copernicus. Copernicus refers to this briefly, near the end of his letter of dedication.

[11] Contrary to Edward Rosen, I do not believe Copernicus differentiates his position from the Pythagorean belief in a moving earth. What he says is, "Let no one suppose that I have *gratuitously* (*i.e.*, rashly or without reasons) assumed, with the Pythagoreans, the motion of the earth . . . . " That this is not a disassociation of his belief from theirs, as Rosen believes, is supported by the places in the *Revolutions* where he clearly states his indebtedness to them for their ideas of terrestrial mobility, and not just for their policy of private disclosure – for example, the passage in the middle of the dedication and two separate references in Bk. I, Chap. 5.

[12] Rosen, in *Three Copernican Treatises*, 3rd. ed., has convincingly documented the beginning of Copernicus's doubts about Ptolemy as early as 1496–1501 in Bologna, when, as a student, he assisted the astronomer Novara, who publicly demonstrated inaccuracies in latitudinal figures in Ptolemy's *Geography*, indicating that the earth may not be motionless. He was also familiar with Regiomontanus's *Epitome* (1496) and his criticisms of Ptolemy's lunar theory, criticisms which are reflected in Copernicus's earliest astronomical writing, the *Commentariolus*.

[13] These references to divine revelation are found in two places in the letter of dedication and three times in Book I, Chaps. 9–10, but the entire introduction to Book I (which was deleted from the first published text in 1543) conveys that theological point. Whether the theological justifications contained in Rheticus's *Narratio Prima* were stimulated by Copernicus cannot be known; but they are consistent with his views.

[14] After all, 'Metaphysics' was not the title Aristotle gave to the work; it acquired that title because it came right *after* his *Physics*; hence, 'after-Physics'. *He* called it 'Wisdom' or 'Theology', which, for him was First Philosophy.

[15] The tradition of Natural Theology which these beliefs reflect came to be expressed in western Christendom by the concept of nature as a 'secondary revelation' of God and is also to be found in numerous references to the 'Book of Nature' in the literature of the sixteenth century. The first explicit reference to this is in the writings of John Scotus Erigena in the fourteenth century, although the concept goes back to Augustine.

[16] In this respect, it is potentially superior to Ptolemy in that it explains the anomaly of retrograde motion, as well as the order and periods of the planets.

[17] That Copernicus is dissembling is shown by the fact that he expressed such fears in a letter to Andreas Osiander. This letter has vanished, but we know of its contents from Kepler's citation of portions of Osiander's reply to Copernicus. If Kepler's report is accurate, then Copernicus was not as confident as he sounded in his dedication. For that reason, I suspect, he tactfully but pointedly refrained from making any comment about biblical interpretation which could be used against him.

[18] It should be noted that, in one sense, this *is* in accordance with Aristotle's basic distinction between celestial and terrestrial motion (in I:8). It is by placing the earth itself among the planets that it acquires rotational motion while leaving the physics of the earth intact.

[19] The clarity of Copernicus's assertions of the immobility of the stellar sphere and the sun make it all the more surprising to read in O. Neugebauer's *The Transmission of Planetary Theories in Ancient and Medieval Astronomy*, New York, 1955, p. 27: "The question as to which body is 'at rest' is of course without any interest, particularly when no such physical body existed in the whole Copernican system."

[20] In I:7, while refuting Ptolemy's arguments against a moving earth, Copernicus attributes an argument to Ptolemy that if the earth rotated on its axis, it would have disintegrated long ago. In I:8 he refutes this argument by reference to his doctrine that "rotation is natural to a sphere". The mystery here is that, as far as I can determine, Ptolemy never argued specifically against the diurnal rotation of the earth nor ever wrote in terms that could even vaguely be construed as a concept of centrifugal force, which Copernicus ascribes to him. It is true that he argued against 'those' who advocated terrestrial motion on the grounds that such motion would "leave animals and other objects hanging in the air", and even that such motion would cause the earth to "fall out of the cosmos". But, and this fact seems to have escaped Copernicus, Ptolemy did not differentiate between diurnal rotation and annual revolution and did not argue that the earth would 'dissipate' under diurnal rotation. To date, I have found no reference in contemporary literature calling attention to this curious misstatement on the part of Copernicus. It could be that it was then a popular impression that Ptolemy had taught some concept of centrifugal force, considering the way Ptolemy was taught (usually at third-hand) in the universities. If that were so, Copernicus would naturally have felt the necessity of rebutting this argument. It would have been more advantageous for him to have cited Ptolemy's precise words, however. This would have showed his opponents that even his greatest predecessor had not mentioned any possible 'dissipation' of the earth, thus strengthening his own physical arguments.

[21] *Physics*, IV:4 (212a, 20–21), trans. by P. H. Wickstead and F. M. Cornford, Loeb Classical Library, 1929, Vol. I.

[22] His *Letter Against Werner* (1524) contains his early arguments against the motion of the eighth sphere. It is not often noticed that, inasmuch as it was also Aristotle's belief that immobility is more noble and divine, the fact that the earth as a whole is immobile in his system represents a serious inconsistency, which Copernicus must have been pleased to eliminate!

[23] In the Christian tradition, God is pre-eminently the *creator* of the universe, a concept entirely absent in Aristotle. For him, matter was co-eternal with the Unmoved Mover and was thus uncreated. Moreover, the biblical conception of God is everywhere of a God who *acts* in continual creative and redemptive fashion. Aristotle's God had no motion. As Prime Mover, God was defined as pure Actuality and, therefore, unmoved. As Aristotle argued in his *Metaphysics*, the Prime Mover moves others by being their final cause, that is, by simply being the object of their love and desire. As a Christian, Copernicus had to modify the theology of Aristotle's system to be consistent with biblical conceptions of God. In this, he is following Thomas Aquinas, whose first argument for the existence of God is a carbon copy of Aristotle's *Physics*, Book 8, and the *Metaphysics*, Books Lambda and Beta. Their respective concepts of deity thus share some functions; for both, deity defines 'place' by providing a final cause and limit to account for change and motion. Thus, for both Aristotle and Copernicus, God is the First Principle of Being and Becoming and thereby establishes the rationality of the universe, which makes knowledge possible.

[24] Examples may be found in *Deuteronomy* 33:27; " . . . the place where I cause my Name to dwell"; *Psalm* 90:1; "Lord, you have been our dwelling place . . . " (*cf. Psalm* 132:5, 7). In the *Targums* (Aramaic paraphrases of the Bible), the targum of *Exodus* 25:8 (which reads; "Let them make me a sanctuary that I may dwell among them") is rendered by; " . . . that I may let my Shekhina dwell among them." For a further development of these themes, see Max Jammer, *Concepts of Space*. Harvard Univ. Press, Cambridge, 1954.

[25] Clear statements of the immobility of the sun can be found in I:10 of the *Revolutions*: "I also say that the sun remains forever immobile . . . "; and later, "In the center of all rests the sun."

[26] The associations of the sun and light with deity in the biblical tradition are numerous, beginning with the first creation account in *Genesis* 1 where the first thing to be created is light, signifying the visible manifestation of God, who was present before the ordering of the cosmos began.

[27] Gerald Holton has documented in his study of Kepler that what was implicit in Copernicus becomes explicit in Kepler, for whom the sun fulfills three functions: it functions as mathematical reference point, as physical mover, and as theological center. See 'Johannes Kepler's Universe: Its Physics and Metaphysics', in R. M. Palter (ed.), *Toward Modern Science*, Farrar, Strauss, and Cudahy, New York, 1961, Vol. II.

[28] Among the few who have noted that the basis of Copernicus's confidence is primarily theological are: E. A. Burtt, *The Metaphysical Foundations of Modern Science*, Humanities Press, New York, 1952, Ch. II; A. O. Lovejoy, *The Great Chain of Being*, New York, 1960, p. 111; Werner Heisenberg, 'Tradition in Science', in O. Gingerich (ed.), *The Nature of Scientific Discovery*, Smithsonian Institution, Washington D. C., 1975; C. F. von Weizsäcker, *Die Einheit Der Natur*, Munich, 1971.

Once could go further and argue for the importance of 'meditation' to Copernicus, a word Copernicus twice uses in his letter of dedication. On its significance for Copernicus, see Karol Gorski, *Mikotaj Kopernik Srodowisko Spoteczne I Samotnosc*, Ossolineum, Warsaw, 1973. An English precis will soon be published under the title, 'The Social Background of Copernicus and His Solitude'.

[29] Thomas Kuhn, whose earliest writings were about the phenomenon of discovery, has had something helpful to say about that in his 'Objectivity, Value Judgment and Theory Choice', reprinted in *The Essential Tension*, Univ. of Chicago Press, Chicago, 1977.

[30] I. Bernard Cohen, *Franklin and Newton*, The American Philosophical Society, Philadelphia, 1956, Chap. XXVI, pp. 190, 657.

BRUCE T. MORAN

# WILHELM IV OF HESSE-KASSEL: INFORMAL COMMUNICATION AND THE ARISTOCRATIC CONTEXT OF DISCOVERY

## I. INTRODUCTION

That a period of scientific revolution coincided with a period of theological, social and economic change in Europe is an obvious commonplace in literature dealing with the history of early modern science. The question of how specific elements in sixteenth-century society affected the actual process of discovery and the diffusion of new ideas is less clearly understood. In particular, the structure of informal methods of scientific interaction which made possible the transmission of new procedures, discoveries, and technical innovations among mathematicians, astronomers, and naturalists prior to the development of formal scientific organizations, has been left, for the most part, unexplored. One of the aims of this study, then, is to delineate an important mode of scientific interaction during the late Renaissance by examining the role of princely courts in the construction of informal patterns of scientific and technical information exchange. Among various forms of aristocratic patronage, it is possible to reconstruct a special type of courtly involvement characterized by the direct participation of princes in scientific and technical projects. Such courts not only provided an environment for the development of technical proficiency and innovation but also were capable of initiating vast networks of scientific correspondence based upon pre-existing religious and political avenues of communication.

This discussion focuses primarily upon the organization of collective and observational projects at the German court of Wilhelm IV of Hesse-Kassel (1532–1592). An examination of Wilhelm's botanical, technical, and astronomical programs illustrates the implementation of several qualitative features of the experimental component of modern science, emphasizing the use of precision, collaboration, and the systematic collection of information. On a broader level, the present analysis seeks to establish a relation between social context and the origin of methodological values within the limits of a particular historical case study. Aside from developing a critical appraisal of the court's role in the creation of methods and ideas significant to the history of science, a further task will be to align the projects of the Kassel court with the social and intellectual conditions which influenced their construction.

67

*T. Nickles (ed.), Scientific Discovery: Case Studies,* 67–96.
*Copyright* © 1980 *by D. Reidel Publishing Company.*

Thus Wilhelm's unique religious and political position among German princes and his understanding of the idea of *reform* within the context of humanist philological criticism and the procedures of early sixteenth-century naturalists are viewed as essential components in explaining both the conceptual origin of the Kassel projects and the social structure upon which they developed.

## II. THE SOCIAL CONTEXT OF INFORMAL INTERACTION AT HESSE-KASSEL

Two religious and political issues which threatened the stability of Protestantism in Germany during the latter part of the sixteenth century established Wilhelm's central position in a network of communication among Protestant princes: (1) continued discord between Lutherans and Calvinists and (2) the perpetuation of doctrinal differences between those Lutherans who followed the moderate and conciliatory views of Philip Melanchthon (1497–1560) (*Philippists*) and those who preferred strict adherence to the teachings of Luther (*Gnesio-Lutheraner*).

The dogmatic framework of this latter conflict developed as a result of a theological controversy which erupted between Melanchthon at the University of Wittenberg and Flacius Illyricius (1520–1575) at the University of Jena.[1] Although the history of the controversy is complex, the dispute centered initially upon what Flacius identified as insidious Roman elements within Melanchthon's theological programs. Flacius's insistence upon the condemnation of adiaphoristic practices and the outright repudiation of the Leipzig Interim and all other suspected Roman formulae continued to place him at variance with Melanchthon and produced a hopeless theological split with Melanchthon's followers at Wittenberg.

The controversy grew increasingly bitter after Melanchthon's death in 1560. Indeed, the threat to evangelical unity brought about by the dispute greatly concerned Wilhelm IV, who, with Duke Julius of Braunschweig and the Elector August of Saxony, promoted the efforts of Jacob Andreae (1528–1590), chancellor of the University of Tübingen, to establish religious unity out of persisting intra-Lutheran discord.

Wilhelm's own theological views, more sympathetic with the moderate constructions of Melanchthon and other first-generation reformers, left no doubt of his anti-Flacian disposition. With the Saxon Elector the Flacian party fared no better. However, where August grew concerned over the extent to which Flacian Lutheranism had gained the upper hand among theologians in Saxony, the Elector's suspicions also increased concerning some Wittenberg theologians, several of whose positions, especially those pertaining to the Eucharist, seemed to have much in common with corresponding Calvinist teachings.[2] Doctrinal similarities shared by particular Wittenberg theologians and Calvinists were underscored by Jacob Andreae in 1570. As a result, rather than healing the schism, Andreae recognized the choice to be between blatant Lutheranism and 'crypto' Calvinism and opted for Flacian rigor. Following a courtly purge of suspected Calvinist infiltrators, August pursued the construction of a general doctrinal code which, he believed, would put an end to Flacian-Philippist dissension. In accordance with the Elector's plan, the theologians who assembled at Torgau in 1576 concluded a 'formula of concord' known as the Torgau Book. Rather than compromise, however, the Torgau Book systematically erased Melanchthon's influence and maintained a definition of the Eucharist in accordance with the teachings of ubiquity professed by Andreae and other Swabian theologians.

Neither the Torgau formula nor its refined successor, the Bergen Book, gained Wilhelm's approval. Aside from his dislike of the ubiquity notion, a doctrine which he would not allow to be taught at the University of Marburg, Wilhelm emphatically rejected the authority of a few theologians to undermine the moderate views of Melanchthon and the amended Augsburg Confession. Moreover, Wilhelm's desire for a Christian reconciliation with the Calvinists required a more liberal atmosphere than afforded by either formula. By not subscribing to the Torgau Concord, Wilhelm retained his intermediary position between Lutheran and Calvinist princes, interacting as easily with Heidelberg professors and the Calvinist Johann Casimir as with Wittenberg mathematicians and the Elector August. This position, in which both Calvinists and Lutherans sought Hessian sympathies and support, expanded Wilhelm's social footing upon which communicative links with princely courts and universities, equally antagonistic to one another, might be individually established.

Hesse's geographical position along major trading routes also helped maintain regularly frequented avenues of communication between Kassel and other important cities and princely courts. The actual transportation of commercial goods took shape as one of the most important features of the economy of Hesse-Kassel in the latter part of the sixteenth century, as the interior German markets of Cologne, Frankfurt, and Leipzig usurped more and more the traditional role of the Hanseatic league. With the development of this upper German economic circle, Hesse's routes of transit became important factors in the passage of commercial traffic. Dealers from Cologne regularly journeyed through Hesse on their way to the Leipzig fair. From Frankfurt the transport of goods passed overland to Kassel and from Kassel

by water through Bremen to the Dutch states. Transportation through Hesse
was especially important to Frankfurt merchants who acquired economic
parity with their Cologne competitors through a direct route of exchange
with the Netherlands.[3] Situated between the economic system of the Hanse,
prosperous Dutch trading centers, growing central German cities, and Italian
markets, the Kassel court stood at the crossway of commercial routes which
spanned the empire. Upon these varied and far reaching avenues of communication, Wilhelm initiated a network of correspondence with other courts,
universities, and private individuals which not only supplied the Kassel court
with religious and political reports, but also, when directed by the Landgraf's
botanical, technical, and astronomical interests, functioned as an important
informal mode of scientific interaction.

### III. BOTANICAL PROJECTS AND THE CONTEXT OF REFORM

With many Renaissance princes, Wilhelm shared an interest in the curious and
the rare. Yet to these diffuse entertainments the Landgraf added a more
personal, focused concern for the ordering and observation of nature. This
interest reflects a different intellectual context — a movement characterized
by attempts to remedy apparent inaccuracies in ancient and medieval texts
by means of diligent, direct observation of natural phenomena and events.
The critical roots of this methodology lay in humanist soil. The partial decay
of humanist confidence in ancient sources, illustrated by futile attempts to
reconcile philological inconsistencies among classical authors, nourished the
re-emergence of a view of knowledge dependent upon meticulous observation,
systematic collection, and detailed description of all manner of natural
*particularia*. This aspect of the Landgraf's approach to nature and the social
network employed in its pursuit came first to be developed in the construction of Wilhelm's botanical catalogs and in the organization of the Kassel
botanical gardens.

    The encyclopedic accumulation of natural objects and amassing of botan-
ical observations and descriptions begun by early sixteenth-century naturalists,
directed Wilhelm's interest in plants and provided him with methods designed
to restore the accuracy of botanical understanding: collecting, cataloging,
and carefully observing subjects received from a group of interacting adepts
scattered throughout Europe. A major source of new botanical information
at Kassel originated from an acquaintance with the works of the leading
naturalists of the mid and later sixteenth century. Yet Wilhelm's botanical
knowledge did not depend entirely upon published accounts. To increase the

number and variety of plants within his gardens, and to acquire information of the most recent botanical discoveries Wilhelm brought to bear the full weight of his informal, courtly network of communication. With well known naturalists, fellow Protestant princes, family relations, as well as professors at the Hessian University of Marburg, Wilhelm exchanged seeds and plants and utilized the direct services of political agents, civil servants, personal representatives, and religious acquaintances in other countries to purchase specimens for his court. A list of such interactions would be tedious at best.[4] Information derived from neighboring courts tended usually toward the description of curious plants discovered accidentally in local regions. However, letters which passed between Wilhelm and his brother Ludwig, the Landgraf of Hesse-Marburg, extended beyond the simple exchange of unusual specimens. The correspondence of each prince reflects a mutual interest, reminiscent of earlier humanist criticism, in the proper naming of subjects whose identity neither Wilhelm nor Ludwig had been able to establish from ancient or contemporary sources.[5] Living plants – and when these seemed too fragile to transport, botanical drawings – circulated between the Kassel and Marburg courts so that Wilhelm's judgments concerning their identity could be compared with those of his brother, whose opinions were guided by the medical faculty at the University of Marburg.

The enthusiastic transmission of descriptions, illustrations, and actual subjects among naturalists should be viewed within the context of humanist epistolography and marked against political and economic conditions which allowed for the maintenance of convenient and extensive means of correspondence. Moreover, two assumptions, pertaining to the acquisition of knowledge developed throughout the Renaissance, were essential to meaningful interaction: (1) that a thorough understanding of the natural world could not be obtained simply by means of glossing ancient texts and (2) that the correction of the inaccuracies and omissions within the works of the ancients could not be done apart from nature, as an isolated enterprise, but required a new role for the naturalist, based upon the sharing of discoveries, experiences, and information.

Naturalists like Conrad Gesner (1516–1565) depended heavily upon wide circles of friends and acquaintances, through which knowledge of new or unusual discoveries might be derived. A list of 227 autographs of friends and visitors inscribed over a ten year period in a small book entitled *Liber Amicorum* suggests the breadth and variety of Gesner's potential correspondence network.[6] From such contacts Gesner received numerous offerings of rare animals, exotic plants, and minerals and acknowledged those who augmented his collections, listing the names of 52 contributors in the preface of his *Historia Animalium.*[7]

In the same manner, two naturalists closely associated with Wilhelm of Hesse, Joachim Camerarius (1534–1598) and Carolus Clusius (1526–1609), obtained much information concerning plants and animals from their many friends and colleagues and communicated their findings regularly to the Kassel court. As with humanist collectors of the *quatrocento* who journeyed frequently between universities and the courts of humanist patrons in the process of recovering and translating ancient manuscripts, Camerarius and Clusius traveled extensively, establishing informal ties with both university and courtly circles.

After studying philosophy and medicine at Leipzig, Wittenberg, and Bologna, Camerarius traveled with the Imperial physician Johannes Crato (1519–1585) in 1563 from Breslau to Padua and thereafter into Hungary. The association with Crato brought Camerarius into contact with the imperial courts at Vienna and Prague and with the circle of humanist scholars and natural philosophers installed there by the Emperors Ferdinand and Rudolf II. By 1564 Camerarius had established himself as a physician in his native city of Nürnberg and erected there a well known medicinal garden in 1569.[8] Although earlier contact with Wilhelm's court is likely, Camerarius first visited the Kassel gardens in 1578. Regular correspondence between Camerarius and the Landgraf began in the following year, becoming especially frequent throughout the 1580's and continuing to the year of Wilhelm's death in 1592.[9] The correspondence suggests an active and continuous exchange of botanical specimens between Kassel and Nürnberg. Many letters contain lists of plants which had matured in either locality and which in turn could provide each correspondent with seeds and cuttings. Camerarius especially benefited from the collaboration. From Wilhelm, Camerarius received substantial financial assistance which allowed the Nürnberg physician to purchase the library of Conrad Gesner in 1581.[10] The importance of the Kassel gardens in the development of Camerarius's own collections, and an appraisal of Wilhelm's botanical interests, appears in the preface of Camerarius's *Hortus Medicus et Philosophicus* (Frankfurt, 1588):

There are now present in Germany as elsewhere the most praiseworthy princes who do not refrain from themselves practicing this art [*i.e.* botany] ... [And I must point out how much] the Landgraf Wilhelm IV of Hesse ... holds a singular love of all the arts and so brings a certain natural inclination to this study. He therefore did not hesitate to provide for himself an exact knowledge of plants and not only cultivates in his gardens a number of rare and select specimens, which Germany has never had before, but also [nurtures there] various fruit bearing trees which it has not fallen to my lot to see elsewhere . . . . Moreover, he most graciously embraces others who study botany even helping and promoting their undertakings by exceptional generosity. Indeed, I particularly am indebted to him and I openly admit having received not a few specimens, brought to me by his highness before all others.[11]

Both Wilhelm and Camerarius corresponded individually with the famous Dutch naturalist, Carolus Clusius. Following a period of study in the Netherlands, Germany, and France, Clusius conducted numerous botanical inquiries throughout central Europe and, in the company of the son of Anton Fugger, journeyed on a botanizing expedition in 1564 through Spain and Portugal.[12] In the autumn of 1573, Clusius arrived at the imperial court in Vienna, at the invitation of Maximilian II. Although formally charged with the organization and care of the court gardens, Clusius continued his own botanical travels, making observations throughout Austria and Hungary. The accumulated observations and collections derived from these and other travels, formed the basis of Clusius's best known botanical studies, to which were added Latin translations of the works of Dodonäus, Belon, Garcia de Orta, and other naturalists.

By 1576 Wilhelm had made contact with Clusius at Vienna and had already received shipments of seeds and plants from the imperial court.[13] Clearly, Wilhelm regarded the imperial gardener as an important botanical agent, eagerly supplying him with lists of specimens most desired for his gardens and awaiting word of Clusius's most recent discoveries. Through a variety of incentives, Wilhelm attracted Clusius several times to Kassel and finally awarded the naturalist with a stipend in 1588, on the condition that Clusius would live close to the court in nearby Frankfurt.[14]

Both Camerarius and Clusius gained for the Kassel court access to important circles of botanical correspondence. Through contact with the focal members of such circles, the observations and discoveries of persons not directly connected with the court were nevertheless filtered through to Kassel. Knowing Wilhelm's interests, Camerarius and Clusius each screened incoming correspondence and 'selectively switched'[15] relevant information to the Landgraf. In a tradition long established among humanist writers, whereby letters apparently designed for a specific correspondent were shared with wide circles of interested scholars, copies of letters pertaining to botany, or actual letters themselves, were regularly gathered together and sent to Hesse-Kassel. Thus, Camerarius often appends letters received from Clusius and other naturalists to his own correspondence and in one instance adjoins a collection of 'Italian letters' received at Nürnberg and recognized as being of interest to Wilhelm.[16]

## IV. PRINCES AND MACHINES

The practical application of mathematics through the construction of useful machines and instruments is an important corollary to the expanded general interest in mathematics during the fifteenth and sixteenth centuries. In Italy, the involvement of humanist scholars in the collection of ancient manuscripts not only brought about the translation of ancient literary and philosophical works but prompted as well the recovery of Greek mathematical texts. Many princely patrons, especially the dukes of Urbino, the Medici and Farnese families, the distant kings of Poland and Hungary, as well as church officials such as the scholar-Cardinal Bessarion (1403–1472) and Pope Nicholas V, enthusiastically supported the discovery and translation of unknown manuscripts and introduced, at their courts, mathematicians and technicians into larger humanist retinues of scholars, poets, and artists.[17]

At the same time, practical needs stemming from dynastic consolidation, exploration, and territorial and commercial expansion emphasized skills pertaining to surveying, cartography, navigation, and fortification and contributed to the increasing presence at court of the mathematician practitioner. While Italy possessed many skilled practitioners, who invented and constructed mathematical devices and who could be called upon for judgments relating to a variety of military, architectural, and artistic problems,[18] the principal centers of instrument making in the late fifteenth and sixteenth centuries were found across the Alps, in Germany and the Low Countries.

Above all, the German cities of Augsburg and Nürnberg gained a recognized preeminence among centers of instrument production. Both cities were important economic centers as well; Augsburg regarded as the headquarters of Fugger banking interests, Nürnberg positioned along an important trade route connecting Italy with the Netherlands. The ease of communication from Nürnberg and the availability of astronomical instruments brought the well known mathematician and astronomer Regiomontanus (1436–1476) to the city in 1471.[19] Regiomontanus's own constructions of observational instruments continued the tradition of instrument making in Nürnberg. The renown of Nürnberg technicians lasted well into the following century, a period best represented there by the globes and measuring instruments of Johann Schöner (1477–1547) and Georg Hartmann (1489–1564) and the clockwork mechanisms of Peter Henlein (1480–1542) and Christian Heyden (1526–1576).[20]

Programs linking instrument making with astronomical observation at Nürnberg provided a major influence in the organization of Wilhelm's technical and astronomical projects. From Nürnberg Wilhelm engaged Andreas Schöner, the son of Johann Schöner, during two years 1558–1560, in several observational projects leading to the construction of planetary and lunar tables. Aside from his observational work, Andreas also completed at Wilhelm's court the first book of a large treatise concerning the design of sundials and astrolabes, *Gnomonice . . . libri tres* (Nürnberg, 1562), which, in accordance with Wilhelm's advice, was dedicated to the Emperor Maximilian II.

The preface of the work is interesting and furnishes evidence of princes in the sixteenth century who not only patronize mathematical studies, but who have become themselves capable mathematical practitioners. After praising the Emperor's own mathematical interests, Schöner extols the merits of several German princes who have become occupied with the design and manufacture of scientific and practical instruments. However, "the greatest in these arts," he writes, "is the illustrious prince Wilhelm, the Landgraf of Hesse." [21]

At Kassel, princely interest in fine mechanical engineering led to the design of projects closely aligning scientific and technical pursuits. The variety of instruments constructed at the Kassel court attracted the attention of the French philosopher, Peter Ramus (1515–1572), who praised the Landgraf as one of several German princes who had developed interests in mathematical studies. "The Landgraf Wilhelm appears to have transported Kassel to Alexandria," he writes, "he has so furnished Kassel with the makers of instruments necessary for observing the stars." [22]

Aside from its observational instruments, Wilhelm's court was perhaps best known as a center for the construction of clockwork-driven celestial spheres and astronomical clocks. One of the best examples from Wilhelm's collection is the famous *Wilhelmsuhr* manufactured by the Marburg mechanician, Eberhart Baldewein (1525–1592), for the Landgraf in 1561. [23] Baldewein's clocks and the clockwork globes of his successor at court, Jost Bürgi (1552–1632), each represented the changing position of celestial bodies over particular time intervals and thus were valued as automated illustrations of celestial order. In certain instances designs could be made to represent novel theoretical positions as well. A small stationary model of the geoheliocentric planetary scheme, commissioned from Jost Bürgi by the Landgraf in 1588, [24] and a later astronomical clock depicting both geocentric and heliocentric planetary configurations, [25] also manufactured by Bürgi, presumably at the court of the Emperor Rudolph II, demonstrate that such celestial machines might also fulfill a pedagogic purpose.

The Kassel court, although a major producer of celestial spheres and clocks, was not alone in the production of astronomical *automata*. The imperial courts of Charles V, Ferdinand, Maximilian II, and Rudolph II as well as smaller courts in Saxony, the Palatinate, Bavaria, Württemberg, and Prussia also commissioned similar machines. Medieval examples, such as the *horologia* of Richard of Wallingford (*c.* 1292–1335) and Giovanni de Dondi (1318–1389), suggest a long tradition of technical skill and ingenuity in the construction of astronomical clocks. Nevertheless, Renaissance princes pursued the construction of celestial *automata* far more intensely than medieval patrons, emphasizing the expression of technical precision in depicting celestial positions as a more important patronage value than artistic design.[26]

Princes who participated directly in scientific and technical projects followed models of involvement already established among Hapsburg patrons. The brother of Charles V, Ferdinand (1507–1564), in addition to employing clockmakers from Innsbruck, Munich, Prague, Pressburg, and Vienna at his court,[27] observed several comets with Paul Fabricius (1519 or 1529–1588), his mathematician and personal physician, and corresponded and conversed with Georg Hartmann in Nürnberg concerning the inclination of compass needles.[28] Through the use of rotating paper constructions designed by Peter Apianus (1495–1552) in an elegant volume known as the *Astronomicum Caesareum* (Ingolstadt, 1540), both Charles and his brother were provided with the means to predict motions of the heavenly bodies solely on the basis of a calculating machine. Apianus's work served as a prototype for many astronomical clocks developed in the later sixteenth century. A metal imitation of one of Apianus's models, moved by gearwork and set in motion by a hand crank, was prepared by Wilhelm of Hesse prior to the construction of the Kassel clocks.[29] Later, assisted by Andreas Schöner, Wilhelm helped compile observational tables, recorded within the Landgraf's copy of Apianus's treatise, which underlay the manufacture of the *Wilhelmsuhr.*[30]

Throughout the construction of a second astronomical clock in 1569 for the Elector August of Saxony, Wilhelm maintained an active association with the workshop of Eberhart Baldewein, the instrument's chief architect. Instructions and suggestions concerning the clock's preparation, especially ideas pertaining to the instrument's driving mechanism, passed frequently between the Landgraf and his mechanician. Baldewein's own ideas concerning the instrument's design were sent to Wilhelm in the form of sketches, while each component of the instrument, when completed, was submitted to Wilhelm for his approval.[31]

Wilhelm's involvement in technical projects places him among several German prince-practitioners. In the Palatinate, the Elector Ott-Heinrich not only patronized the construction of precision mathematical instruments and artifacts but also occupied himself with their design. At least two extant sundials, signed and dated by Ott-Heinrich,[32] confirm his mathematical interests and attest to the increasing aristocratic acceptability of the personal development of mathematical and technical skills. To the many odometers and pedometers commissioned by August of Saxony, the Elector added his own improvements and designed at least four compass dials for use in the Saxon mines.[33] In Braunschweig, the interests of Duke Julius in mining and metallurgy led the prince to design several machines, described in a remarkable manuscript named by Julius the *Instrumentenbuch*, for mining operations in the Harz.[34]

Princely involvement in new technology proceeded normally from the consideration of practical problems in surveying, cartography, and mining. Such activities obtained added significance in the expansion and centralization of early modern states. The specific determination of the extent of state and judicial boundaries, and the independent acquisition of wealth through trade and the development of natural resources, stood at the foundation of personal authority and attached important political and economic operations to the functions of court mathematicians and artisans. The use of precise methods of observation and the systematic collection of information at Kassel extended as much to the court's interest in the compilation of statistical surveys as to the composition of botanical and stellar catalogs. Armed with explicit instructions concerning the precise methods by which information was to be collected and ordered, officials were dispatched from Kassel in order to ascertain, by personal inquiry, the population, social stratification, possessions, resources, and taxes of each village, city, and estate throughout the principality.[35] New land measurements were also made, with an eye to increased accuracy in distinguishing boundaries; and an index of the principal forests, forest rights, and landholders was prepared. The results, compiled in 1585, represented an inventory of current administrative statistics which provided the Landgraf with an exact knowledge of the state's economic character and internal structure.

While the practical utility of precision instruments in the implementation of projects economically and politically beneficial to the state resulted in close collaboration between princely and artisan practitioners, the intellectual component of critical reform, which motivated the organization of Wilhelm's botanical and astronomical projects, is also responsible at Hesse-Kassel for the court's interest in precise mechanical engineering. Since the restoration of observational discrepancies between recorded and firsthand accounts of

natural phenomena depended upon an investigative method embracing exact
measurement, the success of Wilhelm's astronomical projects came to rest, in
the first instance, upon the development of a precision technology capable of
refining the observational capabilities of the court's astronomical instruments.
Utilizing techniques already employed by Wilhelm's close astronomical correspondent, Tycho Brahe, and informally, although only partially, transmitted
to Kassel in 1584, the Kassel court clockmaker Jost Bürgi improved the
precision of the Landgraf's observational instruments so that measurements
which earlier could be made to ± two minutes of arc could thereafter be
recorded to within ± thirty seconds, or even to within fifteen seconds.[36]
Improvements in sighting and in the subdivision of degrees of arc were adopted
to the court's well known azimuthquadrants,[37] among the first of their type
in Europe. Also, a new instrument, the sextant, was introduced to the Kassel
observatory. Since Tycho always viewed the sextant as his own invention,
Wilhelm's measurements with the instrument served to demonstrate its utility
and to confirm at Uraniborg the sextant's precision capabilities. Consequently,
Tycho notes in his *Astronomiae Instauratae Mechanica* (1598):

The value of this instrument [i.e., the sextant] has been proved in a splendid way by the
fact that the distances found by its aid in Kassel agree within a minute, indeed within
one half minute with those found by us in Denmark with our sextants.[38]

Wilhelm had already made numerous astronomical observations prior to
the refinement of the Kassel instruments in 1584. Yet the increased degree of
precision now attainable by the quadrants and sextants at his court once
again stirred the Landgraf's interest in an astronomical project oriented toward the renewed observation of the stars.

## V. THE KASSEL STELLAR PROJECTS

The need for more accurate measurement of the stars and planets is a recurring
theme in the astronomical literature of the later fifteenth and sixteenth
centuries. In the preface of a work concerning planetary theory, dedicated to
Wilhelm of Hesse in 1571, Caspar Peucer, rector of the University of Wittenberg and personal physician of August of Saxony, advised that the Alphonsine
tables and the tables of Copernicus (*i.e.*, the Prutenic tables) were no longer
certain enough and that new and better observations were necessary for
the correct description of planetary movement.[39] Peucer's criticisms were
confirmed by experience at the Kassel court. After observing a lunar eclipse
in the latter part of January 1580, Wilhelm reported his measurements of

the eclipse's duration to the professor of mathematics at the University of Marburg, Victorinus Schönfeld (1525–1591), and instructed Schönfeld to calculate the exact time and duration of the event 'ex fontibus Copernici,' i.e., from the Prutenic tables of Erasmus Reinhold (1511–1553). In compliance with the Landgraf's wishes, Schönfeld calculated the time of the total eclipse from both Reinhold's tables and from the ephemerides of Cyprianus Leovitius. In both, Schönfeld announced, there appeared wide variations between prediction and the Landgraf's actual observations.[40]

Reference to the lack of predictive certainty within even the most trusted astronomical catalogs is frequent in Wilhelm's astronomical correspondence. Beyond disagreements between his own observations and those positions predicted in classical texts, Wilhelm also found that "the tables of Copernicus, Schöner, Ptolemy, and others disagree many times among themselves."[41] Writing to Schönfeld in 1586, Wilhelm revealed his plan for the renewed observation of the fixed stars, "for we find in the tables such great diversities of not only 5 – 6, but often 10 – 11 degrees difference."[42] In the same year Wilhelm's mathematician, Christoph Rothmann (c. 1550–c. 1605), recorded that because of the considerable difference in longitude and latitude within the tables of the ancients, Wilhelm had concluded that one could understand nothing from them and that astronomical learning depended entirely upon new astronomical measurements.[43]

Various conditions contributed to Wilhelm's interest in the formation of programs of precise stellar observation. Problems relating to the Landgraf's technical projects, particularly the necessity of calibrating the court's astronomical clocks; the persisting debates between Protestants and Catholics concerning the Gregorian reform of the calendar (to which Wilhelm remained both theologically and mathematically opposed); and the court's interest in Peter Ramus's Scholarum Mathematicarum (Basil, 1569), in which theoretical descriptions of celestial motions were ruled out in favor of diligent observation and calculation in the acquisition of astronomical understanding, each underscored a general, although urgent, need for exact re-measurement of the stars and planets.

Beyond these, however, two further influences were most important in directing the Kassel stellar programs: (1) Wilhelm's association, through Andreas Schöner, with the tradition of observational reform already well established by Regiomontanus, Bernard Walther, and Johann Schöner in Nürnberg and (2) the recognition of observational problems arising as a result of the court's astrological projects.

With his patron, Bernard Walther (1430–1504), Regiomontanus established in Nürnberg an astronomical observatory, printing press, and workshop for manufacturing mathematical instruments, in an attempt to correct the observational errors which, so he considered, had arisen in ancient texts due to the corruptions of medieval translators and commentators.[44] The reform program at Nürnberg entailed projects of exact observation complemented by the re-editing of classical texts and the publication of contemporary works, including Regiomontanus's own treatises. Walther himself completed 746 solar and 615 planetary measurements while attempting to correct easily recognizable disagreements between direct measurement and the Ptolemaic and Alphonsine tables.[45] However, while the observational aspect of Regiomontanus's intended reform thus continued in Nürnberg, Walther, who purchased the manuscripts and instruments of his associate following Regiomontanus's death in 1476, did little to further the mathematician's program of publication. Many of Regiomontanus's writings consequently remained sequestered in Walther's library until purchased by the city of Nürnberg in 1522.[46]

The reform efforts of Regiomontanus and Bernard Walther resumed once again at Nürnberg with the work of Johann Schöner (1477–1547), globe-maker and, after 1526, professor of mathematics at the Nürnberg Gymnasium. Schöner, who viewed the Nürnberg observations as an essential part in the recovery of astronomical learning in Germany, edited several of the unpublished manuscripts of Regiomontanus after 1531 and published his own astronomical tables, the *Tabulae Resolutae*, in 1536. These tables circulated widely, with additional printings appearing in 1551 and 1561 as part of a collection of the mathematician's works, followed by a separate edition in 1587–1588. At Wittenberg the tables were enthusiastically endorsed by Melanchthon whose preface to the 1536 edition emphasized their practical utility as an instructional aid in astronomical studies.[47]

Direct contact between Wilhelm and the Nürnberg tradition resulted in 1558, when Andreas Schöner began his two year residency at the Kassel court. Little is known of Andreas's mathematical education although at least an introduction to mathematics and astronomy was probably provided him by his father.[48] Nevertheless, following the death of Johann Schöner in 1547, the Schöner library, which by now included several of the works listed by Regiomontanus in his reform program, came into the possession of Andreas. Continuing Regiomontanus's original design, Andreas added to the collection by seeing through the press the collected works of his father[49] and a work pertaining to the computational foundation of one of Regiomontanus's important catalogs of observation, intended for comparison with classical sources, *Tabulae Primi mobilis* (1514).[50]

Certainly Andreas was in a position to discuss the observational work of his father and to comment upon the attempt to restore the accuracy of astronomical observations made earlier at Nürnberg. Yet, while Wilhelm in this way encountered the Nürnberg tradition of observational reform, the idea of what needed reforming by renewed measurement was interpreted differently at Kassel. Unlike the measurements of Regiomontanus, Wilhelm's observations were not intended solely as an attack upon medieval scholasticism. Rather, the observational projects at Wilhelm's court were influenced and directed by noticeable disagreements as much within current catalogs, including those of Regiomontanus and Schöner in Nürnberg, as within classical tables. Wilhelm thus broke with a strictly humanist understanding of observational reform. Regardless of the errors in ancient sources (whatever their origin), if contemporary observations did not agree among themselves, there was continued need of exact measurement based upon refinements in procedure and technology.

Examinations of Wilhelm's astronomical interests have frequently focused upon the Kassel observations as precursors of modern scientific measurement.[51] While the procedures in use at Wilhelm's court and the methods of modern science both value exact description, placing the Landgraf in such a forward-looking posture avoids the entire question of the role of esoteric pursuits in the formation of the court's astronomical projects. Yet such interests contributed a major incentive toward exact observation at Kassel.

The direct influence of stars and planets upon human action was plainly rejected by Wilhelm. The Landgraf objected, therefore, to the assertions of divinatory astrology, which employs the determinist belief that celestial bodies direct human inclinations and necessitate occurrences. By rejecting this, however, Wilhelm still maintained a wide spectrum of astrological options. Natural objects and events, distinct from human behavior, might be affected by secondary (i.e., physical) causes produced by alignments in the heavens. Moreover, celestial manifestations could be understood as revelations of divine order and therefore looked upon as presaging changes in pre-established historical epochs. Wilhelm adopted both beliefs, viewing as portents both unexplained occurrences, such as the new star of 1572, and predictable major celestial alignments, the most important of which was the conjunction of Saturn and Jupiter.[52]

Although earlier conjunctions received the attention of the Kassel court, the prediction of a maximum conjunction in 1584 inspired an astrologically oriented program of precise measurement. The scope of the endeavor was outlined to August of Saxony as early as 1576.

I have been occupied for some time with the maximum conjunction of Jupiter and Saturn which will occur in a few years almost in the same point of the heavens where the last comet appeared. This will undoubtedly bring about such a change . . . that I wish to extract and calculate accurately the times of convergence as well as the times of mean and real conjunction and, therefore, to note the moments of historical change [which coincide] with it. 'This also will allow one to conjecture more accurately since the result of these changes and the corresponding time of the occurrence in the heavens [is known]. Although this will be a great labor, it will nevertheless be such a work the likes of which has never been seen before. [53]

   Wilhelm was well acquainted with contemporary astrological literature dealing with major conjunctions. Discussions of the astrological importance of conjunctions appeared in Peter Apianus's *Astronomicum Caesareum*, a work well known to the Landgraf. Other works concerned with the historical changes announced by major conjunctions, were read at Kassel.[54] Predictions of the religious and political changes to occur as a result of the maximum conjunction of 1563 were prepared by Victorinus Schönfeld at Marburg and sent to Wilhelm. However, in predicting the time of this conjunction, neither the Alphonsine nor the Prutenic tables proved trustworthy. These were no slight deviations. Predictions based on the Alphonsine tables erred by more than a month, while the Prutenic tables were unable to predict the actual day of the occurrence. The lack of predictive dependability within frequently used tables must have been as obvious to Wilhelm as it was to Tycho Brahe, who recalled the discrepancies with alarm in his *De Nova Stella* (1573).[55]

   The maximum conjunctions of 1563 and 1584 were observational events important to the Kassel court. Yet in fixing the time and positions of the convergence of Saturn and Jupiter, Wilhelm confronted an underlying observational problem. To establish the position of Saturn, Wilhelm employed a procedure similar to Bernard Walther's method of deriving planetary positions from two fixed stars. However, in this operation Wilhelm found that the traditional astronomical tables only confused his attempts to determine the correct place of the planet. Accordingly, Rothmann writes that

When [the Landgraf] by means of the left shoulder of Orion [Betelgeuse] which he derived from the tables, had established the true position of Saturn and had used yet another fixed star, also derived from the tables to verify the same position of Saturn, he found essentially differing results as if the star of the left shoulder of Orion had deviated several degrees from the positions given in the tables. He [the Landgraf] concluded at that point to restore [*restituit*] accurately the positions of the fixed stars.[56]

   Wilhelm's motives for observing Saturn may not have been entirely astrological. In fact, Wilhelm had discovered discrepancies between tabular and observational measurements long before his concern with the conjunction of 1584. Yet planetary observations at Kassel are comparatively rare. Venus and Jupiter gained the greater part of Wilhelm's planetary measurements for reasons, to be discussed hereafter, connected with the court's stellar projects. An awareness of Saturn's astrological significance in maximum conjunctions must have played a substantial role in establishing its observational value. The need for observational reform was therefore the greater, since both astronomical and astrological understanding depended upon the ability to predict major alignments, which predictions rarely proved exact when derived from traditional tables.

Although it is impossible to determine the specific number of observations compiled by the Kassel observers, their general range and orientation can be presented through a survey of extant manuscript references and printed sources. Since it is assumed that many more observations were made than appear within available catalogs, the information provided here is intended only as a partial index of observational direction and incidence at the Kassel court.

The most consistent, long-term set of observations relate to the position of the sun. Observers at Kassel measured solar meridians each year almost without interruption from 1561 to 1596, with a preponderance of measurements (342 recorded observations) falling between 1584 and 1590, the years of the court's most active stellar observations. In all, the Kassel lists document over 650 meridian measurements of the sun, to which must be added the observation of solstices and Christoph Rothmann's observations of the sun at 'twilight', for which there exists no direct numerical account.

The observation of solar meridians constituted an important prelude to the organization of the court's stellar programs. By determining the position of the sun at the meridian, observers could establish the obliquity of the ecliptic, which in turn provided a frame of reference for the determination of stellar positions by laying the basis for all measurements in celestial longitude and latitude. Moreover, by observing the right ascension of the sun, the Kassel astronomers were able to measure the right ascension of a star by connecting it to the sun through an intermediary planet.

Three distinct periods of stellar observation stand out from the Kassel manuscripts. An initial interval, roughly 1559–1561, coinciding with the residence of Andreas Schöner at Wilhelm's court, actually combined several observational efforts, including the measurement of an unspecified number of fixed stars by the Landgraf.[57] During the two years 1566–1567, Wilhelm again involved himself in the observation of the stars and recorded at that time 58 stellar positions.[58] These were cataloged according to right ascension and declination as well as by longitude and latitude and were combined ultimately by Rothmann in the court's stellar catalog, the so-called *Hessischen Sternverzeichnis*. Following the arrival of Christoph Rothmann (1577) and Jost Bürgi (1579) at the Kassel court, and with the refinement of the Kassel observational instruments in 1584, began the third and most intensive period of stellar measurement. Over the next five years, particularly during 1585–1586, Rothmann carried out hundreds of stellar observations measuring meridian altitudes, culminations, and angular distances and arranged these into several extant catalogs.[59]

As such evidence indicates, the Kassel measurements were not scattered but focused primarily upon positions of the sun and stars. With the notable exception of Jost Bürgi's systematic observation of the planets, undertaken for the most part after the Landgraf's death, planetary observations at Kassel were of secondary concern unless relating to astrological events or to the court's projects of stellar measurement. Thus observations of Jupiter and Venus, whose transits across the meridian were used to measure stellar right ascension, comprised an essential part of the Kassel stellar projects.

According to Rothmann, the basis of the court's stellar program rested upon the accurate measurement of the right ascension of two stars, *Oculus Tauri* (Aldeberan) and *Canem Minorem*.[60] Initially the Kassel astronomers determined right ascension by calculating first the angular distance between the planet Jupiter and the sun and then observing the distance between Jupiter and a designated star.[61] Since the right ascension of the sun could be derived from prepared solar tables, the absolute right ascension of the star was established directly through the observation. The fact that Jupiter possessed a smaller diameter than the moon (used since the time of Hipparchus to link the positions of the fixed stars to the sun), a slower motion and an extremely small parallax undoubtedly entered into the decision at Kassel to use the planet. However, since Jupiter and the sun never appeared together above the horizon, it was not possible to measure the distance between them directly. The problem was less severe at Kassel than elsewhere, since the development of an innovative type of timekeeping escapement by the court's clockmaker Jost Bürgi (in which two balances were made to act in steady opposition, beating out nearly regular seconds) made it possible to relate the two bodies through temporal rather than angular measurement.

The use of Jupiter involved trusting the Kassel clocks for several hours, ample time for even Bürgi's 'cross beat' escapement to fall several seconds short of precise measurement. By 1587, therefore, a technique employing the planet Venus, which could be observed while the sun was still above the horizon, was adopted at Kassel.[62]

To measure distances between stars, Rothmann initially employed the Kassel sextant. Later, however, the Kassel observers chose to use one of Bürgi's second-beating clocks with the quadrant set at a fixed azimuth.[63] Since the apparent motion of celestial bodies traverses fifteen minutes of arc in one minute of time, the angular distance between the stars in an east-west direction was determined by timing the interval of two stars crossing the same azimuth point.

Astronomical observation by means of time measurement reached its fullest and most consistent expression in the sixteenth century at the Kassel court. The observations of Wilhelm and Rothmann of azimuths as well as altitudes and times and meridian altitudes and times are the first programatic attempts to establish the positions of heavenly bodies by such a procedure. Nevertheless, the underlying method of relating time and angular measurement did not originate with Wilhelm and his assistants, nor were the Kassel observers the sole agents of the technique. In Nürnberg, Bernard Walther had measured the risings of Mercury and the sun as well as a lunar eclipse by means of a weight-driven clock.[64] Tycho also used weight-driven clocks to determine the right ascension of twelve stars observed with the comet of 1577 and, in the description of his mural quadrant, refers to four clocks contained in the observatory at Hveen. Yet such clocks were never capable of the precision afforded by Bürgi's escapement. Without this technical advantage, Tycho found observation by means of clocks far from satisfactory and recognized himself the underlying difficulty with the clocks at Uraniborg: that the weight of the cord suspending the weight, when rolled and unrolled, actually added to the irregularity of the total weight involved in the operation.[65]

## VI. COMMUNICATION AND COSMOLOGY

Collaboration among observers is an essential element in the confirmation of exact scientific measurement. At Kassel, correspondence through princely channels allowed Wilhelm to corroborate measurements through an exchange of information within a well defined courtly circle. Through this network, Wilhelm summoned the judgments and reports of court physicians, mathematicians, and university professors as part of an on going comparative analysis of observations and opinions concerning unusual celestial appearances, comets, and the sudden presence of the new star of 1572.

The new star gained much attention at Hesse-Kassel. It was, however, through princely correspondence that Wilhelm first noticed the phenomenon. To Friedrich of the Palatinate, who sent to the Landgraf observations of the new star obtained at the University of Heidelberg, Wilhelm confessed that since assuming full responsibility for the government of Hesse-Kassel (1567), the affairs of state had left him almost no time to pursue his mathematical and astronomical interest. "For that reason," Wilhelm continued, "we did not know of this star until we were reminded of it by our dear cousin and brother, the Elector of Saxony [i.e., August of Saxony]."[66]

Wilhelm first observed the new star on December 3, 1572, and on the following day communicated his early measurements to the Saxon court. At this time at least, although warning that detailed measurements needed still to be made, Wilhelm considered the phenomenon to be situated remotely from the earth, yet still within the circle of the moon. From the Saxon court Wilhelm requested in return the observations of the Wittenberg mathematician Caspar Peucer. Comparison of Peucer's measurements with those obtained at Kassel however proved disappointing. Nor were renewed observations able to decrease discrepancies of $1°$ + in longitude and one-half degree in latitude, leading Wilhelm to seek resolution of the disagreement by submitting both the Kassel and Wittenberg measurements to the judgment of Victorinus Schönfeld at Marburg, where each could be compared with Schönfeld's own sightings.[67]

While the Wittenberg measurements thus failed to confirm the Kassel observations directly, Wilhelm did not doubt the care given the measurement of the new star in Saxony, but considered that the Wittenbergers operated at a disadvantage because of technical deficiencies in the Saxon instruments.[68] A further problem at Wittenberg, the Landgraf noted, may have arisen since the observations in Saxony were based upon the "common reckoning from the tables of fixed stars" — which tables Wilhelm knew to be incorrect, resting his own measurements upon observations of the stars observed anew from Kassel.[69] Such obstacles could also account for the difference in the star's parallax calculated at Kassel and Wittenberg. Whereas Wilhelm finally established a parallax not exceeding three minutes, Peucer, even with repeated observations, was unable to calculate a parallax of less than nineteen minutes. Finally, in an attempt to obtain observational parity at both Kassel and Wittenberg, Wilhelm decided to transmit both the technology and procedures employed at his court to Saxony, sending Peucer and his fellow mathematicians a beating clock (*Horologium pulsatite*) and a Kassel quadrant.[70]

Completed calculations at Kassel placed the appearance beyond the moon, between Venus and the sun. The problems posed by such a position within the distinct realms of an Aristotelian cosmology were obvious to the Landgraf.

And thus it is clear that the comet is placed in the upper part of the sphere of Venus ... whence it is certainly consistent that it is not situated in the elementary region. Since the *physici* allow no generation or corruption in that place it is the more amazing.[71]

From other courts as well, Wilhelm received reports and opinions of the new star. Through contact with Ludwig of Württemberg (1554–1593), Wilhelm requested the measurements of the mathematicians at the University of Tübingen and thereafter received the opinions and observations of Philip Apianus (1531–1589).[72] From the Palatinate the measurements and judgments of Cyprianus Leovitius were also sent into Hesse.[73] In the evaluation of these reports, Wilhelm employed his own university mathematician, sharing the observations of university professors in Württemberg, Saxony, and the Palatinate with Victorinus Schönfeld, who appraised each description in comparison with the Kassel and Marburg measurements.[74]

Most observers of the new star also observed and commented upon the comet of 1577. Arguments concerning the physical position of the new star carried over to discussions of the comet and focused again upon the determination, or lack of determination, of parallax. Several astronomers, including Wilhelm, Tycho Brahe, Michael Maestlin, the Württemberg physician Helisaeus Roeslin, and Cornelius Gemma, concluded from their observations that the comet must describe a path above the lunar sphere.[75] Others, notably Thaddeus Hegecius at the imperial court, Bartholomew Scultetus, professor of mathematics at Leipzig, Andreas Nolthius, and the Nürnberg astronomer, Georg Busch (who dedicated his treatise concerning the comet to Wilhelm IV), found by comparison a large parallax for the comet and placed it therefore below the moon.[76] In all, Wilhelm recorded 67 observations of the event, specifying the time of the observation and denoting the comet's occidental azimuth and altitudes. These were later published by Tycho in Chapter Ten of his *De Mundi Aetheri . . . Phaenomenis* (1586).

As with the new star, the calculation of a superlunary path for the comet of 1577 carried with it important theoretical consequences and cosmological implications. From his own measurements of the comet of 1577, and with the support of Christoph Rothmann's observations of the comet of 1585, for which again no parallax was determined, Wilhelm found sufficient incentive to break a customary philosophical silence. "The principle of the philosophers," he concluded, "is destroyed, that comets should be generated in the upper region of the air under the circle of the moon."[77]

It would be interesting to know exactly how Wilhelm stood regarding the question of the material composition of the heavens. One can, however, only approximate an answer. As is generally understood, sixteenth-century Aristotelians were of the same mind neither in their theory of celestial matter nor in the degree to which they insisted upon a strict distinction between the heavens and the earth. Moreover, the placement of comets above the moon did not in itself ruin every interpretation of Aristotelian world order. Lack of comet parallax could be made consistent with the basic Aristotelian distinction between sublunary and superlunary regions by sacrificing solid spheres for an eternal and perfect fluid, from which material comets might also be formed.[78]

Unfortunately, we possess no statement from Wilhelm himself as to how his observations influenced his overall comprehension of heavenly matter. Lacking evidence to the contrary, we may assume that while the Landgraf questioned the conservative mode which insisted upon a strict distinction between celestial substances and the elements of the sublunary region, Wilhelm still accepted some type of Aristotelian explanation of celestial and terrestrial differences. However, a more far reaching, and indeed solidly anti-peripatetic, interpretation of Wilhelm's physical views seems evident from an oration prepared by the Italian natural philosopher, Giordano Bruno (*c*. 1548–1600), given at the University of Wittenberg in 1588.

Among the Germans we find not only princes who care enough to foster the study of astronomy in general ... but also, above all, we find the rescuer of [astronomical learning] the great Landgraf Wilhelm of Hesse. Rather than alienating his senses and the intelligence of his eyes [this prince] learns astronomy by experience [*callet astronomiam*] and not through the exorbitant peripatetic philosophy attached to the Ptolemaic theory. He rejects the adoption of orbs and spheres with affixed or imbedded stars. He knows that comets are of the same substance as other stars and that they pervade through the extent of the aetherial region, bearing witness, for that reason, to one continuous heaven without a level of air and a level of aether. He has also observed that the newly appearing stars ... penetrate and pass through those unsuitable spheres. From whence it follows that such a region with its inpenetrable, indivisible and unalterable *Quinta Essentia* is a chimera.[79]

Bruno's description must be regarded with suspicion. Like Peter Ramus and Andreas Schöner, Bruno found much to praise in the mathematical and astronomical interests of German princes. The measurements of comets at Kassel, to which he again refers in Book VI, Chapter XX, of his *De Innumerabilibus, immenso et infigurabili* (1591), provided observational evidence consistent with Bruno's own theoretical ideas concerning an infinite universe with no distinct or privileged parts,[80] and such views undoubtedly colored his interpretation of Wilhelm's physical conception of the heavens. Indeed, Bruno's descriptions seem better suited to the physical conceptions of Wilhelm's assistant, Christoph Rothmann, who openly rejected both solid spheres and all distinctions separating celestial and terrestrial regions.[81]

To whatever degree the observations of the new star of 1572 and the measurement of the comets of 1577, 1580, and 1585 altered Wilhelm's Aristotelian physical views of the universe, the traditional mathematical structure of heavenly movement remained unaffected by their investigation. Throughout his observations Wilhelm remained committed to the Ptolemaic description of planetary motion. Yet, by preferring this model to others, Wilhelm was not intolerant of alternative systems. As a result, the Kassel court offered an open intellectual environment for the discussion of both Copernican and geoheliocentric theories. The prevalence of such an atmosphere at court suggests that Wilhelm himself recognized the applicability of the Ptolemaic, Copernican, and Tychonic models at least for purposes of calculation. Indeed, theoretical pluralism of this sort had already allowed for the acceptance of the Copernican hypothesis within the mathematics curriculum at Protestant universities in Germany, notably Wittenberg, Altdorf, and Königsberg during the last half of the sixteenth century.[82]

As we have seen, political and religious links between Kassel and Wittenberg had been well formed since the earliest years of the Reformation. Theologically and personally, Wilhelm adopted the moderate stance of Melanchthon, many of whose doctrinal essays were collected at the Hessian court. Moreover, contact with several university mathematicians, such as the Landgraf's own court mathematician Christoph Rothmann, Johannes Praetorius at Altdorf, and Melanchthon's son-in-law Caspar Peucer, nurtured the Wittenberg connection. Through the mediation of Caspar Peucer, Melanchthon's copy of Ptolemy's *Almagest* came to Kassel for use in a more clearly humanist attempt to reform textual errors by comparing the original text with later translations.[83] In emphasizing the study of mathematics and astronomy at Wittenberg and in introducing the Copernican model within the mathematics curriculum, Melanchthon played a central role. Following Melanchthon's personal reappraisal of the heliocentric model, Wittenberg mathematicians adopted an interpretation of the Copernican theory as a mathematical or heuristic device while rejecting its physical implications.[84] As with the Wittenbergers, it seems certain that Wilhelm remained receptive to new theories which possessed a high degree of observational correspondence for predictive purposes, while preferring the Ptolemaic model overall, largely for reasons of convenience and from an inclination to the physical connotations of a stationary earth.

Wilhelm's theoretical preferences were not those of his assistant, Christoph Rothmann. Within a correspondence involving Rothmann, Wilhelm, and Tycho Brahe, beginning in 1585 and continuing to the year of Wilhelm's death,[85] Rothmann takes the position of a confident Copernican, after describing his own attempt to construct a geo-heliocentric planetary model.

Rothmann conceived of the main elements of a geo-heliocentric cosmology around 1583 and illustrated them within a manuscript, extant at Kassel, the *Astronomia*.[86] As with Tycho's model, the essential design placed the sun and moon in motion around the earth while all the other planets centered upon the sun. Although it is not possible to discuss the details of Rothmann's system here, at least a partial understanding of his intentions in its construction arise from a letter to Tycho written in 1587.

I also wrote ... the *Astronomia* in which I carried over the hypotheses of Copernicus to the mobility of the sun ... and made manifest by the Prutenic calculations the single schemes of the theorics. But since truly I found through observation that the Prutenic tables were remarkably distant from the truth of the sky, I hesitated until now in its publication lest I should entrust to students something full of errors of calculation ... I was incited to this thing by the most bitter sorrow, because I saw, not without great indignation, that in the universities, where these studies especially ought to flower, almost nothing is known concerning this [*i.e.*, the Copernican] theory. Moreover the Prutenic calculations are in the hands of everyone, yet many less draw upon their original precepts, nor are they to be found who understand his [Copernicus's] hypotheses, not even among those who profess astronomy and think themselves great mathematicians.[87]

Rothmann's indignation at not finding the Copernican hypothesis well represented at German universities is interesting in light of the theory's frequent acceptance within mathematics curricula. Through Erasmus Reinhold (1511–1533) and Caspar Peucer, students at Wittenberg continued to encounter the Copernican theory, if only as a means to facilitate calculation, throughout the latter part of the sixteenth century. Rothmann must have been aware of the extent to which the Copernican theory had been accomodated within German schools. Yet, like Praetorius, Michael Maestlin, and Tycho Brahe, Rothmann may have considered that by emphasizing Copernicus the mathematician, the Wittenberg tradition had neglected long enough the important cosmological elements of the Copernican system.

Rothmann planned to publish his *Astronomia* after substituting Tycho's observations for those recorded in the Prutenic tables. Although employing Tycho's measurements, the intended heuristic purpose to be served by the model remained unchanged. Before its publication, therefore, Rothmann decided to relinquish his 'crypto' Copernican design, preferring to argue thereafter with Tycho as a defender of the heliocentric view.

The conditions which led Rothmann to defend the Copernican theory rest both upon his early education at Wittenberg and his experiences at the Kassel court. At Wittenberg, Rothmann was one of many students in the 1560's and 1570's to explore the mathematical utility of the Copernican theory. Observational data obtained at Kassel through the use of the court's instruments and procedures led him to consider more intensely the cosmological aspects of the heliocentric scheme. With other court mathematicians like Jost Bürgi, Rothmann discussed his thoughts concerning the reality of the Copernican universe, while informal communication links between Kassel and Tycho's observatory provided him with an opportunity to confront his Copernican views through friendly polemical correspondence. Within this correspondence the Kassel mathematician constructed his formal defense of the Copernican theory, arguing not only on the basis of mathematical economy and harmony, but answering also to the physical and religious objections encountered by the supposition of a moving earth.

## VII. CONCLUSION

By suggesting the extent to which the Kassel court functioned as an institutional node of scientific and technological activity, the preceding study has set forward several features of an important aristocratic component of discovery within Renaissance science. The role of the court in the establishment of informal networks of communication contributed a valuable social aspect to the diffusion of ideas in the sixteenth century and expanded the opportunities for collaboration among mathematicians, technicians, and naturalists. While the importance of printed sources in the dissemination of formalized ideas and general knowledge about the world cannot be underestimated, contact among researchers at the actual threshold of discovery and experience takes place, as much today as in the Renaissance,[88] most frequently within informal social modes. As with printed material, the communication of observations and research experience according to specific patterns of informal interaction could bring important information to a large number of interested readers who themselves held no direct contact with the communication source.

The study of the Kassel court has led also to the identification of a specific type of Renaissance prince who both recognized the social and political importance of science and technology and engaged himself in scientific and technical pursuits. The influence of such princes, I suggest, is at least partially responsible for the development of new methodological values among artisans and mathematical practitioners in the sixteenth century. Amid projects of political and economic interest to the court, procedures which emphasized technical and descriptive precision and the systematic collection of information, attained not only practical significance but also aristocratic respectability.

In the organization of botanical and stellar projects at Kassel, Wilhelm of Hesse takes a place beside many in the Renaissance who recognized the necessity of direct observation and the precise description of natural phenomena as initial procedures in the recovery of knowledge following the perception of apparent inaccuracies within ancient texts. Thus Wilhelm's court affords one of the clearest alignments of procedural values characteristic of the projects of prince practitioners and illustrative as well of an important and extensive intellectual tradition in the Renaissance. Such a confluence of utilitarian and intellectual procedures plays a significant role in forming the context of a developing empirical-experimental methodology in early modern science, the essential operational features of which are themselves discovered and later refined as parts of the seventeenth century's experimental philosophy.

*Department of History*
*University of Nevada, Reno*

## NOTES AND REFERENCES

[1] Concerning the Flacian-Philippist controversy see: Wilhelm Prager, *Matthias Flacius und Seine Zeit* (Erlangen, 1859–1861); Lauri Haikolu, *Gesetz und Evangelium bei Matthias Illyricus: eine Untersuchung zur Lutherischen Theologie vor der Konkordienformel* (Lund, 1952); Hans-Werner Gensichen, *We Condemn: How Luther and 16th-Century Lutheranism Condemned False Doctrine*, trans. by Herbert J. A. Bouman (St. Louis, 1967).
[2] Robert Calinich, *Kampf und Untergang des Melanchthonismus in Kursachsen in den Jahren 1570–1574* (Leipzig, 1866).
[3] Ludwig Zimmermann, *Der Ökonomische Staat Landgraf Wilhelm IV* (Marburg, 1933), vol. I, pp. 167–168.
[4] Most of Wilhelm's botanical correspondence with other princes must be consulted in manuscript: Hessisches Staatsarchiv, Marburg: 4A. 32. 28; 4A. 32. 18; 4A. 32. 19; 4B. 40. 20; 4B. 40. 25; 4B. 40. 44; 4B. 40. 60; 4B. 5. 3.
[5] Hessisches Staatsarchiv, Marburg: 4B. 40. 24; letters 1578–1580. See also Hermann

Friedrich Kessler, *Landgraf Wilhelm IV von Hessen als Botaniker* (Cassel, 1859), p. 18.

[6] Richard J. Durling, 'Conrad Gesner's Liber amicorum: 1555–1565', *Gesnerus: Vierteljahrschrift für Geschichte der Medizin und der Naturwissenschaften* 22 (1965), 134–159. See also the section 'Einfuhrung zum Bilderteil', by Rudolf Steiger in *Conrad Gesner 1516–1565: Universalgelehrter, Naturforscher, Arzt* (Zurich, 1967), pp. 129–130.

[7] *Conradi Gesneri . . . historiae animalium lib. I. de quadrupedibus viviparis . . .* (Tiguri, 1551).

[8] Melchior Adam, *Vitae Germanorum Medicorum*, in: *Dignorum lauda Vivorum . . .* (Frankfurt a. M., 1706), pp. 154–159.

[9] Hessisches Staatsarchiv, Marburg: 4A. 31. 36. See also Maximilian Reess, *Über die Pflege der Botanik in Franken von der Mitte des 16. bis zur Mitte des 19. Jahrhunderts* (Erlangen, 1884).

[10] Christian Bay, 'Conrad Gesner (1516–1565): The Father of Bibliography', *The Papers of the Bibliographical Society of America* 10 (April, 1916), 53–86. Hessisches Staatsarchiv, Marburg: 4A. 31. 36; Wilhelm to Camerarius, April 14, 1581.

[11] Joachim Camerarius, *Hortus Medicus et Philosophicus: In quo plurimarum stirpium breves descriptione, Novae Icones . . .* (Francofurti ad Maenum, 1588), *ad lectorum.*

[12] F. W. T. Hunger, 'Charles de L'Escluse 1526–1609', *Janus* 31 (1927), 139–151. Idem, *Charles de L'Escluse, Nederlansch Kruidkundige: 1526–1609*, Vol. I (S-Gravenhage, 1927).

[13] Kessler, p. 15. Hessisches Staatsarchiv, Marburg: 4B. 40. 23; 4A. 32. 12; 4A. 31. 37. See also Gy. Istranffi, *Études et commentaires sur le Code de L'Escluse, Augmentés de quelques Notices Biographiques* (Budapest, 1900).

[14] Hunger, *Charles de L'Escluse, Nederlandsch Kruidkundige*, Vol. I, p. 168.

[15] The term is used by Herbert Menzel, 'Planning the Consequences of Unplanned Action in Scientific Communication', in *Communication in Science: Documentation and Automation*, ed. De Reuck *et al.* (Boston, 1967), 57–71. *Idem.*, 'Informal Communication in Science: Its Advantages and Its Formal Analogues', in *The Foundations of Access to Knowledge*, ed. Edward B. Montgomery (New York, 1968), pp. 153–163.

[16] Hessisches Staatsarchiv, Marburg: 4A. 31. 36; Camerarius to Wilhelm, March 13, 1585; May 27, 1587; March 2, 1588; March 14, 1591.

[17] Paul Lawrence Rose, *The Italian Renaissance of Mathematics: Studies on Humanists and Mathematicians from Petrarch to Galileo* (Genève, 1975).

[18] See: A. G. Keller, 'Mathematicians, Mechanics and Experimental Machines in Northern Italy in the Sixteenth Century', in *The Emergence of Science in Western Europe*, ed. Maurice Crosland (New York, 1976), pp. 15–34.

[19] Derek J. Price, 'Precision Instruments: To 1500', in *A History of Technology*, ed. Charles Singer *et al.* (New York, 1957), Vol. III, pp. 582–619.

[20] Paul Melchger, 'Nürnberg und seine Uhrmacher vor dem 17. Jahrhundert', *Schriften des Fachkreises 'Freunde Alter Uhren'in der Deutschen Gesellschaft für Chronometrie* 8 (Ulm, 1968/69), 6–13.

[21] Andreas Schöner, *Gnomonice . . . hoc est: de descriptionibus horologiarum scioteri-corum omnis generis . . . libri tres* (Noribergae, 1562), *epistolae nuncupatoria.*

[22] *P. Rami Scholarum mathematicarum libri unus et triginta* (Basilae, 1569), p. 67.

[23] See. H. Allen Lloyd, *Some Outstanding Clocks over Seven Hundred Years 1250–1950* (London, 1958), pp. 46–60.

[24] *Cf. Tychonis Brahe Dani Opera Omnia*, ed. J. L. E. Dreyer (Hauniae, 1919), Vol. VI, p. 157.

[25] Henry C. King, *Geared to the Stars* (Toronto, 1978), pp. 78–84; esp. p. 83.

[26] Bruce T. Moran, 'Princes, Machines and the Valuation of Precision in the Sixteenth Century', *Sudhoffs Archiv* 61 (1977), 209–228.

[27*] Erwin Neumann, *Der Königliche Uhrmacher Moritz Behain und seine Tischuhr von 1559* (Luzern, 1967), pp. 8–9.

[28] Ernst Zinner, *Deutsche und Niederländische Astronomische Instrumente des 11.–18. Jahrhunderts* (München, 1956), p. 385.

[29] Bernard Sticker, 'Landgraf Wilhelm IV und die Anfänge der Modernen astronomischen Messkunst', *Sudhoffs Archiv* 40 (1956), 15–25; p. 18. Reprinted in Sticker, *Erfahrung und Erkenntnis* (Hildesheim, 1976), pp. 234–240.

[30] These obervations were recorded by Schöner in the Landgraf's copy of Peter Apianus's *Astronomicum Caesareum* (Ingolstadt, 1540), Murhardsche Bibliothek der Stadt Kassel und Landesbibliothek, Kassel: MS astron. 2° 16.

[31] Susanne Voight, 'Ebert Baldewein, der Baumeister Landgraf Ludwigs IV von Hessen-Marburg, 1567–1592', Dissertation at the University of Marburg, 1942, p. 32.

[32] Max Engelmann, *Sammlung Mensing: Altwissenschaftliche Instrumente* (Amsterdam, 1924), p. 34, number 237. A second sundial signed 'O. H. P.' (Ott-Heinrich Pfalzgraf) was discovered in 1970 by the curators of the Mensing collection, Adler Planetarium, Chicago, as part of a private collection in Amsterdam.

[33] Max Engelmann, 'Die Wegmesser der Kurfürsten August von Sachsen', *Mitteilungen aus den Sachsischen Kunstsammlungen* 6 (1915), 11–43. Also, Ad. Drechsler, *Mathematisch-Physikalischer Salon im Westlichen Flügel des Zwingers* (Dresden, 1874), p. 5.

[34] Niedersachsisches Staatsarchiv, Wolfenbüttel: 2 Alt. 5228. See Gerd Spies, 'Werkzeuge, Geräte und Maschinen in Braunschweigischen Steinbruch', *Museum und Kulturgeschichte: Festschrift für Wilhelm Hansen*, ed. Martha Bringmeier *et al.* (Münster, 1978), pp. 233–244.

[35] Zimmermann, *Ökonomische Staat . . .*, Vol. I, pp. 111–288.

[36] Tycho Brahe, *Opera Omnia*, Vol. VI, p. 31.

[37] *Ibid.*, Vol. III, p. 127.

[38] Tycho Brahe, *Tycho Brahe's Description of His Instruments and Scientific Work as Given in Astronomiae Instauratae Mechanica*, trans. and ed. by Hans Raeder *et al.* (Kobenhaven, 1946), p. 79.

[39] Caspar Peucer, *Hypotheses Astronomiae, seu Theoriae Planetarum ex Ptolemaei et aliorum . . .* (Wittenberg, 1571).

[40] Hessisches Staatsarchiv, Marburg: 4A. 32. 2.

[41] Hessisches Staatsarchiv, Marburg: 4F Kursachsen 82; November 17, 1586.

[42] Hessisches Staatsarchiv, Marburg: 4A. 31. 8.

[43] 'Tabulae Observationum Stellarum Fixarum . . . anno 1586', Murhardsche Bibliothek, Kassel: MS astron. 2° 6. Partially transcribed by Rudolf Wolf, *Astronomische Mitteilungen* 45 (Zurich, 1856–1896), pp. 126 ff.

[44] Rose, *The Italian Renaissance of Mathematics*, pp. 90–117.

[45] Donald deB Beaver, 'Bernard Walther: Innovator in Astronomical Observation', *Journal for the History of Astronomy* 1, pt. 1 (February, 1970), 39–43. Lynn Thorndike, *A History of Magic and Experimental Science* (New York, 1923), Vol. V, p. 366.

[46] Thorndike, Vol. V, p. 339.

[47] Philip Melanchthon, *Corpus Reformatorum*, ed. Carlos Gottlieb Bretschneider (Halle, 1834), Vol. III, p. 118.

[48] Johann Gabriel Doppelmayr, *Historische Nachricht von dem Nürnbergischen Mathematicis und Kunstlern* . . . (Nürnberg, 1730), pp. 79–81.

[49] Johann Schöner, *Opera Mathematica* (Nürnberg, 1561).

[50] *Joannis Regiomontani . . . fundamentum operationum . . .* Neuberg a.d. Donau, 1557).

[51] Cf. Bernard Sticker, 'Die Wissenschaftlichen Bestrebung des Landgrafen Wilhelm IV', *Zeitschrift des Vereins für Hessische Geschichte und Landeskunde* 67 (1956), 130–137. *Idem.*, 'Landgraf Wilhelm IV und die Anfänge der Modernen astronomischen Messkunst', *loc. cit.*

[52] Hessisches Staatsarchiv, Marburg: 4A. 31. 26; 4F Kursachsen 71. Wilhelm to August, October 24, 1580.

[53] Hessisches Staatsarchiv, Marburg: 4F Kursachsen 60. Wilhelm to August, November 2, 1576.

[54] Among these are the *Praedictiones Astrologiae*, 1558, of Nicolas Caesareus Leucopetraeus and Cyprianus Leovitius's *De Conjunctiones Magnis*. Murhardsche Bibliothek, Kassel: 4° MS astron. 10. Hessisches Staatsarchiv, Marburg: 4A, 31. 13; 4A. 31. 8.

[55] *Tychonis Brahe, Dani, De Nova et Nullius Aevi Memoria Prius Visa Stella* . . . (Hafniae, 1573).

[56] Murhardsche Bibliothek, Kassel: MS 2° astron. 5, number 7, fol. 8r. Wolf, *Astronomische Mitteilungen* 45, pp. 133–134.

[57] Hessisches Staatsarchiv, Marburg: 4A. 31. 17.

[58] Murhardsche Bibliothek, Kassel: MS 2° astron. 5, number 19.

[59] Murhardsche Bibliothek, Kassel: MS 2° astron. 5, number 4; MS 2° astron. 6; MS 2° astron. 7.

[60] Murhardsche Bibliothek, Kassel: MS 2° astron. 6. Wolf, *Astronomische Mitteilungen* 45, p. 126.

[61] Murhardsche Bibliothek, Kassel: MS 2° astron. 5, number 8, 48r.

[62] The use of Venus had already been employed at Nürnberg and was further refined by Tycho. *Ibid.*

[63] Sticker, 'Landgraf Wilhelm IV und die Anfänge der Modernen Astronomischen Messkunst', *loc. cit.*

[64] Donald deB Beaver, *loc. cit.*

[65] H. von Bertele, 'Precision Timekeeping in the Pre-Huygens Era', *Horological Journal* (December, 1953), 794–816.

[66] Hessisches Staatsarchiv, Marburg: 4A. 31. 29.

[67] Hessisches Staatsarchiv, Marburg: 4A. 31. 12. Also, Tycho Brahe, *Opera Omnia*, Vol. III, pp. 114 ff.

[68] Hessisches Staatsarchiv, Marburg: 4F Kursachsen 51; December 21, 1572.

[69] *Ibid.*

[70] Hessisches Staatsarchiv, Marburg: 4F Kursachsen 51.

[71] Hessisches Staatsarchiv, Marburg: 4A. 31. 12. Also, Tycho Brahe, *Opera Omnia*, Vol. III, p. 114.

[72] Apianus's letter was later published by Tycho, *Opera Omnia*, Vol. III, pp. 158–161.

[73] Hessisches Staatsarchiv, Marburg: 4A. 31. 13.

[74] Hessisches Staatsarchiv, Marburg: 4A. 31. 8.

[75] C. Doris Hellman, *The Comet of 1577: Its Place in the History of Astronomy*, 1944 (rept. New York, 1971), Chapter III.

[76] *Ibid.*, Chapter IV.

[77] Tycho Brahe, *Opera Omnia*, Vol. VI, p. 49.

[78] William H. Donahue, 'The Solid Planetary Spheres in Post Copernican Natural Philosophy', in *The Copernican Achievement*, ed. Robert S. Westman (Berkeley, Los Angeles, 1975), pp. 244–275.

[79] *Oratio Valedictoria a Jordano Bruno Nolano D. Habita, ad amplissimos et clarissimos professores, atque auditores in Academia Witebergensi anno 1588, 8 Marti* in *Jordani Bruni Nolani Opera Latine conscripta* (Neapoli, 1879), Vol. I, pars 1, pp. 18 f.

[80] See: Robert S. Westman, 'Magical Reform and Astronomical Reform: The Yates Thesis Reconsidered', in Westman and McGuire, *Hermeticism and the Scientific Revolution* (Los Angeles, 1977), pp. 1–91.

[81] Murhardsche Bibliothek, Kassel: MS 2° astron. 5, number 8, fol. 36r–v. Wolf, *Astronomische Mitteilungen* 45, p. 147. Also, *Willebrord Snellii Descriptio Cometae qui anno 1618 ... effulsit. Huc accessit Christophori Rothmanni ... descriptio accurata cometae anni 1585* (Lugduni Batavorum, 1619), pp. 146–155.

[82] Robert S. Westman, 'The Melanchthon Circle, Rheticus and the Wittenberg Interpretation of the Copernican Theory', *Isis* 66 (1975), 165–193. Zofia Wardeska, 'Die Universität Altdorf als Zentrum der Copernicus-Rezeption um die Wende vom 16. zum 17. Jahrhundert', *Sudhoffs Archiv* 61 (1977), 156–164. Götz von Selle, *Geschichte der Albertus – Universität zu Königsberg in Preussen* (Würzburg, 1956), pp. 72–73.

[83] Hessisches Staatsarchiv, Marburg: 4F Kursachsen 82.

[84] Westman, 'The Melanchthon Circle ... ,' *loc. cit.* Students at Wittenberg most likely encountered the ideas of Copernicus initially through Erasmus Reinhold's commentary on Peurbach's *Theoricae Novae Planetarum* (1542), a basic text in astronomy. See: Owen Gingerich, 'The Role of Erasmus Reinhold and the Prutenic Tables in the Dissemination of the Copernican Theory', *Colloquia Copernicana* 2 (Warsaw, 1973), 43–62.

[85] This correspondence is published by Tycho Brahe in his *Epistolae Astronomicarum* (Uraniburgi, 1596).

[86] Murhardsche Bibliothek, Kassel: MS 4° astron. 11.

[87] Tycho Brahe, *Opera Omnia*, Vol. VI, pp. 118–119. For a discussion of Rothmann's model see Christine Schofield, 'The Geoheliocentric Hypothesis in Sixteenth-Century Planetary Theory', *The British Journal for the History of Science* 2 (1965), 291–296.

[88] See Derek J. Price, 'Is Technology Historically Independent of Science? A Study in Statistical Historiography', *Technology and Culture* 6 (1965), 553–568; *Idem.*, 'Structures and Publication in Science and Technology', in *Factors in the Transfer of Technology*, ed. by William Gruber *et al.* (Cambridge, Mass., 1964), pp. 91–104.

PAUL McREYNOLDS

# THE CLOCK METAPHOR IN THE HISTORY
# OF PSYCHOLOGY

I

The major breakthrough in methods of inquiry that eventuated in science as
we know it today began in the thirteenth century and culminated in the work
of Galileo, Huygens, Newton and others in the sixteenth and seventeenth
centuries. Parallel with this development of science, and closely related to it,
was a relatively sudden spurt in the development, construction, and use of
various kinds of machines. A machine can be defined as a dynamic physical
device with a number of functionally interrelated parts. The major complex
machine to be developed in this period, and the first to be widely distributed
in the history of civilization, was the mechanical clock. The important role of
the clock in the development of Western culture (Mumford, 1934, 1967) in
general, and of modern science in particular (Laudan, 1966; Moran, 1977;
Price, 1975, Chap. 2), has been noted by others, but the relation of the
mechanical clock to the emergence of scientific psychology is a story that has
not previously been told.

If we assume that models and metaphors play a key role in scientific
inquiry,[1] and that a scientific discipline having more questions than answers
is likely to attempt the utilization of new kinds of models wherever they
become available in the general culture, then we might expect that shortly
after the development and popularization of the mechanical clock, clock-like
metaphors would begin to appear in psychology. Such was in fact the case,
as we will see presently: indeed, in the latter seventeenth and early eighteenth
centuries, clock analogies were as popular and important for psychological
theorizing as computer analogies are today.

Before looking at some of the instances of the clock as a metaphor of the
mind, it will be helpful to briefly review the history of the mechanical clock.
Though devices for measuring time — chiefly the clepsydra, or water clock —
were well-known in ancient times, the mechanical clock in Europe is believed
to date from the latter thirteenth century. The origin of the device remains
obscure: formerly it was believed that the mechanical clock was invented by
some unknown European craftsmen, probably living in Northern Italy, and
that the invention represented a sudden innovation 'out of the blue'. There is,

97

T. Nickles (ed.), Scientific Discovery: Case Studies, 97–112.
Copyright © 1980 by D. Reidel Publishing Company.

however, now good reason to doubt this, and to suspect that the European development of the clock owed a great deal to ancient Greek and later Islamic technology; and there is even a possibility that a key part of the mechanical clock — the escapement mechanism — was derived in part from Chinese models (Needham and Price, 1960).

Be all this as it may, it is known that during the fourteenth century large mechanical clocks, driven by weights, began to appear all over western Europe. These devices, elaborate and often beautiful, typically stood in the medieval marketplace, or perhaps in a tower. At first, the movements of the clock possibly signified only the times when someone was to strike a bell to mark the passage of the hours. Later, mechanisms operated by the clockwork itself struck the bells, often through the device of automata — or 'jacks': some of these structures are still seen in Europe, as, *e.g.*, in the square at Venice. It is important to emphasize that many of the clocks of the period we are discussing were highly complex devices that revealed not only the time of day, but a variety of astronomical data as well. They were, in terms of the technical standards of the era, exceedingly impressive instruments.

The early European mechanical clocks were powered — as just noted — by weights. During the sixteenth century — and possibly a little before (Joy, 1967, p. 17) — a different form of motive power — in the form of springs that could be compressed or wound — made its appearance. The great advantage of springs was that the clock could be made portable, and also much smaller — indeed, down to the smaller versions that came to be known as 'watches', after their use during 'watches' on shipboard. The next development was the use of the pendulum — the first pendulum clock was probably constructed in 1657 (Tait, 1968, p. 44) — which for the first time made possible the fabrication of highly precise clocks. Though a functioning pendulum cannot be moved and hence is inappropriate in a watch, it was discovered that a fine spring — a *hairspring* — could be made to oscillate regularly, and thus to guarantee accuracy in small instruments (the larger spring, which provides the power for the watch, being the *mainspring*).

The early history of mechanical clocks is exceedingly fascinating, and the importance of clock technology in the course of civilization can scarcely be exaggerated. Clocks were the first great technical triumph, the paradigm for all later mechanical breakthroughs. For psychology the influence of clocks as models was both profound and lasting. Most importantly, the clock-model strongly suggested for psychology, as for all sciences, the themes of regular, recurring, predictable, automatically controlled events, and hence broadened and hastened the acceptance of the conception of natural law. Further — and of special significance for psychology – the amazing precision and complexity of clockwork suggested the possibility that living organisms might usefully be conceptualized as machines.

II

Apparently the first philosopher-psychologist to employ the clock as an analogue of psychological functioning was John Amos Comenius, the brilliant seventeenth century Moravian theologian and educator. In his classic work, *The Great Didactic* (Comenius, 1657), which was completed in 1632 but not published till some years later, Comenius wrote as follows:

... Just as the great world itself is like an immense piece of clockwork put together with many wheels and bells, and arranged with such art that throughout the whole structure one part depends on the other, and the movements are perpetuated and harmonised; thus it is with man .... The weight, the efficient cause of motion, is the brain, which by the help of the nerves, as of ropes, attracts and repels the other wheels or limbs, while the variety of operations within and without depends on the commensurate proportion of the movements.

In the movements of the soul the most important wheel is the will; while the weights are the desires and affections which incline the will this way or that. The escapement is the reason, which measures and determines what, where, and how far anything should be sought after or avoided. The other movements of the soul resemble the less important wheels which depend on the principal one. Wherefore, if too much weight be not given to the desires and affections, and if the escapement, reason, select and exclude properly, it is impossible that the harmony and agreement of virtues should not follow, and this evidently consists of a proper blending of the active and the passive elements.

Man, then, is in himself nothing but a harmony ... as in the case of a clock or of a musical instrument which a skilled artificer has constructed ... (pp. 47–48).

Something of the combined sense of amazement and pride with which men of that era viewed the mechanical clock, not unlike the feeling of awe we have toward the computer, is revealed in the following words from Comenius:

... How is it that such an instrument [i.e., the clock] can wake a man out of sleep at a given hour, and can strike a light to enable him to see? How is it that it can indicate the quarters of the moon, the positions of the planets, and the eclipses? Is it not a truly marvellous thing that a machine, a soulless thing, can move in such a life-like, continuous, and regular manner? Before clocks were invented would not the existence of such things have seemed as impossible as that trees could walk or stones speak? Yet every one can see that they exist now. (p. 96)

The two major pioneers in the beginnings of modern psychology in the seventeenth century were Descartes and Hobbes. Both were contemporaneous with Galileo, though younger than he, and both were strongly influenced by the rising currents of science and the new emphasis on mechanics, including the prime example of a machine, the clock. Descartes, in both of his two major works on psychology – *The Treatise of Man* (1662) and *Passions of the Soul* (1649) – employed the clock metaphor to convey his model of behavior in machine-like terms. Thus, in the former book, in discussing the movements of the limbs in his machine conception of man, Descartes states that these occur naturally "entirely from the disposition of the organs – no more nor less than do the movements of a clock or other automaton,[2] from the arrangement of its counterweights and wheels" (1662, p. 113). Similarly, in *The Passions of the Soul*, in referring to the movements of both brutes and men, he comments that these are determined by "the brain, nerves and muscles, just as the movements of a watch are produced simply by the strength of the springs and the form of the wheels" (1649, p. 340).

Hobbes, a more thoroughgoing mechanist than Descartes, also found the clock, or watch, metaphor a helpful one. Here are the opening lines of his Introduction to his classic work, *Leviathan* (1651):

Nature (the Art whereby God hath made and governes the world) is by the *Art* of man, as in many other things, so in this also imitated, that it can make an Artificial Animal. For seeing life is but a motion of Limbs, the beginning whereof is in some principall part within; why may we not say, that all *Automata* (Engines that move themselves by springs and wheeles as doth a watch) have an artificiall life. For what is the *Heart*, but a *Spring*, and the *Nerves*, but so many *Strings*; and the *Joynts*, but so many *Wheeles* giving motion to the whole Body, such as was intended by the Artificer? (p. 3)

Psychological theorists of the seventeenth and eighteenth centuries differed in the extent to which they believed that behavior could be rationalized by a mechanical interpretation of organisms – as, indeed, they do today – and they used the clock metaphor, interpreted in different ways, to make their points. Thus Descartes conceived that all animal and most human behavior could be understood in machine-like terms, but that the higher mental functions of man could not be. Hobbes, in contrast, made no such exception. A later, and even stronger and more explicit exponent of a machine analogy of behavior – for both animals and humans – was Julien de La Mettrie, a French physician whose book, *Man a Machine*, appeared in 1748. This book, which is highly important in the history of mechanistic psychology, includes a number of allusions to a watch – or parts of a watch, such as the springs or wheels – as an analogue of man as a machine. These metaphorical references are well summarized in the following lines from La Mettrie:

I am right! The human body is a watch, a large watch constructed with such skill and ingenuity, that if the wheel which marks the seconds happens to stop, the minute wheel turns and keeps on going its round, and in the same way the quarter-hour wheel, and all the others go on running when the first wheels have stopped because rusty or, for any reason, out of order. (1748, p. 141)

Descartes, Hobbes, and La Mettrie all employed clock metaphors not only in order to illustrate and communicate their conceptions of man, but also as a means of helping to render their mechanical models of behavior more plausible. However, writers of the period who were not in sympathy with a mechanical orientation — and they were a majority — also sometimes utilized the clock metaphor to make their points. An example would be the third Earl of Shaftesbury (Anthony Ashley Cooper), whose psychological conception emphasized the role of affections and passions in determining behavior. In his popular book, *Characteristics of Men, Manners, Opinions and Times* (1711), Shaftesbury indicated his rejection of the mechanical approach in the following words:

It has been shown before, that no animal can be said properly to act otherwise than through affections or passions, such as are proper to an animal. For in convulsive fits, where a creature strikes either himself or others, 'tis a simple mechanism, an engine, or piece of clockwork, which acts, and not the animal.

Whatsoever therefore is done or acted by any animal as such, is done only through some affection or passion, as of fear, love, or hatred moving him. (p. 285)

In 1747 a book titled *An Enquiry Into the Origin of the Human Appetites and Affections* (reprinted in McReynolds, 1969) was published anonymously in Lincoln, England. The author of this book is unknown, but the book has sometimes been attributed to a James Long. In any event, the book presented an associationistic theory that was, for that day, quite advanced. Since associationism is itself rather mechanistic, one might expect the author of this book to accept the machine analogy, as exemplified by the clock metaphor; in fact, however, he did so only to a limited extent, as indicated in the following quotation:

... though many ... Associations arise mechanically ... yet have we it very much in our Power, either to strengthen and confirm, or to impair and eradicate, them. This is, I believe, Fact. Tho' it must be own'd, 'tis the Opinion of some that we have no Freedom, no Principle of Agency, but are like Machines, a Piece of Clock-work, for instance, wholy passive, etc. But I would appeal to those Gentlemen, whether they do not feel within themselves a Power both of determining and acting independently on the Objects which solicit their Choice! They must allow it to be so. (pp. 325–326)

A major theme in German psychological thought, since the time of Leibniz, has been an emphasis on dynamic, wholistic, and organismic factors in conceptualizing mental processes. Because of this essentially nonmechanistic tradition, one might expect that the early German philosophers would have found the clock metaphor of the mind somewhat less than apt. This seems to have been the case. Thus Leibniz (1898, p. 321), who was in part responsible for the organismic orientation in psychology, stated that:

The unity of a clock ... is in my view quite other than that of an animal; for an animal may be a substance possessing a genuine unity, like what is called ego [*moi*] in us; while a clock is nothing but an aggregate [*assemblage*].[3]

### III

So far I have examined the role of the clock metaphor in the development of a machine model of the mind, and in discussions of the adequacy of such an approach. There were, however, other analogies utilizing the clock in seventeenth and eighteenth century psychological thought, and I turn now to a brief consideration of these.

Perhaps the best known early use of clocks in a metaphorical way is the famous simile of two clocks, employed first by Geulincx (reported in Latta, 1898, pp. 331–332) and later by Leibniz (1898, pp. 331–334). It will be recalled that Descartes had proposed that the mind influences the body through the mediation of the pineal gland. This position was followed by the doctrine of occasionalism, which accepts the duality of mind and body but holds that they do not interact; rather, the apparent correlation between the mental and physical is due to the direct intervention of God. Geulincx, in order to illustrate this position, used the analogy of two clocks that keep the same time even though neither affects the other. A similar illustration was employed by Leibniz – apparently without awareness of Geulincx's prior usage – in portraying his conception of pre-established harmony, which holds that the mental and physical operate in harmony because they were originally synchronized by God.

An important figure in British psychology in the early eighteenth century, though he himself was not a psychologist, was Joseph Butler, Bishop of Durham. In his Preface to a collection of his sermons (1729), Butler employed the example of a watch to illustrate and support his modern-appearing view that in order to understand the nature of man one cannot merely examine the different aspects of man – passions, appetites, affections, and reflection – separately, but must conceive of these as parts of an overall system, in order to point up the relations among them. Here is the relevant passage:

Every work both of nature and of art is a system; and as every particular thing, both natural and artificial, is for some use or purpose out of and beyond itself, one may add, to what has been already brought into the idea of a system, its conduciveness to this one or more ends. Let us instance in a watch – Suppose the several parts of it taken to pieces, and placed apart from each other: let a man have ever so exact a notion of these several parts, unless he considers the respects and relations which they have to each other, he will not have anything like the idea of a watch. Suppose these several parts brought together and any how united: neither will he yet, be the union ever so close, have an idea which will bear any resemblance to that of a watch. But let him view those several parts put together, or consider them as to be put together in the manner of a watch; let him form a notion of the relations which those several parts have to each other – all conducive in their respective ways to this purpose, shewing the hour of the day; and then he has the idea of a watch. Thus it is with regard to the inward frame of man. Appetites, passions, affections, and the principle of reflection, considered merely as the several parts of our inward nature, do not at all give us an idea of the system or constitution of this nature; because the constitution is formed by somewhat not yet taken into consideration, namely, by the relations which these several parts have to each other; the chief of which is the authority of reflection or conscience. It is from considering the relations which the several appetites and passions in the inward frame have to each other, and above all, the supremacy of reflection or conscience, that we get the idea of the system or constitution of human nature. And from the idea itself it will as fully appear, that this our nature, i.e., constitution, is adapted to virtue, as from the idea of a watch it appears, that its nature, i.e. constitution or system, is adapted to measure time. (1729, pp. ix–x)

Butler was one of a number of writers of the period who held that human beings have an inherent motive toward virtue. This assumption is evident in the last sentence above.

Another interesting use of the clock metaphor appears in a little book by Benjamin Franklin published in 1725. Franklin – then only nineteen years old – had arrived in London the year before, and had eagerly immersed himself in the intellectual currents of the day. In his treatise, titled *A Dissertation on Liberty and Necessity, Pleasure and Pain,*[4] Franklin argued against the existence of free will. In the following he utilized the analogy of a clock to show how confused – and thus unlikely – the universe would be if each individual had complete free will.

All the heavenly Bodies, the Stars and Planets, are regulated with the utmost Wisdom! And can we suppose less Care to be taken in the Order of the *moral* than in the *natural* System? It is as if an ingenious Artificer, having fram'd a curious Machine or Clock, and put its many intricate Wheels and Powers in such a Dependance on one another, that the whole might move in the most exact Order and Regularity, had nevertheless plac'd in it several other Wheels endu'd with an independent *Self-Motion*, but ignorant of the general Interest of the Clock; and these would every now and then be moving wrong, disordering the true Movement, and making continual Work for the Mender, which might better be prevented, by depriving them of that Power of Self-Motion, and placing them in a Dependance on the regular Part of the Clock. (1725, pp. 12–13)

A fourth instance of a clock metaphor in seventeenth century thought that is of interest to the history of psychology is that discussed by Laudan (1966), though he does not note its specific relevance to psychology. In contemporary psychological theorizing there is what may be called the 'black box' question. If one thinks of the human mind as a kind of black box, then one point of view, espoused by strict behaviorism, is that it is possible to understand and predict behavior without knowing what is going on in the box, simply by analyzing inputs and outputs;[5] a contrasting point of view, on the other hand, holds that it is necessary, in order to make sense of behavior, to formulate meaningful hypotheses as to what is going on within the black box. A somewhat similar methodological problem was discussed by Descartes (in *Principles of Philosophy*, Principle *CCIV*; 1644, pp. 300–301), using the analogy of a clock whose face we can see but whose inner workings are hidden. Descartes conceived that a like situation holds with respect to many natural phenomena and argued that, under these circumstances, it is appropriate to formulate hypothesized mechanisms to explain the phenomena, and that if these hypotheses are compatible with the actual occurrences, then that is as far as one can go. This position, including the metaphor of the clock, was utilized – as Laudan shows – by Boyle and others to support a probabilistic approach to knowledge.[6] We can conclude that the use of a clock metaphor in the context just described helped to advance the practice of developing hypotheses and constructing models in scientific methodology.

IV

It is evident from the various examples that we have surveyed to this point that the clock metaphor played a significant and interesting role in the development of modern scientific psychology. It is, of course, not possible to definitively separate the extent to which the existence of the mechanical clock actually contributed to the birth of mechanistic psychology from the extent to which, in contrast, it merely served as a convenient metaphor for a machine interpretation of behavior that was developing independently.[7] Probably both possibilities are valid in part. The ways of scientific discovery are diverse and complex, and not easily fathomed, but assuredly they involve scientists seeing old problems in new contexts, and – on occasion – perceiving new problems, new questions. In this perspective, it seems likely that the wide presence of the mechanical clock in the general culture of the seventeenth and eighteenth centuries provided a handy cognitive schema in terms of which newer thinkers could readily conceive of man in machine-like terms. To be sure, this influence was mainly in the sense of suggesting a new way of looking at old questions — i.e., nobody proposed literal translations of clock mechanisms in understanding the human brain — but this is the usual way of scientific progress.

There is one respect, however, in which I believe that the clock analogy contributed more directly to scientific advance in psychology. This effect, which was more in the nature of setting a new question for psychology than of providing a possible approach to older questions, concerned the area of psychology that we now call motivation, which deals with the internal factors that cause an organism to behave in certain ways. Though this field, in terms of its substance, was by no means new in the period we are discussing — it had been richly developed by classical Greek thinkers — an important new thread entered during this period. This new feature was the fuller awareness that there are forces within the organism — motive forces — that energize the movements of the organism. It will be recalled that it was in this period that physics — in the work of Galileo, Newton and others — concentrated on the analysis of mechanical forces. It was also in this period that the term 'motive' — in the writings of Hutcheson, Hume, Helvétius, Bentham, and others — began to be systematically used in psychology to designate the internal forces that power and impel behavior.

My thesis is that the example of the mechanical clock was responsible to a significant degree — I do not claim that it was the only factor — for mediating the entrance into psychology of the concept of motive. A clock or watch, it may be noted, consists of two aspects — first, the mechanism itself, and second, a motive force to keep it operating. This is clearly put in this passage from the latter eighteenth century philosopher-theologian, William Paley (1802):

... when we see the watch *going*, we see proof ... that there is a power somewhere, and somehow or other, applied to it; a power in action; — that there is more in the subject than the mere wheels of the machine; — that there is a secret spring, or a gravitating plummet; — in a word, that there is force, and energy, as well as mechanism. (p. 525)

It is this conception of a motive force in clocks that I am suggesting was carried over to the theory of organisms when the clock, or machine model was applied to man. Just as clocks need a motive force to keep them going, so do humans. The perceived relevance of the motive force in clocks as an analogue for driving human behavior is nicely illustrated in these lines from Franklin's (1725) monograph [Franklin's view was that behavior ('Motion') is determined by the need to be relieved of unpleasant sensations ('Uneasinesses')] : "As fast as we have excluded one Uneasiness another appears, otherwise the Motion would cease. If a continual weight is not apply'd, the Clock will stop" (p. 15). Similarly, La Mettrie (1748) noted that the functioning of the body can be likened to that of a pendulum, and that it "can not keep up forever", but that "It must be renewed, as it loses strength" and "invigorated when it is tired . . . " (p. 135).

As we noted earlier, the first motive force used in driving clocks was waterpower; this was succeeded by the use of falling weights (as in a contemporary Grandfather's clock); and finally, except in the larger instruments, by coiled springs. The word 'springs' — and, more specifically, 'mainspring' — then entered the language as ways of conveying the idea of motive forces within a machine, or — by analogy — within an organism.

The earliest instance of this usage of which I am aware is in the following lines from Michel de Montaigne (1580) " . . . a sound intellect will refuse to judge men simply by their outward actions; we must probe the inside and discover what springs set men in motion"[8] (p. 244). In a somewhat more concrete fashion, La Mettrie (1748) stated of the brain that it "can be regarded, without fear of error, as the mainspring of the whole machine, having a visible influence on all the parts" (p. 135). And the influential French philosopher Claude Helvétius (1758) proposed, in a passage prescient of Freud — that among civilized men, "the love of women is . . . the main spring by which they are moved" (p. 261).

The term which gradually came to be used to refer to the motive forces within man, and which we today term simply 'motives', was 'springs of action', adapted directly from the mainsprings of clocks and watches. Thus, we find Francis Hutcheson (1742) in his *An Essay on the Nature and Conduct of the Passions and Affections* writing that:

If we examine the true *Springs* of human Action, we shall seldom find their Motives worse than *self-Love*. Men are often subject to *Anger*, and upon sudden *Provocations* do Injuries to each other, and that only from Self-Love, without Malice; but the greatest part of their Lives is employed in Offices of *natural Affection, Friendship, innocent Self-Love*, or *Love* of a *Country*. (p. 109)

Franklin's work, referred to earlier, includes the statement that "Uneasiness [is] the first Spring and Cause of all Action" (p. 16). David Hume, in making the point in his *An Inquiry Concerning Human Understanding* (1748) that human nature has always been pretty much the same, comments that the chief use of history is

> to discover the constant and universal principles of human nature by showing men in all varieties of circumstances and situations, and furnishing us with materials from which we may form our observations and become acquainted with the regular springs of human action and behavior. (pp. 94–95)

In 1815 Jeremy Bentham published a book titled *A Table of the Springs of Action* (reprinted in McReynolds, 1969). This small volume, which had probably been written considerably earlier, and which was the first book devoted exclusively to the topic of human motivation, consists of a large fold-out table presenting a taxonomy of motives organized around the conception that individuals seek pleasures and avoid pains. The fact that the phrase 'springs of action' was selected as part of the title of this book indicates that by this time the expression had entered the common language as a meaningful term for human motivation.

After this time – and to some extent up to the present day – one finds the term 'springs of action' increasingly employed to denote the inner forces powering behavior. William James, in his *Principles of Psychology* (1890, Vol. II, p. 549), headed a section 'Pleasure and Pains as Springs of Action', and William McDougall, the great motivational psychologist earlier in the present century, also employed the term:

> The human mind has certain innate or inherited tendencies which are the essential springs or motive powers of all thought and action .... (1909, p. 19). [And further] ... It must be said that in the developed human mind there are springs of action of another class, namely, acquired habits of thought and action. (p. 43)

(McDougall disagreed with Bentham, James, and others in his insistence that pleasure and pains do not constitute springs of action). In 1927 the first textbook in the field of motivational psychology appeared; it was by M. K. Thomson, and was titled *The Springs of Human Action*. In the following year the British psychologist Charles Spearman published a paper on 'A new method for investigating the springs of action'.

To restate and summarize my present point – I am suggesting that the concept of psychological motive, in the sense of an inner force driving and energizing behavior, arose in psychology in part through the example and intellectual stimulation of the spring mechanisms that became popular in the sixteenth century for providing the motive power in clocks and watches. Though it is easiest to defend this argument if we think of the process I have described as being only the development of a verbal metaphor, in which the term 'springs of action' becomes attached, for reasons of felicity of expression, to a concept already central in psychology, I would maintain that something more substantive was involved. The fact is that prior to the period under review there was no concept of motive in psychology in the sense noted above.[9] Rather, the prevailing theory of behavior was that the individual decided – on the basis of Reason – what he wanted to do, and then did it, circumstances permitting. It is therefore plausible to conceive that the *idea* of motives forces in man, as well as the label, is in part due to the clock analogy.

## V

The general aim of this paper has been to develop the thesis that the advent of the clock as a machine in wide usage in the sixteenth, seventeenth and eighteenth centuries contributed significantly to the birth and growth of the machine model in psychology. In this sense the clock played somewhat the same role in psychological theorizing in the earlier period that the computer does today – *i.e.*, as a complex machine in a different field which by its metaphorical suggestiveness stimulates a new look at psychological processes.

This similarity should not be surprising. A clock is, after all, a simple computer – programmed to yield temporal data – and is the direct ancestor of the large present-day computers. The legacy of the clock as a metaphor of the mind is, however, more directly represented in contemporary culture by cartoons which still occasionally appear showing a person having clockwork for a brain,[10] and by the continued use in common speech – when we don't understand a man's behavior – of the query, "I wonder what makes him tick?"

In concluding, I wish to comment on the implications of the history of the clock metaphor in psychology for our understanding of the process of scientific discovery. Typically, analyses of this process focus on the individual scientist and inquire how he came upon the themes that permitted him to see an old problem in an innovative way. For example, in the case of Darwin it is often noted that his thinking was stimulated by Malthus's views on population growth. In the present paper we have looked at this whole problem from essentially the opposite perspective; *i.e.*, we have identified a theme – the mechanical clock – and followed its influence throughout a scientific field.

This influence, we have seen, was quite pervasive, though it is also note-worthy that not all psychological thinkers of the period found the metaphor apt. We may suppose that those who invented the analogy between the clock and the mind, and who used it and profited from it, were those who were mentally prepared for it, *i.e.*, that in terms of the nature of the psychological questions they were asking, and of their particular world views and habits of thought, the example of the clock, stated as a metaphor, helped them to make sense of behavior.

This interpretation suggests an interaction, in the process of scientific discovery, between the minds of individual scientists and the availability in the current culture of new sources of creative metaphors. Thus, in the present instance the existence of mechanical clocks would not have contributed to scientific advance had there not been receptive minds ready to perceive the relevant metaphor in a creative way; but at the same time, and regardless of how well prepared their minds were, the theorists involved could not have developed the clock metaphor, with all of its suggestive implications, had there not been clocks.[11]

*Department of Psychology*
*University of Nevada, Reno*

## NOTES

[1] The role of metaphors and analogies is especially prominent in psychological discourse and psychological theorizing. This fact, which has been known, if not fully appreciated, for some time (*cf.* McReynolds, 1970, 1971), follows in part from the circumstance that a person's inner, mental processes can be conceptualized in terms meaningful to other persons only through the mediation of mutually observable entities and events. For example, the sensation of pain is described in terms such as sharp, dull, and stabbing. Similarly, psychological terms such as drive, repression, and tension derive from obvious physical analogues.

[2] Descartes employed a number of mechanical metaphors other than the clock. Most prominently, he thought of his model of man as analogous to the hydraulically operated automata – mechanical figures simulating animals and human forms – that were popular in his day.

[3] Leibniz's general opposition to artificial machines as models of living things is elaborated in his *Monadology* (paragraph 64; Leibniz, 1898, pp. 254–255). Leibniz writes that " . . . The organic body of each living being is a kind of divine machine or natural automaton, which infinitely surpasses all artificial automata. For a machine made by the skill of man is not a machine in each of its parts. For instance, the tooth of a brass wheel has parts or fragments which for us are not artificial products, and which do not have the special characteristics of the machine, for they give no indication of the use for which the wheel was intended. But the machines of nature, namely, living bodies, are still machines in their smallest parts *ad infinitum*. It is this that constitutes the difference between nature and art, that is to say, between the divine art and ours."

[4] This brief tract – it is only 32 pages in length – was set in print by Franklin himself in the shop in which he was employed. Only a hundred copies were printed, and most of these were later destroyed by Franklin, who came to regret certain points in the work. Though an incisive treatise, it cannot have had much influence; its terminology does, however, reflect the usages current in its time.

[5] Behaviorists differ among themselves in the extent to which they consider it appropriate to frame hypotheses concerning internal, unobservable psychological processes. Thus, Clark Hull and E. C. Tolman considered a limited number of such hypotheses to be essential, whereas B. F. Skinner – a stricter behaviorist – strongly avoids such hypotheses.

[6] Laudan points out that Descartes's espousal of a hypothesis-forming approach was in fact incongruent with his more explicit *a prioristic* methodological orientation. Laudan also notes that Henry Power turned the watch metaphor against Descartes by holding, in effect, that it is possible to look inside the watch to see how it works. This orientation is exemplified in contemporary psychology by psychophysiologists and neuropsychologists who attempt to study directly brain activities associated with mental processes.

[7] Clocks were of course not the only machines that were constructed during this period and which contributed to the gradual development of a mechanical conception of man; they were, however, the most technically sophisticated machines then present, and they were widely distributed throughout the culture. (For general treatments of the history of mechanistic models in psychology consult Campbell, 1970; Jaynes, 1970; McReynolds, 1971; and – especially – Lowry, 1971.) With respect to the contemporary psychological scene, behaviorism is perhaps the prime example of a mechanistic orientation. Such an orientation, however, is by no means limited to the different varieties of behaviorism. Thus, all current perspectives which employ the $S-R$ (stimulus – response) paradigm are basically mechanistic. Further, modern cognitive psychology is to a large degree mechanistic in outlook, as is neuropsychology.

[8] I am indebted to Professor David McClelland (1951, p. 383, and Personal Communication) for awareness of this quotation.

[9] To be sure, there had long been a thriving theory of passions – conceived as strong impulses within a person which may take over certain of his or her behaviors, and in the face of which he or she remains passive (hence the name 'passions') – and this topic covered much of the same territory that the concept of motive was to deal with in the newly emerging psychology. But while the concepts of passion and motive are related, they are by no means the same. 'Passion', as contrasted with 'motive', was used in a more limited way, referring to only that minority of behaviors that was believed not to be subject to control by reason. Further, 'passion' did not carry the connotation of an inner force energizing all behavior. Finally, 'passions' were more or less limited to those impulses that we today call 'emotional'.

[10] I will give two examples. One is an interesting book by Dean E. Wooldridge (1963) titled *The Machinery of the Brain*; the cover of this book depicts a face dawn in profile, with the inside of a watch occupying the place of the brain. The second example is the cover of the magazine *Psychology Today* (April, 1969), which illustrates a woman with clockwork for a brain. Both illustrations convey quite directly the notion of the mind as a machine modeled after the clock.

[11] I wish to thank the editor, Thomas Nickles, for his helpful comments on this paper.

## BIBLIOGRAPHY

*All references are to the indicated translations or reprints.*

Anonymous: 1747, *An Enquiry into the Origin of the Human Appetites and Affections*, Lincoln, England, reprinted in McReynolds (1969).

Bentham, J.: 1815, *A Table of the Springs of Action*, Richard and Arthur Taylor, London.

Butler, J.: 1729, 'Preface' to *Sermons*, 2nd ed., reprinted in *J. Butler's Works*, Vol. II, Oxford, pp. iii–xxvii, 1836.

Campbell, B.: 1970, 'La Mettrie: The Robot and the Automaton', *Journal of the History of Ideas* 31, 555–572.

Comenius, J. A.: 1657, *The Great Didactic*, translated from the Latin and edited by M. W. Keatinge, Adam and Charles Black, London, 1910.

Descartes, R.: 1644, *Principles of Philosophy*, translated from the Latin and edited by E. S. Haldane and G. R. T. Ross, *The Philosophical Works of Descartes*, Cambridge Univ. Press, Cambridge, 1967, Vol. I, pp. 201–302.

Descartes, R.: 1649, *Passions of the Soul*, translated from the French and edited by E. S. Haldane and G. R. T. Ross, *The Philosophical Works of Descartes*, Cambridge Univ. Press, Cambridge, 1967, Vol. I, pp. 329–427.

Descartes, R.: 1662, *The Treatise of Man*, translation from the French and commentary by Thomas S. Hall, Harvard Univ. Press, Cambridge, 1972.

Franklin, B.: 1725, *A Dissertation on Liberty and Necessity, Pleasure, and Pain*, London.

Helvétius, C.: 1757, *Essays on the Mind*, translated from the French, Albion, London, 1810.

Hobbes, T.: 1651, *Leviathan*, republished by Dutton, New York, 1950.

Hume, D.: 1748, *An Enquiry Concerning Human Understanding*, reprinted in A. Flew (ed.), *On Human Nature and Understanding*, Collier Books, New York, 1962.

Hutcheson, F.: 1742, *An Essay on the Nature and Conduct of the Passions and Affections*, 3rd ed., London.

James, W.: 1890, *Principles of Psychology*, Vol. II, Henry Holt, New York.

Jaynes, J.: 1970, 'The Problem of Animate Motion in the Seventeenth Century', *Journal of the History of Ideas* 31, 219–234.

Joy, E. T.: 1967, *The Country Life Book of Clocks*, Hamlyn House, Feltham, Middlesex.

La Mettrie, J. O.: 1748, *Man a Machine*, translation from the French, Open Court, Chicago, 1912.

Latta, R.: 1898, 'Introduction and Notes', in G. Leibniz, *The Monadology and Other Philosophical Works*, translated by R. Latta, Oxford Univ. Press, London.

Laudan, L.: 1966, 'The Clock Metaphor and Probabilism: The Impact of Descartes on English Methodological Thought, 1650–65', *Annals of Science* 22, 73–104.

Leibniz, G.: 1898, *The Monadology and Other Philosophical Works*, translated by R. Latta, Oxford Univ. Press, London.

Lowry, R.: 1971, *The Evolution of Psychological Theory: 1650 to the Present*, Aldine-Atherton, Chicago.

McClelland, D.: 1951, *Personality*, Dryden, New York.

McDougall, W.: 1909, *An Introduction to Social Psychology*, 2nd ed., Methuen, London.

McReynolds, P. (ed.): 1969, *Four Early Works on Motivation*, Scholars Facsimiles and Reprints, Gainsville, Florida.

McReynolds, P.: 1970, 'Jeremy Bentham and the Nature of Psychological Concepts', *The Journal of General Psychology* 82, 113–127.

McReynolds, P.: 1971, 'Statues, Clocks, and Computers: On the History of Models in Psychology', *Proceedings, 79th Annual Convention*, American Psychological Assoc., pp. 715–716.

Montaigne, M.: 1580, *Essays*, translated from the French by D. M. Frame in *The Complete Works of Montaigne*, Stanford Univ. Press, Stanford, California, 1967, pp. 1–857.

Moran, B.: 'Princes, Machines and the Valuation of Precision in the 16th Century', *Sudhoffs Archive* 61, 209–228.

Mumford, L.: 1934, *Technics and Civilization*, Harcourt, Brace, New York.

Mumford, L.: 1967, *The Myth of the Machine*, Harcourt, Brace and World, New York.

Needham, J., and D. Price: 1960, *Heavenly Clockwork, the Great Astronomical Clocks of Medieval China*, Cambridge Univ. Press, Cambridge.

Paley, W.: 1802, *Natural Theology*, reprinted in *The Works of William Paley, D. D.*, Peter Brown, Edinburgh, 1825, pp. 435–554.

Price, D.: 1975, *Science Since Babylon*, enlarged edition, Yale Univ. Press, New Haven.

Shaftesbury, 3rd Earl of: 1711, *Characteristics of Men, Manners, Opinions, Times*, reprinted by J. M. Robertson (ed.), Bobbs-Merrill, Indianapolis, 1964.

Spearman, C. E.: 1928, 'A New Method for Investigating the Springs of Action', in M. L. Reymert (ed), *Feelings and Emotions: The Wittenberg Symposium*, Clark Univ. Press, Worcester, Massachusetts, pp. 39–48.

Tait, H.: 1968, *Clocks in the British Museum*; London.

Thomason, M. K.: 1927, *The Springs of Human Action*, Appleton, New York.

Wooldridge, D. E.: 1963, *The Machinery of the Brain*, McGraw-Hill, New York.

HOWARD E. GRUBER

# THE EVOLVING SYSTEMS APPROACH TO CREATIVE SCIENTIFIC WORK: CHARLES DARWIN'S EARLY THOUGHT[1]

Scientific work and thought is not a single process but a complex group of activities organized and orchestrated toward certain ends. The diversity of approaches taken by participants in this conference — by philosophers, historians, sociologists, and psychologists interested in scientific inquiry — bears witness not only to controversy but to the multifaceted complexity of the scientific enterprise.

Unfortunately, much theoretical and empirical work in this area has been piecemeal. Each investigation focuses on the operation of one or a few processes, or on the development of one or a few concepts. What is badly needed is a *systemic view* of the whole thinking person engaged in scientific work. This requires a great increase in attention to case studies. Only through the arduous work of following the growth of a thought process over extended periods of time can we see how its many facets are interrelated. The aim of such studies is not merely to pile up interesting narratives. Rather, the hope is that in studying and conceptualizing the individual creative scientist as a working system, we will shed light on very general issues connected with scientific discovery.

Two conceptual shifts are in order and under way. First, there is a general discomfiture with the dichotomy between the context of justification and the context of discovery. In my view, this means that some way must be sought of understanding scientific work and thought that treats its many aspects as part-processes in one indissociable system of intellectual struggle and growth. Second, renewed interest in discovery processes means that we must get beyond 'Aha!' Once we free ourselves from the view that discovery is summed up and wrapped up in the mysteries of the 'Eureka Experience', the direction to which we naturally turn is the conceptualization of scientific thought as protracted, purposeful, constructive work.

These conceptual shifts suggest two related methodological changes. First, as stated above, the systemic conceptual goal demands close examination of cases. Second, the magnitude and Protean character of this task will probably require new forms of interdisciplinary collaboration.

Although such collaborations have not yet occurred in any important degree, there is one moderately close approximation: the Centre International

113

*T. Nickles (ed.), Scientific Discovery: Case Studies*, 113–130.
*Copyright © 1980 by D. Reidel Publishing Company.*

d'Epistémologie Génétique in Geneva, under the leadership of Jean Piaget. There, the empirical focus has been mainly on the development of children's thinking. But the theoretical focus is on the slow processes of construction in which ideas and systems of ideas are generated. This constructivist view of intellectual growth provides a valuable theoretical springboard for the study of scientific creativity.

Precisely because of its systemic nature, such an effort must touch upon many questions: all the components of a very complex process. In the present essay I sketch out what my students and I have been calling an *evolving systems approach* to the study of creative thinking. Issues to be dealt with include: intentionality, the relations between emotions and thought, scientific thinking as a series of structural transformations, metaphoric thought as part of the process of abstraction, differential uptake of complex ideational structures, and the place of insight in an evolving structure of ideas.

The material I will draw upon includes Darwin's writings before the autumn of 1838, that is, before he arrived at a clear idea of the theory of evolution through natural selection.[2]

## I. AIMS AND LIMITS OF THE EVOLVING SYSTEMS APPROACH

In July 1837, when Darwin began his First Notebook on evolution, he seems to have presupposed that a theory of evolution must include some conception of the origin of life itself. His first theory, the *monad theory of evolution*, which I shall describe below, reflects this view. After a few months, however, he recognized that a theory could be constructed that would be useful, while remaining silent about the origin of life. Considering the state of biochemical knowledge in those days, almost a century before Oparin[3] began his work on the origin of life, Darwin made a wise and creative decision. Every theoretical effort operates within certain boundary conditions. Darwin chose to describe nature as an ongoing system in which organized beings interact with each other and with their milieu, this manifold interaction producing the entire ever-changing panorama of life evolving.

While he put aside the search for the beginnings of life in the appearance of some simple organism, Darwin's theoretical intention was certainly to include the notion of improvement and continuous progress (past, present, and future) as necessary consequences of whatever natural laws might be found to explain evolution. This can be seen both in his First Notebook and in the final paragraph of the *Origin of Species*, written 22 years later (1859). Thus, both in its earliest stages and in its mature form, Darwin's theory displays two aspects, entailing the same natural laws. On the one hand, the theory attempts to account for the *progressive* evolution of complex forms from simpler ones; on the other hand, the theory gives an account of organic nature as a stable system undergoing perpetual transformations. These two aspects correspond to what J. B. S. Haldane has referred to as the *dynamic* phase and the *static* phase of evolutionary theory. In a complete theory, they are mutually supportive and indispensable to each other.

In this respect, our work on creative scientific thinking resembles Darwin's theoretical efforts. On the one hand, we would like to press our own reconstructions back as far as possible toward beginnings, and to bring out the progressive, constructive nature of creative work. On the other hand, both the present state of knowledge about the psychology of thinking and the documentary evidence available for any particular case study, *e.g.* Darwin, requires that we put our main emphasis on later phases. In these later phases, prominent aspects of thinking resemble a stable system undergoing transformations. Thus we will see both growth and transformation as moments of Darwin's thinking.

We have found it useful to conceive of the creative person as an evolving system comprised of three great subsystems: an organization of knowledge, an organization of purpose, and an organization of affect.

There is a peculiar difference between historians of science and scientists themselves in their accounts of scientific work. The former focus almost entirely on the growth of ideas. The latter are prone to stress the relation of this growth of ideas to the individual's plans and intentions, and to a variety of aesthetic and affective experiences. In my own work on Darwin's thinking, I too began by putting the emphasis on the growth of ideas, but not in the way that historians usually do. I was interested in using Darwin's notebooks to reconstruct his thought processes during the relatively brief period of less than two years in which he first worked purposefully to construct a plausible theory of evolution. Moreover, insofar as I was able, I tried to consider Darwin's thinking as a whole. To be more specific, I treated the ideas in Darwin's notebooks on evolution (the First, Second, Third, and Fourth Notebooks) in their complex interaction with his ideas on man, mind, and materialism (the M and N Notebooks). Most important of all, my efforts were devoted to treating Darwin's thinking at a given moment as an organized system composed of interacting substructures. This system exhibited interesting interactions with his contemporaries, and reflected his own recently past experiences during the voyage of the Beagle (1831–1836). Resulting from all this were the inner tensions in the system of ideas, tensions giving rise to the evolution of that system.

## II. SCIENTIFIC THINKING AND THE AFFECTIVE
## TRANSFORMATIONS

But as Piaget and Inhelder have pointed out, the growth of cognitive structures is intimately connected with what they have called 'the affective transformations'.[4]

In the childhood period of concrete operations, thought is centered upon things. In the adolescent period, formal operations develop and the person can now treat thought itself as the object of thought. New abilities of hypothetical thinking, reflectivity, and abstractness emerge. These can, under favorable circumstances, be enormously liberating: the real becomes a special case of the possible; reflecting upon the structure of possibilities transforms the meaning of reality.

In Darwin's life, this relationship took a rather special form. In childhood he had formed what may be called an unusually strong *cathexis with nature*. As his thinking became more abstract, he never lost this connection. In later life he may have lost his ability to enjoy human aesthetic productions – the music, art, and poetry which had in early manhood given him so much pleasure. But there is no evidence that he ever lost his voracious enjoyment of nature itself.[5]

His appetite for the concrete objects of natural history is well illustrated by his account of an entomological incident during his university days at Cambridge. Overturning a log, in quick succession he captured two beetles desirable for his collection. While still holding one in each hand, he spied a third, even more desirable. To free one hand for the new capture, he popped one beetle into his mouth. It squirted an acrid liquid and the pain forced him to drop his specimens. The story illustrates not only his deep and spontaneous cathexis with nature, but also a kind of *courage* that had to survive and evolve considerably for Darwin to accomplish his life's task.

Thus, when we learn that Darwin loved the writings of Alexander von Humboldt and read them aloud to his university friends, the point is not that he absorbed his cathexis with nature from Humboldt. Rather, his encounter with Humboldt was among those that helped him to *transform* his love of nature into the scientific quest that became his life. Darwin could appreciate more deeply the feeling for nature of this great 'scientific traveller' (Darwin's phrase in later life), and for awhile seek to emulate him, precisely because he already had an affective structure into which Humboldt's views could be assimilated.

Among the earlier experiences that shaped this affective structure were his knowledge of the work of his grandfather, Dr. Erasmus Darwin. Dr. Darwin exhibited a "rare union of poetry with science" (Charles Darwin's phrase. See below.). Not only did he write the well known evolutionary essay *Zoonomia*, he also wrote poetical works such as *The Temple of Nature, or the Origin of Society* and *The Loves of the Plants*. These works covered a great deal of ground. A prominent topic was the sexual nature of plant reproduction, treated in a high-flown emotional style, with abundant use of the poetic device of personification.

Not to be forgotten also are the several natural theologians he knew, the Reverend Professors he studied with at Cambridge, and some who were not professors. Not only did they love and revere nature as the handiwork of God, but as men of science they studied it. Not only did they imbue Darwin with his lifelong fascination with biological adaptation, they also showed him a valuable part of the life style he needed, a certain unity of love and thought.

Of course, Darwin went far beyond these origins and devoted himself to explaining adaptation without recourse to God. In this way his life is an illuminating example of the *differential uptake of ideas*. It is not enough to say that the creative person was 'influenced' by certain experiences; this describes him far too passively. We have to show how he assimilates ideas selectively into his own growing intellectual structures, and shapes them to his own purposes. We glimpse Darwin moving along this path in a passage he wrote in his *Diary*, early in the voyage of the Beagle:

But these beauties are as nothing compared to the Vegetation; I believe from what I have seen Humboldt's glorious descriptions are and will for ever be unparalleled: but even he with his dark blue skies and the rare union of poetry with science which he so strongly displays when writing on tropical scenery, with all this falls far short of the truth. The delight one experiences in such times bewilders the mind; if the eye attempts to follow the flight of a gaudy butterfly, it is arrested by some strange tree or fruit; if watching an insect one forgets it in the stranger flower it is crawling over; if turning to admire the splendour of the scenery, the individual character of the foreground fixes the attention; The mind is a chaos of delight, out of which a world of future and more quite pleasure will arise. I am at present fit only to read Humboldt; he like another sun illumines everything I behold.[6]

I have selected this passage to emphasize the high excitement of which Darwin was capable. But it contains another note as well. By this time, Darwin had begun to recognize that he would be more than a scientific traveller, that he would need years of the 'more quiet pleasure' of contemplative thought to make the most of these exuberant experiences.

By now the reader will have noted that I am discussing the relation between emotion and thought, but unlike psychoanalytic writers, I am *not* discussing the role of neurosis and anxiety in creative work. Like more ordinary people, some great thinkers are anxious and neurotic. But the main point is that they must grow beyond their limitations. The positive emotions of which I have been speaking – curiosity, commitment and courage, love of truth, compassion – provide the power necessary for this growth.

Clearly, mere knowledge and intellectual skill, even at a very high level, are not enough to marshall the energies of the person for the longer reaches of time needed for creative work. A strong emotional bond with a chosen subject, and a larger vision, are essential. The creative person must develop a sense of identity *as* a creative person, a sense of his or her own *specialness*. But this cannot be founded on empty fantasies. Tasks must be self-set. A personal point of view must emerge that gives meaning to the choice of tasks. A group of personal allegiances must be formed to provide the mutual support (and sometimes collaboration) that creative work requires. As the individual senses this entire system beginning to function, a new excitement must rise in him and he must be willing and able to assimilate this state of heightened emotionality without retreating into ordinariness.

## III. SCIENTIFIC THINKING AS A SERIES OF STRUCTURAL TRANSFORMATIONS

We can better understand the need for a powerful affective structure, and for a strong sense of purpose and of vision, when we see how very far Darwin had to travel intellectually. This can best be done by treating his ideas as an evolving series of structures.

When he began the voyage of the Beagle, he accepted the views of Natural Theology: the world had been created with a set of organisms, each beautifully adapted to its milieu. During the years of the voyage, through reading and through much geological field work in South America and in the Pacific islands, Darwin abandoned the Catastrophist view of geology and assimilated Charles Lyell's Uniformitarian views: The physical features of the earth are continually undergoing slow transformations governed throughout all time by a uniform set of natural laws. The earth is unimaginably older than the 6,000 years or so suggested in some then contemporary reconstructions of Biblical history. Events which seem cataclysmic from a human perspective, such as earthquakes and floods, when viewed on a geological time scale are only parts of the total process of slow transformation. The young Darwin not only assimilated Lyell's views, he extended them in important ways that are beyond the scope of this paper.

But this change in geological point of view, while it solved certain problems, brought new problems to light. If organisms are perfectly adapted to the milieu for which they were created, what becomes of this adaptation when the milieu changes? The position of Natural Theology, seemingly so perfectly equilibrated, underwent a profound disturbance with the introduction of this question.

During the voyage, Darwin had rich experience with what seemed to be an important part of the answer: If the milieu to which certain organisms are adapted changes sufficiently, those organisms become extinct. In South America, Darwin personally made valuable discoveries of fossil remains of extinct organisms.

But extinction, an answer to one kind of problem, also raised new problems. In a continuously changing world, would not many and eventually all species disappear? Lyell had seen this problem and had proposed the hypothesis of multiple creation: if there are successive depopulations, there must be successive creations of new species, to maintain the earth's living inhabitation in an approximately steady state.

The voyage ended in 1836. By then Darwin was certainly interested in the idea of species mutability to meet changed conditions in the phenomenon of extinction, and in the hypothesis of multiple creations. But it must be stressed that these ideas, even taken together, fall far short of the idea of progressive evolution.

In July 1837, Darwin began his First Notebook on what he then called the 'transmutation of species'. From the first forty pages, we can reconstruct his first theory of evolution. This theory was quite ephemeral, since Darwin began to see flaws in it as he was writing it down. It runs as follows:

From nonliving matter, simple organisms, *monads*, spring into life. At a given time, the monads arising all over the world would be very much the same, due to similar global conditions. Once formed, as different individuals meet with different physical environments, they change, becoming progressively more complicated. The number of species is approximately constant. Since the production and evolution of monads is a more or less continuous process, some species must disappear in order to maintain the species number constant. This 'death of species' occurs in a manner analogous to the death of individuals; all living things must reproduce and all must die. When a given monad has lived its span, it dies and with it die all the species that it has become.

Darwin saw immediately that the monad life-span idea carries with it the implication of the sudden, catastrophic disappearance of many related species. He was sceptical of this conclusion on various grounds: it was a form of the Catastrophism he had struggled to reject during the *Beagle* voyage; the fossil evidence as he read it did not support the idea. Whereas on page 22 of the First Notebook he had written, "If we suppose monad definite existence . . . ", by page 35 we find him writing,

If we . . . grant similarity of animals in one country owing to springing from one branch, and the monucle [used synonymously with *monad*, H. G.] has definite life, then all die at one period, which is not case . . . *Monucle not definite life*. (Darwin's emphasis).

Some of the provisional flavor of Darwin's thinking at this point can be caught in the following passage, which also indicates that the species *man* was included in his thinking about evolution as early as July 1837.

Unknown causes of change. Volcanic island – Electricity. Each species changes. Does it progress?

Man gains ideas.

The simplest cannot help becoming more complicated; and if we look to first origin there must be progress.

If we suppose monads are constantly formed, ? , would they not be pretty similar over whole world under similar climates and as far as world has been uniform, at former epoch? . . .

Every successive animal is branching upwards, different types of organization improving . . .

This tendency to change . . . requires deaths of species to keep numbers of forms equable. But is there any reason for supposing number of forms equable? . . .

Organized beings represent a tree, *irregularly branched*, some branches far more branched, – Hence Genera. As many terminal buds dying, as new ones generated. There is nothing stranger in death of species than individuals.[7]

Although Darwin soon gave up the monad theory, it helped him to generate his *branching model of evolution*. He drew three tree diagrams, each bringing out somewhat different points. Such a diagram, highly formalized, became the only illustration in the *Origin of Species*. It contains within it many ideas: the idea that evolution is a random process; the non-necessity of discovering 'missing links' among contemporary species, since continuity is only 'vertical' in time, not 'horizontal' in space; the idea of divergence and speciation; and the idea of exponential growth which is implicit in any tree diagram.

As already stated, the coupling of the premise of exponential growth with the premise of a constant number of species requires the idea of extinction of some species. In other words, this coupling leads to a *formal* principle of some selection, although it neither proposes any *mechanism* by which selection might be effectuated nor suggests in itself any bias in the selective process.

Among his many pursuits in the year that followed, Darwin searched for an understandable mechanism of selection. Meanwhile, he studied more deeply the problems of hybridization and variation, and consequently came to have some knowledge of plant and animal breeding. But it was not the idea of artificial selection that finally suggested to him that natural selection might be a positive evolutionary force. It was, of course, known to him that many writers had proposed natural selection as a conservative force in eliminating monsters and other deviants, keeping the system of nature stable (*i.e.*, as it had been created). What struck him on September 25th, 1838, on reading the Rev. Thomas Malthus's *Essay on Population* was that *selection coupled with variation* could produce evolutionary change.

It should be noted that there are important *theoretical invariants*: the new theory had important structural carryovers from previous ones. (1) Darwin remained uncertain as to the *cause* of variation. (2) In the monadic theory the plenitude of nature was represented by the production of monads. In the natural selection theory this plenitude was represented in a double way. On the one hand each individual organism has a vast potential to reproduce itself, necessitating selection of only a small proportion for survival and reproduction if population is to remain approximately constant. On the other hand, the amount of variation to be found in nature is much greater than had previously been recognized. (3) The assumption of certain approximate constants in nature, such as the number of species or the size of populations, leading to the inference that exponential growth cannot continue indefinitely. (4) The irregularly branching tree model, present throughout all of Darwin's thinking about evolution from 1837 onward.

The importance of Darwin's reading of Malthus in 1838 should not be exaggerated. As I have pointed out, the branching model had already forced him to search for some principle of selection. He had already been busily acquainting himself with everything known about variation. He had undoubtedly come in contact with Malthus's ideas many times before. But by this time, the structure of Darwin's thinking had finally grown to the point where Darwin could see that the principle of superfecundity might be usefully introduced at a precise point in the theoretical structure.

It should be added that Charles Darwin was no 'Malthusian'. He never suggested that human social policy should be adjusted to prevent population growth. He fathered ten children, of whom six married and four had offspring.

Indeed, if he had addressed himself to this question, it would have been more in keeping with his general way of thinking to suppose that a multitude of small factors operating naturally and spontaneously would be far more efficacious than the deliberate application of a monolithic social policy.

## IV. METAPHORIC THOUGHT AS PART OF THE PROCESS OF ABSTRACTION

Up to this point I have emphasized the interplay between thought and feeling in explaining the growth of ideas. But these are obviously not enough. For science to progress there must be constant exchange between the world of ideas and the world of practical, sensuous experience. But we need to question the form of this exchange. If one stops to reflect upon it, there is really a deep, unsolved problem here. Concrete facts and abstract ideas seem so incommensurable with each other. How can a structure of ideas assimilate a new fact? In a sense, the fact is 'out there' in the world: it must be transformed in some way before it can be assimilated. But how? I wish only to discuss one aspect of the subject, the role of metaphors.

Something must be said here about the problem of definition, which is both terminological and taxonomic. Metaphors, analogies, and models are part of a group of comparison processes by which we use some parts of our knowledge to illuminate others. There are many names for such comparison processes, but there is no adequate taxonomy of them. Indeed, since they have almost always been treated singly, we have no adequate overview of the way in which groups of such comparison processes function in intellectual work. Lacking any such systematic treatment, it is idle to fuss over definitions. We need a large and generous term to cover the whole family of comparison processes. The poet, Howard Nemerov, facing the same problems, has used the term *figures of thought*. [8] In a recent and excellent discussion of Charles Lyell's metaphoric thinking, Martin Rudwick has used the term *analogy* for the same purpose. [9] Rudwick's paper is rare in that he makes a wide-ranging survey of many of Lyell's figures of thought, rather than indulging in the usual penchant for selecting one favorite. In the following discussion I will use *metaphor, image, figure of thought*, and the abbreviated *figure* interchangeably to stand for such comparison processes.

Metaphors are particularly interesting because they can be charged both with sensuous knowledge and with intuitive feeling. They can make specific points clear, and, because their boundaries are not fixed, they can invite the thinker to further exploration. In most discussions of the subject, metaphors are considered one at a time, often in a deliberate quest for the single ruling metaphor that governs a given intellectual enterprise or a whole life, or even a whole discipline.

I do not question the value of such single-metaphor studies. They are a necessary part of the examination of groups and systems of metaphors. In this symposium, Paul McReynolds's discussion of 'The Clock Metaphor in the History of Psychology',[10] is especially valuable and of course has a direct bearing on one of Darwin's intellectual sources, the work of William Paley. In a recent paper of Wise, although his conceptual aim is to trace the growth of a single ruling metaphor in Maxwell's thinking, the author in fact nicely demonstrates the interplay of a number of figures of thought.[11]

Reflecting on Charles Darwin's thinking suggests that metaphors do come in groups, or systems. In Darwin's own case, at least six important figures of thought played a role in his thinking about evolution. Each one plays a special part in the system of his thought. Singling out any one of them as capturing his central idea does serious injustice to the synthetic unity of his thought.

All of these figures appear quite early in Darwin's thinking, in one form or another. I will enumerate them in the chronological order of their occurrence.

*Contrivance*. Darwin speaks of particular organs, or parts of organs, as 'contrivances' or inventions. Of course, Natural Theologians from whom he borrowed this language had in mind the 'Divine Artificer' who had fashioned these contrivances. Darwin's point came to be exactly the opposite: The myriad beautiful adaptations of nature are *not* the inventions of a conscious being, but the result of the 'blind' operation of natural laws. Nevertheless, the idea of a contrivance helped him to preserve his sense of wonder at the remarkable and *almost* perfect fitness of form to function throughout nature.

*Tangled Bank*. In the final paragraphs of the *Origin of Species*, Darwin asks the reader to consider a sort of microcosm of organic nature, a 'tangled bank' in which all sorts of species interact with each other and with their milieu. This scene provides a model of the complex interplay of living things on the grander scale which evolutionary theory is required to explain. It is a metaphor in the sense of a comparison of part with whole. Its particular function in Darwin's system is to bring out the complexity of interactions among contemporaneous organisms. As we have seen above, this image appears early in the *Beagle Diary*.

*Irregularly Branching Tree*. As discussed above, this image appears early in the First Notebook, as a product of Darwin's monadic theory of evolution. Although the monadic theory was soon abandoned, the tree image remained a part of Darwin's thinking. Its particular function is to bring out the *historical* relationships among organisms. Reading horizontally, it depicts contemporaneous organisms; reading vertically, noncontemporaneous ones. This image had explanatory value in a number of ways, appears as the only diagram in the *Origin*, and is referred to throughout the book. Over the years, Darwin thought often about the most appropriate form of the tree image, and there are a number of tree diagrams among his unpublished manuscripts. In fact, the most appropriate form of the tree seems to have varied according to the specific point Darwin was exploring at any given time. For example, in the First Notebook, soon after drawing his first tree diagram, he wrote, "The tree of life should perhaps be called the coral of life, base of branches dead, so that passages cannot be seen" (p. 25). But he adds immediately: "this . . . offers contradiction to constant succession of germs in progress" (p. 26) — indicating that he is still clinging to the idea that 'monads' are continually being produced. Years later he wrote, "Tree not good simile. [It would be better to say:] Piece of seaweed endlessly branching." A faint sketch of such a piece of seawead appears below these words. (Undated manuscript.)

*War*. On September 28, 1838, when Darwin read Malthus's *Essay on Population*, as I have recounted above, he had his first clear insight into the idea of evolution through natural selection. He was struck by the *necessity* of selection as a consequence of the potential for exponential population growth, exceeding the ability of the habitat to support this growth. Malthus's *Essay* is in large part a catalogue of human warfare and of the decimation of human populations. Darwin borrowed the language of warfare for one of his metaphors, dramatizing both the selectiveness and the destructiveness through which the innovative force of evolution may be conceived as working.

*Wedging*. Darwin concluded the passage in his notebook for September 25, 1838, with another metaphor:

One may say there is a force like a hundred thousand wedges trying (to) force every kind of adapted structure into the gaps in the economy of nature, or rather forming gaps by thrusting out weaker ones." (Third Notebook, p. 135).

This image of wedging reappeared in essays he wrote in 1842 and 1844, and in the first edition of the *Origin of Species*, but it was deleted from subsequent editions. It would be interesting to know why Darwin dropped it, since it does convey dramatically the way in which variation and struggle continuously disequilibrate the natural order at almost every point in space and time. It is noteworthy that, in this highly significant passage of Darwin's notes, the wedging image is at least as prominent as the war image.

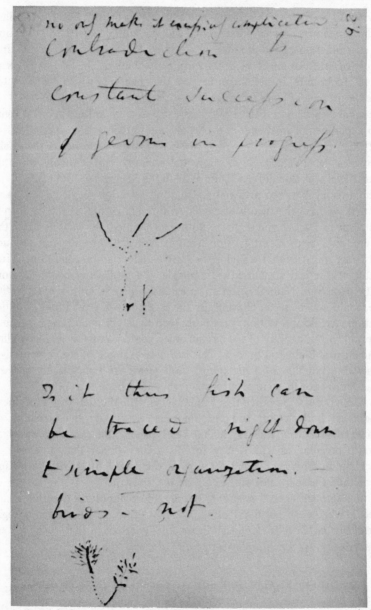

Fig. 1. Darwin's first attempt to sketch the idea of the irregularly branching tree of nature. *Courtesy of Cambridge University Library.*

*Artificial Selection*. This figure was valuable to Darwin because it brings out the *cumulative effects* of long continued selection in a particular direction (*i.e.*, as a function of the duration of a given set of selection pressures). But the role of artificial selection in Darwin's early thinking should not be exaggerated. To be sure, in the *Origin* he gave the chapter in which the subject is treated a prominent place (Chapter I, on variation under domestication). This was probably because artificial selection came as close as Darwin could come to an experimental treatment of evolution, and Darwin was — especially after 1842 when he moved from London to his permanent residence in the village of Down — an indefatigable experimental scientist. But it should be noted that artificial selection did not figure at all in the September 28th insight.

The analogy between artificial and natural selection seems clearer to us now that we have the theory of evolution through natural selection, than it did at first to Darwin and Wallace. Both men pointed out ways in which the two processes seem to differ. For example, domestic breeders can produce 'monstrous' varieties that would never survive under natural conditions. After a stable theoretical structure was available, the similarities could be more readily perceived. Natural selection, previously seen as only a conservative force, could now be seen as having also the possibility of establishing novelties in nature. Artificial selection, previously seen as an ephemeral and capricious exception to the stability of the natural order, could now be seen as modelling a fundamental biological process. This is a nice example of the constructive, synthesizing function of a metaphor: both terms are transformed by their conjunction; once they are seen together, they illuminate each other.[12]

It will of course be noted that I don't agree with Michael Ruse's judgment in assigning both logical and chronological priority to the metaphor of artificial selection.[13] Wallace, in his 1858 paper, still saw the results of domestic breeding as a difficulty for his theory rather than as a way of supporting it. This fact alone shows that the model of artificial selection was neither necessary nor sufficient for the construction of the theory of evolution through natural selection. But the more important question is not to decide whether this or that figure of thought occurs first in a given thought process, but rather to look at each figure as it functions in an evolving system of ideas displaying the interplay (*inter alia*) of many figures.

## V. DIFFERENTIAL UPTAKE OF COMPLEX IDEATIONAL STRUCTURES

Having completed this brief survey of Darwin's system of metaphors, it remains to take note of the remarkable *differential uptake* of Darwin's ideas. In expounding his thought, later writers – including social Darwinists – have put almost the entire emphasis on the images of artificial selection and war. Darwin's own use of the images of contrivance, tree, wedge, and tangled bank have been almost completely neglected. This biased account of Darwin's thinking has had two unfortunate consequences. First, it has given to the idea of struggle the form of *polarized struggle* between two opposed forces (as in human warfare, or as in a contest between Breeder and Nature). In Darwin's own thinking, 'struggle' clearly means something else: the complex interplay of many factors, leading to the differential survival of organisms, depending on their varying adaptation to all the conditions of life.

Second, this biased account of Darwin's thinking puts all the emphasis on forces of selection and destruction in nature, those aspects of the entire process in which stronger organisms flourish and weaker ones die, depicting a world of nothing but winners and losers. But this selective use of Darwin's ideas destroys the dialectical unity of his thought. The creative and explosively productive metaphors of contrivance, tangled bank, and branching tree were equally essential features of Darwin's image of nature.

It would be worth studying the social and ideological sources of this distortion. The differential uptake of Darwin's ideas reflects a rather sour view of nature, as compared with the exuberance of Darwin's feeling and thought about it.[14]

## VI. THE PLACE OF INSIGHT IN AN EVOLVING STRUCTURE OF IDEAS

At several points in this paper I have referred to Darwin's 'insight' about evolution through natural selection, stimulated by his reading of Malthus's *Essay*. The documentary evidence is unmistakeable, that on September 28, 1838, he had such an insight. But the meaning of *insight* is open to discussion. In concluding this paper I want to take up a few points that may help to clarify the evolving systems approach to creative thinking.

*Is 'insight' an explanatory concept?* As I have tried to bring out, creative thinking is a complex system of many inter-related processes. Insight is not an explanation of creativity, it is part of the total set of phenomena that requires explanation. It is important in understanding creative thinking because it reflects and accentuates the occurrence of qualitative leaps in a series of structural transformations.

*Insight as a subjective phenomenon.* The hallmark of insight is its affective

character. Since it has almost always been described in intellectual terms
referring to the cognitive contents of some discovery, this emotional aspect
of insight is not well understood. Our ignorance of this subject is part of a
more general neglect of the positive emotions related to productive life. Pro-
visionally, we may say that insight (sometimes called the 'aha!' or 'Eureka!'
experience) involves feelings of surprise, newly discovered novelty, elation,
a sense of movement toward a goal. In the sense that it is an illumination, we
may think of it as a structural change in cognitive organization, entailing a
feeling of increased clarity. But we should be careful about this point, because
sometimes all that is 'discovered' is a new question.

*Frequency and magnitude of insights.* Insight is often treated as though
the creative person has only one or a few insights in a lifetime. This is probably misleading. In what may be the only empirical study of the question, an
experimental scientist kept a notebook for about ten years. During this time
he made over 30,000 entries of which 680 were marked at the time of entry
as 'sudden illuminations'. This number is approximately one insight per week,
and about two per cent of all events considered worth recording.[15] Although
I have not made a similar count for Darwin, I would estimate that his rate of
production of insights is noticeably higher, perhaps one per day, or even
greater. Even using the lowest plausible figure, the number is very high,
yielding thousands of insights in a creative lifetime. This large number suggests
that moments of insight are expressions of the relatively stable functioning
of a system, rather than of its overthrow. In other words, we are talking
about a series of relatively small transformations representing both purposeful
growth and structural change. This seems to me to be a more satisfying
picture than the view of creativity as a few great leaps into the unknown.

Indeed, the general picture that I have been presenting of Darwin's thought
suggests that any interesting creative process will necessarily entail a group
of transformations in a dense and complex structure composed of many
interacting parts. When evolving structures of ideas are seen in this way, the
hypothesis of sudden great reorganizations of the whole ensemble becomes
less plausible, and the hypothesis of a series of transformations at a number
of critical points becomes more plausible. We should bear in mind that even
when a theory has reached maturity and has been thoroughly examined and
well formulated, no one can think about it all at once. Why should we suppose
that the inventor can?

*Insight and the social nature of thought.* The hypothesis of isolated great
insights corresponds to the idea of the creative person as an isolated genius.
In constructing the evolving systems approach to creative thinking we are

searching for a better understanding of the relation between the occurrence of great individuals and the social, collaborative nature of all human work, including creative thought.

One approach would be to propose that, since many insights are necessary, they can be distributed among many workers, so that important changes are nothing but the cumulative effects of many people's work. This view runs into two difficulties, the inescapable fact of individual greatness, and the corresponding point that new ideas must be somewhere assembled into an effective, working system.

A new configuration has not only a number of new components but many new interactions. Various aspects of a group of problems can be dealt with in different combinations, thus exploiting the special knowledge and preoccupations of different individuals and groups. Meanwhile, an individual with a new and unusual vision and sense of purpose may be re-assembling the whole ensemble in a new configuration. In Darwin's life, during his London years, while he was working out the theory of evolution, he also worked with many specialists, processing the materials brought back from the *Beagle* voyage. In addition, he had ample opportunity to test out some of his ideas on people like Charles Lyell.

Nevertheless, for a long time, he kept his vision of the whole new emerging configuration of ideas quite private. His life exhibited a pattern of purposeful management of the relation between public and private thinking. In one sense, thought is like metabolism and growth: it must take place within an individual organism. In another sense, thought is like reproduction and other biological processes requiring interaction among organisms. It is in the dialectical movement between these two poles that innovations are constructed.[16]

*Institute for Cognitive Studies*
*Rutgers University, Newark, New Jersey*

## NOTES

[1] An earlier version of this paper was given as a series of lectures at the International Symposium on Evolution, Liblice, Czechoslovakia, June 5–9, 1978. I thank Rachel Falmagne for a critical reading and helpful comments on a draft of this paper.
[2] Howard E. Gruber, *Darwin on Man: A Psychological Study of Scientific Creativity*, together with *Darwin's Early and Unpublished Notebooks*, transcribed and edited by Paul H. Barrett, New York: Dutton, 1974; 2nd Edition, University of Chicago Press, 1980.
[3] A. I. Oparin, *The Origin of Life*, New York: Dover Press, 1953 (first published in Russian, 1936).

[4] Jean Piaget and Bärbel Inhelder, *La Psychologie de l'Enfant*, Paris: Presses Universitaires de France, 1975. See also H. E. Gruber and J. J. Vonèche, 'Réflexions sur les opérations formelles de la pensée', *Archives de Psychologie* 44 (1976), 44–45.

[5] Charles Darwin, *The Autobiography of Charles Darwin, 1809–1882*, with original omissions restored, edited by his granddaughter, Nora Barlow, London: Collins, 1958.

[6] Charles Darwin, *Charles Darwin's Diary of the Voyage of H. M. S. 'Beagle'*, edited from the MS by Nora Barlow, Cambridge University Press, 1934. This passage was written on February 28, 1832.

[7] Extracts from first Notebook on Transmutation of Species, pp. 18–22, published in Gruber and Barrett, *op. cit.* I have added some punctuation.

[8] Howard Nemerov, *Figures of Thought*, Boston: David R. Godine, 1979.

[9] Martin J. S. Rudwick, 'Historical analogies in the geological work of Charles Lyell', *Janus* 64 (1977), 89–107.

[10] Paul McReynolds, 'The Clock Metaphor in the History of Psychology'. This volume.

[11] M. Norton Wise, 'The mutual embrace of electricity and magnetism', *Science* 203 (1979), 1310–1318.

[12] For a fuller discussion, see Howard E. Gruber, 'Darwin's tree of nature and other images of wider scope', in J. Wechsler (ed.), *On Aesthetics in Science*, Cambridge, Mass.: MIT Press, 1978.

[13] Michael Ruse, 'Ought philosophers of science consider scientific discovery? A Darwinian case study'. This volume.

[14] Howard E. Gruber, 'The Fortunes of a Basic Darwinian Idea: Chance', in R. W. Rieber and K. Salzinger (eds.), *Psychology: Theoretical-Historical Perspectives*, New York: Academic Press, 1980.

[15] Howard E. Gruber, 'Créativité et fonction constructive de la répétition', *Bulletin de Psychologie de l'Université de Paris* 30 (1976–77), 235–239.

[16] Additional case studies similar in character to the approach outlined here include: Rudolph Arnheim, *The Genesis of a Painting, Picasso's Guernica*, Berkeley: University of California Press, 1962; Frederick L. Holmes, *Claude Bernard and Animal Chemistry*, Harvard University Press, 1974; and Martin Rudwick, 'Darwin and Glen Roy: A "Great Failure" in Scientific Method?', *Studies in the History and Philosophy of Science* 5 (1974), 97–185.

MICHAEL RUSE

## OUGHT PHILOSOPHERS CONSIDER SCIENTIFIC DISCOVERY? A DARWINIAN CASE-STUDY

My concern in this paper will be with Darwin's discovery of his theory of evolution, particularly the part centered on its mechanisms. What I want to know is whether knowledge of Darwin's route to discovery tells us something about the finished theory, say as it is found in the first edition of *The Origin of Species* (1859). Do we, as philosophers, need to know how Darwin got his theory in order to understand his theory? I take it that there is a school of philosophical thought, 'logical empiricism', that would argue that essentially a scientist's route to discovery is irrelevant to his or her finished product. A scientific theory or hypothesis is in some sense intended to be a reflection of reality. Hence, that a scientist may have gotten his ideas after years of painstaking fitting of the data to possible ideas, like Kepler, or in a flash through mystical contemplation of his navel, is of absolutely no concern.[1] Even if Archimedes had never taken a bath in his life, his principle would still have been the same.

I assume that it is this kind of philosophy of scientific discovery (*i.e.*, that there can be no significant philosophy of scientific discovery) that underlies Carl Hempel's quick dismissal of discovery in his (excellent) little textbook, *Philosophy of Natural Science* (1966). A scientist may hit on an idea by the craziest of means, like Kekulé finding the benzine ring through dreaming of a snake swallowing its tail, but the 'real' science has no place for this. Were a herpetologist to complain that benzine cannot be circular because snakes do not swallow their tails, his worries would be dismissed as inappropriate. A similar philosophy seems to be held by Karl Popper, who has the singular distinction of having written a book called *The Logic of Scientific Discovery* (1959), which is not about scientific discovery at all. In a more recent paper (1970, p. 57), Popper states:

[T]o me the idea of turning for enlightenment concerning the aims of science, and its possible progress, to sociology or to psychology (or ... to the history of science) is surprising and disappointing.

And someone like Mario Bunge (1968) seems almost to want us to forget discovery as soon as possible, so that we might not illegitimately read into our theory things which helped us to get to the theory — otherwise we shall

131

*T. Nickles (ed.), Scientific Discovery: Case Studies*, 131–149.
*Copyright © 1980 by D. Reidel Publishing Company.*

start worrying about whether the benzine ring is cold blooded and whether bath salts are necessary for a true application of Archimedes's Principle![2]

I shall argue that this belittling view of scientific discovery is wrong: philosophically castrating in fact. Let us turn at once to history.

## 1. DARWIN'S ROUTE TO DISCOVERY

Charles Darwin published his *Origin of Species* late in 1859. For reasons which are not entirely clear, Darwin had been sitting on his idea for twenty years — it had in fact been fifteen years since he had completed a 230 page draft of his theory[3] — and even when he did publish, it was only because he had been sent a paper by the young naturalist, Alfred Russel Wallace (Darwin and Wallace, 1858), which contained evolutionary speculations uncannily like his own. As is well known, Darwin argued that evolution is chiefly a function of 'natural selection', the differential survival and reproduction of the more adapted over the less adapted, and that this in turn is fuelled by the 'struggle for existence', where animals and plants compete with each other and the environment for limited resources. Then, having produced his mechanism, Darwin applied it to many different areas of biology — geographical distribution (biogeography), instinct, embryology, and others. (See Figure 1.)

The crucial move to natural selection as an evolutionary mechanism was made by Darwin in the fall of 1838: late September—early October, to be more precise. However, controversy exists over precisely how Darwin moved to natural selection. In his *Autobiography* (1969), Darwin claimed that the work of animal and plant breeders using artificial selection gave him the notion of natural selection, and then reading Malthus's *Principle of Population* (1826), with its description of the struggle for existence — a function of geometric population growth potential always outstripping food and space arithmetic growth potential — showed him how to apply selection as a mechanism for evolution. And this route to discovery is confirmed by several other recollections by Darwin of his momentous discovery.

But this account of discovery does not mesh very easily with entries Darwin made in notebooks around the time of the discovery. The importance of Malthus is reinforced, but the key role played by the artificial/natural selection analogy is put in doubt. From comments Darwin made right up to the time that he read Malthus, he seems to have had some doubts about the power of artificial selection and the consequent analogy to natural selection. "It certainly appears in domesticated animals that the amount of variation

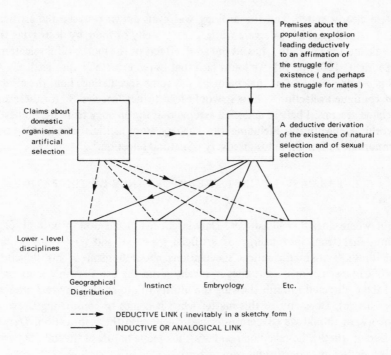

THE STRUCTURE OF DARWIN'S THEORY IN THE <u>ORIGIN</u>

( from Ruse 1975b )

**Fig. 1.**

is soon reached — as in pigeons no new races" (Darwin, 1960, p. D175). And indeed, some scholars have concluded on the basis of this and like passages that Darwin did not really use the analogy from the domestic world in his discovery of natural selection. (See Herbert, 1971, and Limoges, 1970.) Others however are loathe to make a liar out of Darwin. My own position, which I shall state but not really argue for here, is that although Darwin was not as certain of the value of the analogy before Malthus as he became afterwards (and thought that he had been before), the analogy did indeed play an important part in Darwin's discovery. (I do argue my case in Ruse, 1975 and 1979*a*.)

I argue this claim chiefly on the basis of what Darwin read in the months before he read Malthus. We know that in the summer of 1838 Darwin read influential pamphlets on animal breeding, in which the principles of selection were clearly stated, and the analogy was even drawn between the artificial and natural worlds! "A severe winter, or a scarcity of food, by destroying the weak and the unhealthy, has all the good effects of the most skilful selection" (Sebright, 1809, p. 16). We know also that Darwin reacted enthusiastically to this reading and that at that point he did some speculating about the effects of continued selection — how it would lead to new species. So for these and related reasons, I believe and shall assume that the analogy from the domestic world — specifically including the analogy from artificial selection — was important to Darwin in his discovery of natural selection.[4]

## 2. THE CASE AGAINST THE IMPORTANCE OF THE PATH OF DISCOVERY

But where does this all take us? Darwin got his mechanism of natural selection, first from the analogy of artificial selection, and then from reading Malthus's quasi-mathematical speculations about humans in his *Principles* (1826) (speculations, somewhat ironically given the use Darwin was to make of them, directed towards showing the futility of attempting any real progress or change). Does any of this matter when it comes to considering Darwin's theory, or should we concern ourselves solely with the justifications Darwin offered: whether he relied on real laws, the precise nature of the links between his premises and conclusions, and so forth?

Considering matters first at a general level, and recognizing that initial suggestions will probably require some refinement, it should in theory be possible to decide empirically some of the pertinent questions about scientific discovery — at least in a one-way negative manner somewhat akin to the Popperian falsification of scientific hypotheses. Suppose one has a scientist $A$, who gets to theory $T$ by route of discovery $R$. If one now has another scientist $B$, who also gets to theory $T$, but *not* by route of discovery $R$, one can certainly conclude that $R$ was not necessary for getting to $T$, and therefore can hardly be that essential for understanding $T$: it is not going to be embedded in $T$ in any significant way. This line of argument is one-way however, because if $A$ gets to $T$ via $R$ and $B$, not on $R$, does not get to $T$, there is always the logical possibility of a third scientist $C$ who gets to $T$ but not on route $R$.

Now, as I have just said, this is an empirical line of argument. If one has a situation with the right kinds of scientist, one can stop one's *a priori* theorizing and check. And the beautiful thing about the Darwinian case is that one does have just such a situation with the right kinds of scientist: Wallace came to natural selection as well as, but quite independently of, Darwin.

Most or perhaps all the variations from the typical form of a species must have some definite effect, however slight, on the habits or capacities of the individuals. . . . [Consequently, if] "any species should produce a variety having slightly increased powers of preserving existence, that variety must inevitably in time acquire a superiority in numbers. (Darwin and Wallace, 1858, p. 273).

And if this keeps happening long enough, the process "must in the end, produce its full legitimate results" (*Ibid.*, p. 275). What is of crucial importance to us here is that although Wallace was as dependent on Malthus as Darwin for getting to the mechanism, *he did not use the artificial selection analogy*. Indeed, like everyone but Darwin, Wallace looked upon the domestic world as one of the rightful pillars of the case *against* evolutionism! Domestic change is limited; therefore, any analogy is that natural change is limited.

In other words, even if one could legitimately follow Lakatos (1970) in creating history to fit one's philosophical theses, it would seem that in the Darwin-Wallace episode, one could not have more definitive support of the irrelevance of a scientist's route to discovery for understanding that which he discovers. Two men discovered the identical principle of natural selection. For one, the analogy from the domestic world in general and artificial selection in particular was crucial. For the other, the analogy played no role at all; it played an *anti*-role, for he looked upon it as a problem to be surmounted. Obviously therefore, when it comes to understanding a completed theory, discovery is irrelevant.

### 3. SUBJECTIVE AND OBJECTIVE ELEMENTS IN SCIENCE

And yet, I am not sure that this is all that there is to be said, even about the Darwin case. In order to articulate my objections, let me introduce the terms 'objective' and 'subjective', although I am a little hesitant about so doing: too often the terms have meant all things to all people and have been applied indiscriminantly in the most inappropriate of situations. But understanding 'objective' to mean something public, 'out there', with existence independent of the observer, and understanding 'subjective' to mean something which in a very real sense depends on the human mind, which has no reality away from the individual, what I would argue is that science has both objective and subjective elements.[5] More precisely, scientific theorizing contains elements which reflect objective reality, and elements which are more subjective in nature. Moreover, I would claim that whilst the Darwin case certainly shows that science inasmuch as it is objective (*i.e.*, talks about objective things) is independent of discovery, the Darwin case also shows that science inasmuch as it is subjective is dependent on discovery. Furthermore, I am not sure that one can or would want to eliminate the subjective element.

Clearly the Darwin case does point to the fact that independent of Darwin or Wallace there exists a process in organic nature of differential survival and reproduction. This is going on — some animals survive to reproduce, whereas others do not. Insofar as a scientist like Darwin or Wallace grasps this fact, his science is objective, because it is reflecting something independent of the observer. And as we have seen, this objective elements stands in its own right, independently of how anyone comes to grasp it.

But what about the subjective element? Here I am on somewhat shakier ground, if only because, as previously argued, the fact that two scientists taking different routes arrive at different ends does not logically preclude the existence of a third who takes yet another route and ends where the first (or the second) also ends. However, I think that (in theory) some sort of case can be made for the position I am proposing. Even if I cannot make my case logically watertight, I can at least try to make it factually convincing (i.e., if not deductively certain, at least inductively probable). If a scientist builds some element into his theory which there is reason to believe is directly connected with his route to discovery, which for various independent reasons one would think unlikely to be there without that route, which last suspicion is reinforced by the fact that another scientist taking a different route does not put the element into his version of the theory and moreover gives evidence that that is the very thing he would *not* want to do, then one has grounds both for thinking the element in some way subjective and linked with the way of discovery.

Perhaps the kind of case I am trying to establish can best at given plausibility by turning at once again to the Darwin-Wallace episode. I offer the following three suggestions as instances of the importance of a scientist's route to discovery.

## 4. DARWIN'S LANGUAGE

First, take the question of language. Both Darwin and Wallace referred to the struggle for existence, by that name. This was natural enough because the term occurs in Malthus's *Principle of Population* (1826), of key importance to both Darwin and Wallace — and had also been picked up and used by others who were influential on Darwin and Wallace. In particular, the term occurs in the work which probably had more influence than any other on the two evolutionists: Charles Lyell's *Principles of Geology* (1830–1833). On the other hand, only Darwin referred to the process of differential survival and reproduction as 'natural selection'. Darwin claimed that he got the term from breeders, who used it to refer to the natural process of some organisms dying off through the environmental conditions. Breeders were certainly aware of this phenomenon, and Darwin was aware that they were aware, although I have not yet come across a written use of the term.

But this is as it may be. What is of importance to us here is the fact that Darwin obviously got the term in some way from the breeders' use of the term 'selection' to denote their process of picking the organisms they wanted (what we now call 'artificial selection' to distinguish it from natural selection).[7] What I would suggest is that this term 'natural selection' is a subjective element in Darwin's theory, in the sense of subjective that I have characterized. Moreover, I would suggest that without the analogy from the domestic world, Darwin would never have got it or used it, and that the term itself has a status which makes it an integral part of Darwin's theory (i.e., it is not just a symbol but is philosophically interesting and important, as was Darwin's application of natural selection to the fossil record).

That the term is subjective seems fairly obvious. The differential reproduction is public and exists independently of observing theorizers. What we humans call the process is another matter. It could have been called almost anything − 'Twelfth night, or what you will'. Darwin himself suggested that it might have better been called 'natural preservation'. Herbert Spencer suggested, and Darwin later accepted, 'the survival of the fittest' (Darwin and Seward, 1903, Vol. 1, p. 269). Wallace in his paper did not really call the process anything: he just described it.

But would the term have been used without the analogy, and without the prior use of 'selection' by breeders? It seems highly improbable that it would have been, or, to strengthen the claim a little, that if the breeders had used some other term, that term would have been used without the analogy. Logically, monkeys might type Shakespeare; actually, they are not very likely to do so. Take again the (natural!) control experiment of Wallace. He proves that one did not need the analogy to get at the process of selection. But he also suggests strongly that without faith in the analogy, one would be most unlikely to use the term 'selection' to describe the process. As has been mentioned, Wallace followed everyone but Darwin in thinking that if any analogies can be drawn from the domestic world, they *disprove* selection. Wallace wanted as little as possible to do with man's selective power. Consequently, we have the paradoxical situation that whereas Darwin emphasized the analogy, a major theme of Wallace's paper is that there is no significant similarity between the domestic and the natural worlds. Stated Wallace:

It will be observed that this argument [denying evolution on the basis of the domestic world] rests entirely on the assumption, that *varieties* occurring in a state of nature are in all respects analogous to or even identical with those of domestic animals, and are governed by the same laws as regards their permanence or further variation. But it is the object of the present paper to show that this assumption is altogether false . . . (Darwin and Wallace, 1858, p. 269).

In short, Wallace was not looking to the domestic world for guidance, and for him to have adopted the term 'natural selection' would have meant his going against his whole strategy. For all real purposes, therefore, it would seem that Wallace could not have got or employed the term; only one thinking as Darwin thought could have done so. Significantly, it was Wallace who urged Spencer's alternative on Darwin.

At this point, I suspect some readers may be getting impatient. I am labouring to produce a molehill. Let us concede that the use of the term 'natural selection' is subjective. Let us concede that although terms are subjective, a theory must have them. Let us even concede that Darwin could not have got his term 'natural selection' without his route to discovery, and that therefore if we philosophers are to understand Darwin's use of the term, we must make reference to the context of his discovery. So what? The important things Darwin does in the *Origin* are identifying and describing the process of natural selection, and offering evidence for its power, like the evidence of the fossil record and of geographical distribution. None of this required Darwin's route of discovery: the actual words he used are irrelevant. A rose by any other name would smell as sweet.

Obviously there is truth in this objection; but, I contend, not the whole truth. In looking at a theory, particularly as philosophers, we want to consider what is important or essential about the theory. But what is to be counted as important? What is to be counted as essential? One thing which must be included presumably, however one decides, is the elegance of the theory – its simplicity. Another is the extent to which the theory fits with our metaphysical preconceptions – Does it violate accepted beliefs about causality? And yet another is the hard evidence for the theory – What makes it plausible? I would suggest that, judged by precisely these criteria, the particular terms used can be important, and that Darwin's term 'natural selection' is a paradigmatic illustration. After the *Origin* was published, the term natural selection' was as much a matter of controversy as anything else, like the hard evidence. (Ruse 1979a). For instance, some complained that by the use of 'selection' Darwin had unwittingly introduced a theistic concept into his theory – one cannot talk about selection without implying a selector, or rather, a Selector. Others, conversely, agreed with this point, but found no cause for complaint! They felt much happier with Darwin's theory precisely because it necessarily involved God. But whether one deals with supporters or critics, important questions about the acceptability of Darwin's theory revolved about his use of words. It seems clear that the term 'natural selection', coming as it did uniquely from artificial selection, brought with it a certain 'flavour' of artificial selection. People could not think about natural selection any more without thinking about the origin of its name, and this influenced the way they reacted to natural selection itself.[8]

I might add that some of the questions raised by the term 'natural selection' – or at least by the implications of the use of the term – remain with us today. For instance, had Darwin not used 'natural selection' but the more neutral 'natural preservation', he would not then have made the dreadful mistake of allowing Wallace to persuade him to adopt Spencer's 'survival of the fittest' as an alternative. And then we would not have people like Karl Popper still today arguing that evolutionary theory is either second-rate science or disguised metaphysics, because its central notion reduces to the empty tautology that those that survive are those that survive. In short, if we are at all sensitive to the actual historical fate of Darwin's theory – why people accepted or rejected it – we must allow that his use of language was important.

My claims therefore in this section are: Darwin's use of the term 'natural selection' was subjective in the sense specified. It came uniquely from the use of the term 'selection in the domestic world, that is, although its origin was not logically necessary, as a matter of fact it was highly unlikely to come in any other way. Historically the use of the term was important in reactions to Darwin's theory, because it is clear that people could not use the term without in some way thinking of its origins. To us as philosophers, trying to understand Darwin's theory, because apparently people cannot use the term without bringing in origins and because this affects their reactions to Darwin's theorizing, necessarily we must concern ourselves with Darwin's route to discovery.

## 5. ARTIFICIAL SELECTION AS JUSTIFICATION AND HEURISTIC GUIDE

The second way in which Darwin's route to discovery was reflected crucially in his theory centres on the use to which he put the domestic/natural analogy in constructing his theory. Readers of the *Origin* will know that Darwin does not just present the struggle for existence and natural selection right at the beginning. Rather he begins with a detailed discussion of the domestic world and of breeders' successes with artificial selection. Then, having talked about everything from pigeons to sheep, from strawberries to cabbages, Darwin broadens his gaze to the animal and vegetable kingdoms in the wild.

Now, although a number of commentators have been rather inclined to depreciate Darwin's reasoning powers, as he himself admitted the *Origin* is one long argument from beginning to end. In the first chapter, Darwin was not simply wasting time as he summoned up courage to get to the controversial notion of natural selection. Rather Darwin was doing two things (Ruse 1973*b*; 1975*b*). First, he was preparing the way for the *justification* of his overall theory: because artificial selection is so effective in the domestic world, analogously, we should expect it to be effective in the natural world. And later in the *Origin*, Darwin returned repeatedly to this analogy. For instance, in his discussion of embryology, Darwin justified his claim that the differences between embryos and adults is a function of the differences in selective forces, by reference to the domestic world. Second, Darwin was discussing artificial selection as a *heuristic* guide to natural selection. Even if one did not think the analogy offered support, it certainly helped the reader (in Darwin's opinion) to understand what natural selection is all about. (See Figure 1.)

Had Darwin not arrived at natural selection by means of artificial selection, I doubt he would have put the analogy to either of these uses in the *Origin*. One could, I suppose, argue that even if (like Wallace) Darwin had come to natural selection despite artificial selection, he could then later have decided that artificial selection offers support for natural selection and thus have introduced it as justification. This, I take it, was what Wallace's position grew to be (with reservations to be noted), and it is certainly true that it was after he discovered natural selection that Darwin made his most extensive study of the domestic world in his search for evidence for his theory.[9] On the other hand, it seems improbable that Darwin would have sought information from the breeders had he not first thought the analogy of value, and certainly he would not have made the justificatory use of it that he did in the *Origin*. Even more strongly, had Darwin not been led to natural selection via artificial selection himself, he would hardly have offered it in its heuristic guise. For Wallace, it was anything but! Hence, here as before it seems plausible to suggest that Darwin's route of discovery influenced the theory he produced.

I take it that also as before we are dealing with a subjective element in Darwin's theory. What someone finds heuristically valuable does not seem to have the independence of the observer in the way that the brute fact of differential reproduction does. Indeed, it might be felt that everything is so subjective and personal at this point, that although Darwin certainly introduced the analogy into the *Origin*, it was not really part of his theory proper — certainly not inasmuch as it was intended to have a heuristic value. However, I am not sure that this reply is really fair: at least, the reply seems to presuppose an *a priori* view of what constitutes the essence of a theory that one is imposing upon history. Darwin's aim in the *Origin* was to persuade the reader to accept evolution and to accept natural selection as its chief mechanism: that was his 'theory', and the artificial selection discussion/analogy played a key role. Moreover, like the term 'natural selection', artificial selection figured mightily in the controversy in the years after 1859. For instance, although T. H. Huxley was Darwin's most vocal supporter, he could never fully endorse the power of natural selection, because he thought that artificial selection proves the limitations of natural selection! (Hull 1973; Ruse 1979a). Hence, I conclude that unless one prescribes that only formal deductive systems are the 'real' parts of theories, one has no right to exclude artificial selection (and even with such a prescription one has no right).

## 6. THE NATURAL/SEXUAL SELECTION DICHOTOMY

We come to the third and final way in which Darwin's path to natural selection through artificial selection is reflected in the theory of the *Origin*. Although it was always his major mechanism, natural selection was never Darwin's sole putative mechanism of evolutionary change. Darwin was, for example, always a Lamarckian, in the sense of believing in the inheritance of acquired characteristics.[10] But Darwin's major secondary mechanism was *sexual selection*: even in his earliest drafts of his theory, Darwin mentioned this kind of selection (Darwin and Wallace, 1958); in the *Origin* (1859) he spelt it out clearly, albeit without developing it; and then in his seminal work on our species, *The Descent of Man* (1871), Darwin discussed sexual selection in great detail, both as it applies through the animal world and as it applies to *Homo sapiens*. I argue that not only was artificial selection crucial in Darwin's getting to this kind of selection, but it was essential for the place that Darwin gave it in his theorizing. In other words, without understanding Darwin's route to discovery, we cannot understand the structure of Darwin's argument.

Moreover, I claim that whilst sexual selection is not itself subjective, Darwin's treating it as an independent kind of selection is.

First, there is the historical question of how Darwin got to sexual selection. There are hints of sexual selection in the evolutionary meanderings of Darwin's grandfather, Erasmus Darwin, as well as in the writings of others that Darwin read (for example, Sebright, 1809). However, study of what Darwin produced makes it overwhelmingly certain that the key to discovery for Darwin, as well as the conviction that the discovery was important, lay in the analogy from the domestic world and the breeder's power of selection. (Ghiselin, 1969) Breeders select for two things: attributes of animals and plants that are useful to us, like shaggy sheep coats and fleshy root crops, and attributes that are pleasurable to us, like fancy pigeon tails and vicious bull-dogs. It was this division that gave rise to the natural/sexual selection dichotomy. Furthermore, the division that breeders make in the pleasurable attributes was exactly reflected in a division that Darwin always made in sexual selection. Breeders select (*qua* pleasure) for pugnaciousness, as when they breed vicious fighting cocks, and for beauty, as when they breed beautiful birds. For Darwin, in the natural world, these translated into sexual selection through male combat and sexual selection through female choice. We can see therefore that the analogy from artificial selection played a powerful role for Darwin when he came to introduce and justify sexual selection in the *Origin* and in later works. But did he have to have the analogy or metaphor? And what was the status of the natural/sexual selection dichotomy?

Take first the question of the necessity of the analogy from the domestic world. Wallace certainly never discovered sexual selection. Moreover, when he was introduced to the notion, although initially he accepted both forms, before long he became very hostile to sexual selection through female choice. He thought the notion unduly anthropomorphic, imputing human standards of beauty to animals. Instead, Wallace argued that the brightness of males and drabness of females is a function of the need for females, vulnerable as they protect their young, to camouflage themselves, rather than a function of males being chosen by females. In other words, sexual dimorphisms of this kind should be seen as a matter of dowdy females rather than flashy males! (Vorzimmer, 1970, p. 200).

Of course, one might argue that the fact that Wallace had trouble with sexual selection does not exactly prove the necessity of the artificial selection analogy — and this is true. But it is really hard to see how one would come to make the natural/sexual selection division without the analogy, let alone endorse it. There is no other basis than the analogy to make the division, or to separate male combat from female choice. For instance, Darwin talks about sexual selection being less fierce than natural selection, for it does not involve death – but natural selection, as Darwin recognized, does not always involve death either. The key to natural selection, as with sexual selection, is reproduction – survival is incidental. A plant less able to reproduce in the desert than another is selected against, even if both survive. Similarly, the natural/sexual dichotomy is not really based on the relationship of the competitors. Sexual selection can occur only between members of the same species, but Darwin explicitly allows that one can have natural selection between members of the same species also. Nor is the dichotomy based on the animal/plant dichotomy. Sexual selection is restricted to animals, but one has natural selection in that kingdom too.

Similar remarks apply to the division between the kinds of sexual selection. In recent years, particularly with the advent of sociobiology, sexual selection has enjoyed something of a renaissance (Campbell, 1972; Ruse, 1979b). But the male combat/female choice division is not really maintained. One prominent sociobiologist, for instance, divides sexual selective strategies into the 'domestic bliss' strategy and the 'he man' strategy (Dawkins, 1976); but these cut across Darwin's divisions. The 'he man' strategy requires that the female try to mate with the male with the most attractive characteristics: these attractive characteristics include both strength and beauty.

I argue therefore that one cannot grasp Darwin's natural/sexual selection division without recognizing his route to discovery through artificial selection. If one asks what the rationale behind the distinction really is, and as philosophers we surely ought to, we must make mention of the way in which Darwin discovered his mechanisms. Note: I am not saying that Darwin's division of selection into two kinds was wrong. He made good use of it, even though selection through female choice did get surrounded by controversy. I would however suggest that the division is artificial or subjective in the sense discussed earlier. In a typically perceptive passage, William Whewell points out that the difference between a natural and an artificial classification lies in the fact that the former alone can be delimited in two different, logically independent ways:

And the Maxim by which all Systems professing to be natural must be tested is this: – that the *arrangement obtained from one set of characters coincides with the arrangement obtained from another set.* (1840, Vol. 1, p. 521, Whewell's italics)

This very point applies to the case we are discussing. The natural/sexual selection distinction is artificial simply because its only justification is the anology from human selection. (For more on this question of artificial/natural classification see Ruse, 1969 and 1973*a*.)

## 7. CONCLUSION

In three different ways I have tried to suggest that Darwin's route to discovery through the domestic world significantly and uniquely influenced the theory presented in the *Origin*. I have allowed, however, that this influence was at the level I have called 'subjective', and that Wallace's work shows that there was another side, an 'objective' side, to Darwin's theory that lay beyond the route to discovery. I suspect that some of my readers, even if they grant what I have argued so far, will nevertheless feel less than impressed. They will argue that, by my own admission, the 'real' or 'true' part of a theory is understandable without knowledge of discovery. In the case at point, the essential theory of the *Origin* is independent of the precise way that Darwin got to his mechanisms. I have tried to explain why I would not accept this criticism: I believe it to be insensitive to the historical reality of science. But in conclusion, let me hint why I believe that knowledge of a scientist's passage of discovery might be even more crucial to understanding a completed theory than I have so far argued.

As many modern commentators have noted (for instance Laudan, 1977), scientists do not simply set out to give a faithful reflection of reality — any old reality. Rather, they set out with problems that they want to solve. These problems and the strictures that scientists set on themselves in solving the problems — the regulative principles — crucially influence the finished product. Take once again the Darwinian case. First, there is the question of why one should want such a theory as that of the *Origin*, at all? Wherein lies its interest? This may seem like a very odd question indeed, particularly to those of us trained in the logical empiricist tradition. Obviously Darwin's theory is interesting — the plethora of stuff written on it in the past hundred years amply attests to this fact. And in any case, surely there is something logically absurd about asking whether a theory is interesting: at least, inasmuch as one wants to understand the theory philosophically. What one should ask is whether it is true.

However, if we look at history, and *only* if we look at history, we see that the question is not so very odd, nor is it insignificant. The origins of organisms had to be seen as a major problem for Darwin to want to solve it, or for anyone — supporter or critic — to take his suggestions seriously. It was not by chance that the *Origin* appeared mid-way in the nineteenth century or that it appeared in Britain, at that time certainly not the world leader in science (Ruse, 1979*a*). A theory like Darwin's was not going to be important until there was all the unexplained and curious information on geographical distributions, the fossil record, embryology, and so forth; nor would it have been so important had not the British had this rabid desire to mesh their science with their evangelically inspired Biblical speculations. I suspect that most Frenchmen still do not accept Darwin; but to such rational men – Catholic or atheist – the whole matter is really not very important. They are not stunted by the claustrophobic effects of Wesley's Protestant Christianity, and hence to them a theory like Darwin's which challenges these effects does not have an *a priori* morbid fascination. Of course, if you like, this whole question of the importance or interest of Darwin's theory is still subjective, in a way that the objective matter of the theory's truth or falsity is not. But this is a fact I find quite insignificant. There's no point in having a pretty face if no one's going to fall in love with you.

Second, there is the question of the solution of the problem, once it is seen as significant. Darwin bound himself by at least two methodological dicta, or, following the neo-Kantians, what I would call 'regulative principles': norms of what constitute good science, or even more strictly norms to which a scientist's work must adhere if it is to be called 'science' at all (Ruse, 1979*c*).

On the one hand, insofar as possible, Darwin felt he had to be 'Newtonian', where the model for this was the astronomy of the 1830's (Ruse, 1975*c*). This led Darwin to put his central arguments to natural selection in the deductive form as found in Malthus. Also, it was for this reason that the *Origin* is strikingly structured in a fan-like form, with the core of selection explaining in so many different areas: biogeography, paleontology, embryology, and so forth. (Refer back to Figure 1.) This 'consilience of inductions' was what J. F. W. Herschel and William Whewell found so admirable about Newton's theory (as well as the wave-theory of light).

On the other hand, Darwin tried to be 'Teleological'. Darwin is often portrayed as taking God, particularly Paley's divine clockmaker, out of biology; and this is true (Ruse, 1975*d*). All British pre-Darwinians thought that the most distinctive feature about the organic world is the way it shows functions, organization, and ends: the adaptations of the kangaroo are *as if* they were designed, because, in the opinion of virtually all, they *were* designed ("irrefragable evidence of creative forethought": Owen, 1834, p. 348)! Darwin certainly did not follow the herd in thinking that 'as if' implies 'was'. Nevertheless, he agreed fully that the manifestation of organization and ends – 'teleology' – characterizes the organic world, and consequently he deliberately sought a mechanism to explain it. Even after he became an evolutionist, Darwin went on seeking until he found the mechanism of natural selection, which he thought could explain organic teleology. In the *Origin* therefore, Darwin felt quite free — not to say obligated — to use the terms and categories of the Paleyites, like 'adaptation' and 'function'. The concern with ends was no less an obsession for Darwin than it was for his opponents — although the latter did not think that Darwin had done a good enough job.

Both in his Newtonianism and his Teleology, Darwin let his way of discovery influence the theory he produced. It may be argued that any theory has to be axiomatic as Newton's theory was and as Darwin's theory tried to be. This is a moot point. However, not every theory is or tries to be as consilient as Darwin's. Similarly, it would be hard to maintain that a biological theory *has* to be teleological, even though much might be lost if it were not. One could, as one does in physics, stay just at the level of material causation and refuse to ask questions about ends. To ask for the function of the heart would be illegitimate: all one could ask is physiological questions about embryological development, and so forth (Ruse, 1973*a*; 1977).

In these two modes of Newtonianism and Teleology therefore, I would suggest that Darwin incorporated subjective elements into his theory of organic origins, and that these elements were a direct function of his route to discovery. Since these elements were so pervasive and significant in his theory, given also what has been argued previously, I conclude that the attempt to understand Darwin's theory philosophically, without full consideration of the context of discovery, is hopelessly doomed to failure.

*Department of Philosophy*
*University of Guelph, Ontario*

## NOTES

[1] As part of their not-so-subtle denigration of scientific discovery, logical empiricists (*e.g.*, Hempel, 1966) are usually quick to point out that Kepler had some pretty wild neo-Pythagorean ideas about the mystical significance of mathematics.
[2] Honesty compels me to confess that this attitude towards scientific discovery pervades the minor logical empiricist writings, for example Ruse (1973*a*).
[3] Darwin wrote a 35 page sketch of his theory in 1842 and a 230 page essay in 1844. These are reprinted in Darwin and Wallace (1958). Even in the sketch all the essential ideas appear.
[4] Even if I attribute too much to the analogy from the domestic world in Darwin's

actually grasping the rudiments of differential reproduction in September/October 1838, what I shall argue will not thereby fail. My concern, as will become clear, is with Darwin's *theory*. As pointed out (Note 3), the theory of the *Origin* first appeared in outline in the sketch of 1842. Hence, what I shall have to say really depends crucially on all work up to that point. But, between 1839 and 1842 Darwin continued to read extensively in the writings of breeders, definitely by then taking seriously the domestic/wild analogy. Hence it is indubitable than inasmuch as Darwin had a conceptual argument − a theory − the analogy could and did play a vital role.

[5] I am presupposing that the reader, like myself, has a fairly robust sense of reality and is prepared to accept without too much question that chairs, tables, and trees really do exist, whether or not we are around. "The tree is an oak" I take to be about objective fact; "I prefer mahogany to oak" I take to be more in the subjective realm.

[6] Actually, Malthus speaks of struggles for existence.

[7] Note the use of the term 'selection' in the passage from Sebright (1809) quoted above.

[8] Obviously, I am here getting fairly close to the view of metaphor proposed by Max Black (1962) and accepted by Mary Hesse (1966). People viewed natural selection through the 'lens' of artificial selection, and at least as a matter of contingent fact people's understanding of natural selection could not have been the same as it was without the artificial selection metaphor. An extremist might perhaps argue that, *logically*, without the metaphor the understanding could not have been the same. I am hesitant to go quite that far, although it might be argued that even if logically one could have got the term 'natural selection' without the metaphor, the thoughts it would conjure up would not be the same.

[9] In line with the point made in Note 4, what I mean here is that after Darwin had discovered differential reproduction in 1838 and before he wrote down his theory in 1842, he made his most extensive study.

[10] Lamarck's main force of evolutionary change was a kind of teleological drive up the chain of being. This was never part of Darwin's theory.

## BIBLIOGRAPHY

Black, M.: 1962, *Models and Metaphors*, Cornell University Press, Ithaca, N. Y.

Bunge, M.: 1968, 'Analogy in quantum theory: From insight to nonsense', *British Journal for the Philosophy of Science* 18, 265–286.

Campbell, B.: 1972, *Sexual Selection and the Descent of Man*, Aldine Press, Chicago.

Darwin, C.: 1859, *On the Origin of Species by Means of Natural Selection*, Murray, London.

Darwin, C.: 1871, *Descent of Man*, Murray, London.

Darwin, C.: 1960, 'Darwin's notebooks on transmutation of species, Part III' (Notebook 'D'), edited by D. de Beer, *Bulletin of the British Museum (Natural History) Hist. Ser.* 2, 121–150.

Darwin, C.: 1969, *Autobiography*, N. Barlow (ed.), Norton, New York.

Darwin, F., and A. C. Seward: 1903, *More Letters of Charles Darwin*, Murray, London.

Darwin, C., and A. R. Wallace: 1858, 'On the tendency of species to form varieties; and on the perpetuation of varieties and species by natural means of selection', *Proceedings of the Linnean Society, Zoological Journal* 3, 46–62. Reprinted in C.

Darwin and A. R. Wallace, *Evolution by Natural Selection*, Cambridge University Press, Cambridge, 1958.

Dawkins, R.: 1976, *The Selfish Gene*, Oxford University Press, Oxford.

Ghiselin, M.: 1969, *The Triumph of the Darwinian Method*, Univ. of California Press, Berkeley.

Hempel, C.: 1966, *Philosophy of Natural Science*, Prentice Hall, Englewood Cliffs, N. J.

Herbert, S.: 1971, 'Darwin, Malthus, and selection', *Journal of the History of Biology* 4, 209–217.

Hesse, M.: 1966, *Models and Analogies in Science*, Univ. of Notre Dame Press, Notre Dame, Indiana.

Hull, D.: 1973, *Darwin and His Critics*, Harvard Univ. Press, Cambridge.

Kuhn, T. S.: 1962, *The Structure of Scientific Revolutions*, Univ. of Chicago Press, Chicago.

Lakatos, I.: 1970, 'Falsification and the methodology of scientific research programmes', in I. Lakatos and A. Musgrave (eds.), *Criticism and the Growth of Knowledge*, Cambridge Univ. Press, Cambridge.

Laudan, L.: 1977, *Progress and its Problems: Towards a Theory of Scientific Growth*, Univ. of California Press, Berkeley.

Limoges, C.: 1970, *La Sélection Naturelle*, Universitaires de France, Paris.

Lyell, C.: 1830–1833, *Principles of Geology*, Murray, London.

Malthus, T. R.: 1826, *An Essay on the Principle of Population*, 6th ed., London.

Owen, R.: 1834, 'On the generation of the marsupial animals, with a description of the impregnated uterus of the kangaroo', *Philosophical Transactions*, pp. 333–364.

Popper, K. R.: 1959, *The Logic of Scientific Discovery*, Hutchinson, London.

Popper, K. R.: 1970, 'Normal science and its dangers', in I. Lakatos and A. Musgrave (eds.), *Criticism and the Growth of Knowledge*, Cambridge Univ. Press, Cambridge.

Ruse, M.: 1969, 'Definitions of species in biology', *British Journal for the Philosophy of Science* 20, 97–119.

Ruse, M.: 1973a, *The Philosophy of Biology*, Hutchinson, London.

Ruse, M.: 1973b, 'The Value of analogical models in science', *Dialogue* 12, 246–253.

Ruse, M.: 1975a, 'Charles Darwin and artificial selection', *Journal of the History of Ideas* 36, 339–350.

Ruse, M.: 1975b, 'Charles Darwin's theory of evolution: an analysis', *Journal of the History of Biology* 8, 219–241.

Ruse, M.: 1975c, 'Darwin's debt to philosophy: an examination of the influence of the philosophical ideas of John F. W. Herschel and William Whewell on the development of Charles Darwin's theory of evolution', *Studies in History and Philosophy of Science* 6, 159–181.

Ruse, M.: 1975d, 'The relationship between science and religion in Britain, 1830–1870', *Church History* 44, 505–522.

Ruse, M.: 1977, 'Is biology different?' in R. Colodny (ed.), *Laws, Logic, Life*, Pittsburgh Series in the Philosophy of Science, Univ. of Pittsburgh Press, Pittsburgh.

Ruse, M.: 1979a, *The Darwinian Revolution: Science Red in Tooth and Claw*, Univ. of Chicago Press, Chicago.

Ruse, M.: 1979b, *Sociobiology: Sense or Nonsense?* D. Reidel, Dordrecht.

Ruse, M.: 1979c, 'Philosophical factors in the Darwinian Revolution', in F. Wilson (ed.), *Pragmatism and Purpose*, Univ. of Toronto Press, Toronto.

Sebright, J.: 1809, 'The art of improving the breeds of domestic animals in a letter addressed to the Right Hon. Sir Joseph Banks, K. B.', London.

Vorzimmer, P. J.: 1970, *Charles Darwin: The Years of Controversy*, Temple Univ. Press, Philadelphia.

Whewell, W.: 1840, *Philosophy of the Inductive Sciences*, Parker, London.

LINDLEY DARDEN

# THEORY CONSTRUCTION IN GENETICS

Philosophers of science have had relatively little to say about theory construction. Theories were treated by Popper (1960) and the logical empiricists (e.g., Hempel, 1966) as if they arose all at once by a creative leap of the imagination of a scientist, a process whose study was viewed as the province of the psychologist. Only after the creative leap, they agreed, were the philosopher's logical tools useful to evaluate the theory so produced. Even more historical accounts concerned with scientific change, such as Kuhn's (1970), did not discuss the way paradigms or the theories within them were constructed, except that they somehow arose in response to anomalies of their predecessors. Lakatos (1970), who proposed criteria for evaluating progressive research programs, did not discuss how the scientist constructs the program originally, although later additions, he claimed, resulted in some way from the 'positive heuristic'. Even Laudan (1977) who made much of the common place observation that science is a problem-solving activity, focused on the use of solved or unsolved problems to evaluate theories and research traditions; he did not provide an analysis of how a scientist goes about solving a problem. Thus, most of twentieth century philosophy of science, from the logical empiricists to the most recent work, has been within the context of justification, not the context of discovery.

Dichotomizing science into these mutually exclusive contexts and concentrating on justification to the exclusion of discovery distorts the on-going process that characterizes science. A theory rarely, if ever, arises all at once in a complete form. Vague ideas about postulated explanatory factors may take on more form as new data are found and new theoretical components added. A negative result may produce a change in only one part of a theory, with the subsequent modification encorporating new ideas which fit the data better. Connections to empirically confirmed items in another field may be important in constructing part of a theory and at the same time bring a measure of justification. The processes of discovery and justification thus have complex interrelations in the development of a theory over time. Consequently, this paper will use the phrase 'theory construction' which better captures this on-going process than do the mutually exclusive 'context of discovery' and 'context of justification'.

151

T. Nickles (ed.), Scientific Discovery: Case Studies, 151–170.
Copyright © 1980 by D. Reidel Publishing Company.

A central problem in understanding this on-going process of theory construction is to understand the various factors that play roles in it. First of all, the *domain* to be explained, that is, the phenomena to be explained by the theory (Shapere, 1974*a*), will be one factor, which, minimally, acts as a constraint. The theory must be able to account for all or most of the items of the domain. More positively than its constraining role, the domain may provide suggestions for the form the theory may take. Shapere (1974*b*) recently provided an interesting analysis of domain indications for theory construction. But something other than the ideas in the domain must be built into the theory if the theory is to explain that domain. Shapere (1973, p. 22) characterized this difference by saying that theories must introduce *new ideas*, different from those in the domain, in order to account for that domain. Mary Hesse expressed this in linguistic terms: "the essence of a theoretical explanation is the introduction into the explanans of a new vocabulary or even of a new language" (1966, p. 171). Carl Hempel, in discussing the problems of inductive inference said in a similar vein: "scientific hypotheses and theories are usually couched in terms that do not occur at all in the description of the empirical findings on which they rest, and which they serve to explain" (1966, p. 14). He continued by claiming that there could be no mechanical rules for producing the novel concepts found in the theories.[1]

Where do these new ideas, these novel concepts, expressed in a new vocabulary, come from? Even if there are no mechanical rules for their production, are guidelines for search obtainable? What role, exactly, do they play in the construction of a theory? In an attempt to answer these questions, this paper will argue that either analogies or interfield connections may play the role of providing new ideas that are built into a theory according to a schema of theory construction, which will be discussed. Furthermore, a working hypothesis will be proposed: connections to well-developed related fields are likely to be a better source of new ideas than analogies, for reasons to be given, and thus, when interfield connections are available, they should be considered prior to a search for analogies in the steps of constructing a theory. In the specific historical case to be examined, I show that progress occurred in theory construction when scientists shifted from vague analogies to interfield connections as the source of new ideas.

Some scientists and philosophers have discussed the role of analogies in theory construction; fewer have discussed the connections between closely related scientific fields as providing an important source of new ideas. Before turning to a detailed examination of the historical case, we will briefly examine some of these opinions.

The philosopher, Norwood Russell Hanson (1960), in his now classic work on reasoning in discovery, discussed the role of analogies. In retroductive reasoning, Hanson claimed, the scientist is puzzled by some phenomena. Then, according to Hanson, the scientist uses analogy to reason that some type of hypothesis plausibly explains the phenomena. Unfortunately, Hanson did not elaborate on this. Exactly how does the analogy function to suggest a type of hypothesis? What reasons does the scientist have for choosing one analogy rather than another? In what way do analogies make a type of hypothesis plausible?

Scientists have also indicated that analogies play a role in theory construction. G. K. Gilbert, a noted American geologist, said:

To explain the origin of hypotheses I have a hypothesis to present. It is that hypotheses are always suggested through analogy. Consequential relations of nature are infinite in variety and he who is acquainted with the largest number has the broadest basis for the analogic suggestion of hypotheses. (1896, p. 2).

R. J. Blackwell, a philosopher writing in *Discovery in the Physical Sciences* said, in a similar vein:

... the scientist might recognize that the unsolved problematic before him bears resemblances to another problematic which has already been solved. As a result he forms the hypothesis that the same type of explanation also applies to his problem .... This type of transformation leading to theoretical explanation we will call substitution through analogy. (1969, p. 179).

R. Harré, (1960, 1970) provided a more detailed discussion of types of models and the ways they function in theory construction. Although it is not necessary to give all of Harré's baroque terminology in his classification of different types of models, two basic types are of interest to us here, the first to rule it out of our further discussion and the second to focus on. *Homoeomorphs*, according to Harré, have the same thing as both the *source* from which they come and the *subject* which they model. The simplest type are scale models in which the parent situation serves as both the source of information used to construct the model and is the subject modeled. Additionally in the homoeomorph category is a type of model often used by scientists, especially modern biologists. The model is an abstraction from, or an idealization of, a theory; the term 'theory' can often be substituted for this usage of 'model'. This class of homeomorphic models do not play the role of providing new ideas in theory construction.

On the other hand, the second class, which Harré called *paramorphs*, employ analogous relations and are important in constructing theories. The sources of these *analogue models*, as I will call them here, are different from the subjects which they model. Of these models Harré said:

The analogy that is found to hold between certain characteristics of different processes is the basis of the paramorph . . . . We make for ourselves conceptual models, in which something with which we are familiar or which we understand very well is used as an imaginary model of some otherwise obscure process . . . . The more general notion of a conceptual paramorph is a basic element in the analysis of the construction of scientific explanations. (1960, pp. 87–88)

In addition to the role of the analogical model in first constructing a theory, Harré indicated that models play a role in later steps of theory construction. Harré discussed the deployment of a model much as Mary Hesse did elsewhere (Hesse, 1966): one may explore the 'neutral analogy' to see if other unexplored aspects of the model are also similar to the subject and can be used to construct additional postulates. Alternatively, Harré suggested, one may bring in a different analogue model in the further development of the theory. Finally, Harré also discussed a process that will be of interest to us in our further discussion, the use of another science in place of a model:

The reference to another science takes the place of a description of a model . . . . The sciences can be arranged in a hierarchy such that an explanation of facts in one is given in terms of the description of facts in another . . . . (1960, p. 101)

Although my points differ somewhat from those expressed by these philosophers or will develop the analyses further, we agree on the basic point that analogues or other fields of science supply important new ideas in constructing a theory.

In previous work (Darden, 1976), I analyzed the role that analogies played in the construction of Darwin's theory of heredity, and from that case I extracted a general schema for theory construction. In further work done conjointly with Nancy Maull (1977), we examined the generation and function of interfield theories, that is, theories that connect two scientific fields. In the present paper I relate and extend this previous work by arguing that in the case of construction of theories of heredity, specifically, early attempts depended on vague analogies but progress occurred with the replacement of vague analogies by specific interfield connections.

In order to substantiate these claims, we shall now turn to an analysis of an important case from the history of biology, namely attempts to construct a theory of heredity, beginning with Darwin's pangenesis, to the rediscovery of Mendel's laws, to William Bateson's experimental and theoretical work, to the construction of the successful theory of the gene by T. H. Morgan and his associates. The period spans some sixty years, from the 1860s to 1926.

Charles Darwin's writings have proved a rich source for discussing the role of analogies in science. Darwin used many. In previous work (Darden, 1976), I examined Darwin's theory of heredity and will briefly summarize those results here. In 1868 Darwin proposed his provisional hypothesis of pangenesis, the view that hereditary units, called 'gemmules', were produced all over the body, collected in the reproductive areas and were passed on to grow into the embryo. To explain Darwin's use of analogies there, I constructed a general schema for a pattern of reasoning in theory construction.

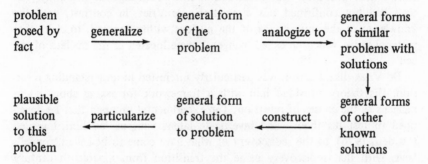

problem posed by fact  →generalize→  general form of the problem  →analogize to→  general forms of similar problems with solutions

↓

plausible solution to this problem  ←particularize←  general form of solution to problem  ←construct←  general forms of other known solutions

Each item in the domain to be explained is considered as a problem to be solved. The next step is the extraction of the general form of problem embodied in that item. For example, Darwin wished to explain reversion, *i.e.*, cases in which an organism has characteristics like one of its ancestors rather than like its immediate parents. This domain item poses the general problem: how can something be present, then disappear, and then reappear in the same form at a later time? The next step, according to the schema, is the search for analogues which embody a problem with the same general form but additionally have solutions to that problem. Darwin said that many things in nature are present, then go into dormancy, then appear again: seeds in plants and dormant buds on plants were given as two analogues. Thus, 'activity, dormancy, and then reactivation' is the general form of the solution found through use of analogies. The general solution is then particularized for the problem posed by the domain item, *e.g.*, gemmules are active, become dormant through one or more generations, and then are reactivated.

The pattern of reasoning may be used in a similar way for each of the domain ˙items. As a result, different analogies play roles in constructing different postulates. At the early stages of theory construction, such as

Darwin's, the analogies ranged widely over numerous processes known to occur in the empirical world. The use of such wide-ranging analogies provides weak plausibility to the postulates constructed using them: such processes are known to occur elsewhere. Testing is immediately in order to see if that process does occur in this case. Another nineteenth century theory of heredity showed improvement that was provided by use of closely related knowledge to construct better postulates. Hugo de Vries, in his *Intracellular Pangenesis* of 1889, explicitly claimed to be developing and improving Darwin's provisional hypothesis of pangenesis. Among the numerous changes that de Vries made was the incorporation of knowledge from cytology, the study of cells. Darwin had proposed that gemmules circulated throughout the body, but tests had not confirmed this circulation. De Vries, in contrast, used the knowledge of the importance of the nucleus within the cell, to claim that material units, which he called 'pangens', were located in the nucleus of the cell.

De Vries, like Darwin, was particularly interested in understanding reversion. His theory provided him with a framework for asking about latent pangens as the causes of reversion. The experimental program that he developed to test his theory and investigate latent pangens led him, I argued (Darden, 1976), to the rediscovery of what have come to be called Mendel's laws. With that rediscovery came the transition from nineteenth century theories of heredity with their wide-ranging domains to Mendelian genetics with its data expressed in numerical ratios of differing hereditary characteristics. The construction of theories to explain those ratios which culminated in the theory of the gene is the historical case to which we now turn.

Genetics emerged as a field in 1900 with the independent rediscovery of Mendel's results by de Vries and Carl Correns. But the theory of the gene was not formulated until some years later by T. H. Morgan and his coworkers. It is important to understand exactly what the new conceptions were that showed such promise that a new field of research emerged to investigate them (Darden, 1977), yet were so limited that it seems inappropriate to call them a theory as of 1900.

Mendel's empirical results are quite familiar. When organisms differing in traits of one character were crossed, one of the traits dominated in the first generation of hybrids. The famous example was the cross between yellow and green peas, with all the $F_1$ generation being yellow peas. When the hybrids were allowed to self-fertilize, the second generation showed the traits in a proportion of 3:1, *e.g.*, three yellow to one green. When plants with two different characters were crossed, *e.g.*, pea color and height of plants, the characters behaved independently, giving a 9:3:3:1 ratio. But Mendel did not stop at merely recording these numerical analyses of the empirical data. Importantly, he also gave a minimal explanation of them by introducing the new idea that the germ cells of the hybrids differed in that each cell had the potential to produce only one of a pair of traits. Furthermore, his other independent assumptions were, one, that the types of germ cells were formed in equal numbers, and, secondly, that the cells combined randomly in fertilization. William Bateson (1902) called the new idea 'purity of the gametes'; de Vries (1900, reprinted in Stern and Sherwood, 1966) expressed his discovery by saying that the germ cells of hybrids are not hybrid but belong to one or the other of the parental types. Such assumptions explained the numerical results, as Mendel's algebraic notation showed: $A$ and $a$ represented the differing traits.

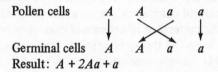

Pollen cells        $A$    $A$    $a$    $a$

Germinal cells    $A$    $A$    $a$    $a$
Result: $A + 2Aa + a$

Since the presence of $A$ produced the dominant appearance, the $A + 2Aa$ combine to give the appearance $3A$ to $1a$ (Mendel, 1866, reprinted in Stern and Sherwood, 1966). Differing characters assort independently to give the $9AB:3Ab:3aB:1ab$ ratio. However, Mendel did not discuss the nature of the '*Elemente*' or '*Merkmal*' in the germ cells; he certainly did not claim that the germ cells carried differing material particles.

At the beginning of the field of genetics, there were good empirical regularities produced by the technique of artificial breeding and the proposal that germ cells of hybrids differed with respect to the characters they later gave rise to. But there was no well-developed theory about these causes of characters. As indicated by the genetic data, the germ cells contained differing black boxes which somehow caused characters. Filling in details about these black boxes and their interrelations became the chief task of genetics.

Although neither were rediscoverers of Mendel's results, William Bateson and T. H. Morgan were important figures in the development of theories in genetics. Bateson's attempts were unsuccessful; Morgan and his associates formulated the theory of the gene. The contrast between Bateson's approach and that of the Morgan school provides us with an excellent case for examining progress in methods of theory construction.

Bateson's sources of ideas about the causes of patterns of inheritance ranged widely over analogies from physics, with a predilection for forces and vibrations. His methods of theory construction resembled Darwin's in the use of numerous analogies from various sources. T. H. Morgan, on the other hand, insisted on tighter ties to empirical evidence. The source of new ideas for much of the construction of the theory of the gene was the neighboring field of cytology. By using interfield connections between postulated genes and observed chromosomes, the Morgan school constructed a theory which quickly surpassed Bateson's in explanatory power and extendability to new data. Thus, progress resulted from abandoning the use of vague analogies and the directed search for interfield connections as guides to theory construction.

A distinction needs to be made explicitly at this point. The theory whose construction is being discussed here is an intrafield theory, completely within the field of genetics. Its successful form, as of 1926, was called by Morgan the 'theory of the gene'. It postulated unobservable genes in the germ cells, whose behavior served to explain the patterns of inheritance of characters determined by artificial breeding. This *intrafield* theory is different from, though closely connected to, the *interfield* chromosome theory of Mendelian heredity, which postulated that genes were in chromosomes. The interfield connections established by the chromosome theory were important in the construction of the theory of the gene; that construction is being discussed in this paper. This discussion represents a shift of emphasis from the work on interfield theories (Darden and Maull, 1977) in which we emphasized the importance for both fields of the interactions between them. Here, the emphasis is on the developments within a single field. Historians and philosophers have often confused the theory of the gene and the chromosome theory because they were so closely related. For the purposes of this paper, the theory whose construction is discussed is the theory of the gene, and the links to cytology important in that construction will be referred to as 'interfield connections' or 'links'.

William Bateson never worked out a well-developed theory of genetics. His own attempts were along quite different lines than the theory of the gene and the chromosome theory, neither of which he ever accepted. Most of his published writings show him as the empirical scientist, doing genetic experiments, coining new terminology (*e.g.*, 'genetics', 'heterozygote', 'homozygote'), criticizing the Morgan interpretations. He appears to be the empiricist, skeptical of theoretical leaps. But unpublished writings and some published work (Bateson, 1913) belie that interpretation. Bateson struggled to formulate his own theory of genetics in terms of vibrations, forces, and vortices. William Coleman (1970), in an excellent article, analyzed these attempts to show

Bateson was using traditions in physics in the late nineteenth century as a source for analogies to construct his theory. Coleman said:

[Bateson's] search for an alternative to strictly atomistic, particulate, or, if you will, material bases for heredity led him to consider a diverting array of models. Of these, the most significant was one derived from popular generalizations of the physical sciences. In this rhythmic or vibratory theory of inheritance is a model whose premises are reasonably clear, whose further development is desperately obscure and whose attractions Bateson could never escape. (1970, p. 267)

Bateson's own discussions of his hypotheses are not numerous. In a letter of 1891 he called an early view the 'Undulatory Hypothesis' and briefly sketched the main idea:

Divisions between segments, petals, etc. are *internodal* lines like those in sand figures made by sound, *i.e.* lines of maximum vibratory strain, while the mid-segmental lines and the petals, etc. are the *nodal* lines, or places of minimum movement. Hence all the *patterns* and *recurrence of patterns* in animals and plants — hence the perfections of symmetry — hence bilaterally symmetrical variation, the completeness of repetition whether of a part repeated in a radial or linear series etc. etc. (B. Bateson, 1928, p. 43)

The cell came to occupy a privileged place in Bateson's attempts at theory construction. But it was not a static structure; instead it was a dynamic vortex:

The cell is a vortex of chemical and molecular change. Matter is continually passing through this system. We press for an answer to the question, How does our vortex spontaneously divide? The study of these vortices is biology, and the place at which we must look for our answer is cell-division. (Bateson, 1907; Quoted in Coleman, 1970, pp. 274–5)
A simple vortex, like a smoke-ring, if projected in a suitable way will twist and form two rings. If each loop as it is formed could grow and then twist again to form more loops, we should have a model representing several of the essential features of living things. (Bateson, 1913, p. 40)

An understanding of living vortices, namely cells, held the key, according to Bateson, for understanding heredity and variation. Heredity was essentially like producing similar cells in cell division or symmetrical parts in a single organism. Variation, on the other hand, was like differentiation in which cells give rise to cells that differ from them. This primacy of the cell and his predilection for dynamic forces were aspects of Bateson's rejection of subcellular material structures, such as chromosomes within the nucleus of the cell, as the important locus of activity. Not only did he reject the chromosome theory, he also was quite skeptical of interpretations of Mendelism as requiring discrete material units. On numerous occasions, Bateson stressed that there was no evidence to show that the Mendelian factors were material substances (Bateson, 1913, p. 268).

We are in the state in which the students of physical science were, in the period when it was open to anyone to believe that heat was a material substance or not, as he chose. (1902, pp. 2–3)

These arguments were meant to leave open the possibility that Bateson's dynamic approach was correct.

Bateson's use of analogies to physics was reflected in the terms he used to report data. An important example on which we will focus was the discovery by Bateson and coworkers of exceptions to the 9:3:3:1 Mendelian ratios (Bateson *et al.*, 1905). These anomalies were called by Bateson either 'coupling', if the dominant traits tended to be inherited together more often than expected, or 'repulsion', if the dominants were found together less often. The 'intensity' of coupling or repulsion varied for different pairs of characters. The example reported in 1905 was for partial coupling in pollen shape and colors of flowers in sweet peas.

By examining this terminology and additional statements that Bateson made, one may infer the form of a theory of heredity that Bateson might have proposed. 'Powers' within the swirling vortex of the germ cells produce characters during development (1913, p. 268). These powers or forces may attract or repel each other with varying intensity in the formation of the germ cells (a denial of independent assortment) with a resultant distortion of the 9:3:3:1 ratio of characters produced. Alternatively, the powers may not influence each other, giving the usual Mendelian ratio.

But this new idea of coupling and repulsing powers did not prove to be fruitful and was abandoned by Bateson in favor of an alternative hypothesis: selective reduplication of certain cells. For instance, instead of equal numbers of germ cells with *AB, aB, Ab, ab* as Mendel proposed, Bateson suggested unequal gametic ratios, *e.g.*, *7AB:1Ab:1aB:7ab*, produced by selective reduplication of the cells containing *AB* and *ab*. Thus, he denied that equal numbers of the different types of germ cells were produced. Instead, a disproportionate number of certain types formed and combined in fertilization to give rise to the anomalous ratios. Bateson and Punnett (1911) called this the 'reduplication hypothesis'.

By abandoning coupling and repulsion in favor of cellular reduplication, Bateson made progress in his method of theory construction. He abandoned vague analogies from the physical sciences which produced a theory difficult, if not impossible, to test. In place of those analogies he substituted a process from the related field of cytology – cell division – as a means of constructing an explanation of the genetic data. Unfortunately, Bateson was not appealing to any specific cytological data about selective reduplication of cells. Ideally, in theory construction one can find empirically confirmed information in a related field with relevant similarities such that it can be used to solve one's problem. But the ideal may not obtain since scientists in the related field may simply not have looked in the right place or at the right things. In that event one can predict phenomena in the related field on the basis of its presumed relation to the theory being constructed. Thus, Bateson's hypothesis predicted selective reduplications. The cytologists found none.

Another way in which the reduplication hypothesis failed was in its extendability. It became entirely unwieldy when extended from the coupling of two characters to the coupling of three or more. The numbers of different types of cells that had to be produced by selective reduplications became quite large. Since extendability to new cases determines the acceptability of a new theory, it should act as a demand in theory construction: choose the analogies or interfield connections which will provide ideas for theory development in the light of additional cases. Bateson failed to take this demand into account and his method of theory construction was the poorer for it.

The successful use of the interfield links between genetics and cytology, begun by Boveri (1904) and Sutton (1903), came to fruit in the hands of T. H. Morgan and his associates between 1910 and 1926. Unlike Bateson, they did not focus on the level of the whole cell and its divisions for sources of new ideas to explain the genetic data. Instead, they made use of knowledge of the subcellular level, specifically knowledge about the chromosomes inside the nucleus of the cell. Their postulations were often made on the basis of known cytological processes or properties and thus their constructions had a measure of empirical support that Bateson's lacked. Also, additional information about chromosomes provided ideas to Morgan and his associates for constructing additional postulates of the theory of the gene; in other words, the interfield links provided ideas for extending the theory.

In 1903 and 1904 Walter Sutton and Theodor Boveri independently postulated the chromosome theory of Mendelian heredity. Both scientists were struck by the similarities in the known properties of chromosomes and (what came to be called) Mendelian genes: (1) they both consisted of pure individuals which did not join or contaminate each other; (2) both genes and chromosomes were found in pairs; (3) in the formation of the sex cells, each set is reduced by one half. On the basis of these similarities Boveri and Sutton proposed that the genes were in or on the chromosomes and made predictions for both cytology and genetics on the basis of knowledge from the other field. Using the knowledge that the chromosome number was much smaller than gene number, they both predicted that some genes would be carried on the same chromosome and would therefore be linked in inheritance. (For more discussion of this case see Darden and Maull, 1977.)

T. H. Morgan began his career with work in embryology and became an outspoken proponent of the use of experiments in biology. Prior to 1910, Morgan was a critic of both Mendelism and the chromosome theory. But his own and others' experimental results caused him to change his views. (For a discussion, though I think probably not the complete explanation of this change, see Manier, 1969, and Allen, 1978.) As has been shown in recent historical writings such as Carlson (1971) and Roll-Hansen (1978), Morgan's students, Sturtevant, Bridges, and Muller, were important contributors to the development of the theory of the gene, often bringing Morgan around to their points of view. Although a detailed examination of the step by step construction of the theory of the gene would be very instructive, such a large task is outside the scope of this paper. Instead, this discussion will concentrate only on the explanation of the deviations from the 9:3:3:1 ratios, namely the postulation of linkage.

One of Morgan's criticisms of the chromosome theory was that the prediction by Sutton and Boveri of characters inherited together had been found in only a few instances (Morgan, 1910a, p. 489). The most striking work for indicating the importance of the chromosomes in heredity, Morgan conceded, was the finding of connections between differing chromosomes and sex determination as discovered by Nettie Stevens and investigated by E. B. Wilson (Brush, 1978a). The experimental work that marked the turning point in Morgan's views about Mendelism and the chromosome theory was Morgan's discovery of what came to be called sex-linked characteristics. As Morgan searched for mutations among *Drosophila* cultures, a white-eyed male appeared among the normal red-eyed flies. Subsequent breeding experiments showed white and red eyes to be linked to sex (Morgan, 1910b). Morgan quickly discovered other characters linked both to sex and to each other. But the linkage was not complete, or, in Bateson's terminology, there were variations in the strengths of coupling. If all the genes were carried on, and remained on, a single intact chromosome, then one expected complete linkage of the characters. A problem thus became to explain the partial linkage.

Morgan made excellent use of recent cytological work to construct on explanation for partial linkage. After discussing Bateson's hypothesis, Morgan said:

In place of attractions, repulsions and orders of precedence, and the elaborate systems of coupling, I venture to suggest a comparatively simple explanation based on results of inheritance of eye color, body color, wing mutations and the sex factor for femaleness in *Drosophila*. (1911, p. 384)

Morgan continued by postulating that the factors (genes) for those characters were located linearly along the chromosome. Citing the work of Janssens (1909) which showed the intertwining of chromosomes during the formation of germ cells, Morgan proposed that some of the homologous chromosomes exchanged parts, thereby producing crossing over of linked factors. Morgan concluded:

Instead of random segregation [independent assortment] in Mendel's sense we find "associations of factors" that are located near together in the chromosomes. Cytology furnishes the mechanism that the experimental evidence demands. (1911, p. 384)

Thus, Morgan, in contrast to Bateson, did not search for analogies from the physical sciences; instead he used empirical knowledge from the related field of cytology and the postulated physical relation of genes to chromosomes to construct an explanation for the genetic data. The data presented the problem: how are factors partially linked to each other? The cytological evidence of linear chromosomes which sometimes intertwine provided the solution: the genes are linked linearly to each other and sometimes parts of the homologous linkage groups cross over.

By 1915 when Morgan, Sturtevant, Muller, and Bridges wrote their now classic *The Mechanism of Mendelian Heredity*, Morgan's explanation of linkage had been markedly confirmed and extended to new cases. Linkage groups had been associated with the other, nonsex chromosomes; the number of linkage groups corresponded to the number of chromosomes. Bridges (1914) had found an abnormal fly which showed corresponding anomalies in both the chromosomes and the patterns of inheritance. Sturtevant (1914) constructed a linear map of the relative positions of genes using the numerical crossing over data. When the data showed less crossing over than predicted, once again the chromosomes provided a way of constructing the explanation: double crossing over occurred or one crossover interfered with the formation of another. (Morgan *et al.*, 1915, pp. 63–64)

In 1926 Morgan gave an elegant presentation of the theory of the gene:

The theory states that the characters of the individual are referable to paired elements (genes) in the germinal material that are held together in a definite number of linkage groups; it states that the members of each pair of genes separate when the germ-cells mature in accordance with Mendel's first law, and in consequence, each germ-cell comes to contain one set only; it states that the members belonging to different linkage groups assort independently in accordance with Mendel's second law; it states that an orderly interchange – crossing over – also takes place, at times, between the elements in corresponding linkage groups; and it states that the frequency of crossing over furnishes evidence of the linear order of the elements in each linkage group and of the relative position of the elements with respect to each other. (1926, p. 25)

Note particularly that nothing is said about chromosomes in this statement of the theory of the gene. The interfield connections to cytology were used in its construction. Furthermore, since the information from cytology was already empirically confirmed, it provided support for the new theory. The theory of the gene, apart from the chromosome theory, contained new ideas, genes associated in linkage groups. The connections to cytology played the same role as analogies in theory construction, but they were a much more powerful source of new ideas since they supplied information about physically related processes.

However, connections to cytology were not sufficient to fill in all the details of the black boxes, the genes. Although the postulation that the genes were in the chromosomes made plausible the assumption that the genes were material particles, it would be the task of new interfield connections established by molecular biology to illuminate the rest of the black box. But we must leave the rest of this case study for another time.

A hypothesis which emerges from this case study and which needs to be tested in others is that progress occurs in theory construction with the abandoning of vague analogies in favor of interfield connections to provide new ideas. At this point a more detailed discussion of 'vague analogies' as opposed to 'interfield connections' is needed. A critic of this hypothesis might ask: What is the difference? In both cases one is postulating similarities to factors in other sciences. In your example, for instance, similarities to electricity are just as good a source of new ideas as similarities to the behavior of chromosomes. How is one to be judged better than the other, the critic continues, except with hindsight as to which produced the successful theory?

My reply concedes something to this objection: analogies and interfield connections do have properties in common. First, a relation of similarity exists between the postulate constructed and either the analogue or the information from another field. Secondly, both may serve as sources of new ideas. Thirdly, in both instances, assuming the analogy has been drawn to a known thing or process, the use of either analogies or interfield connections brings some plausibility to the postulate constructed using it: a similar thing or process is known to occur elsewhere. Finally, the analogy and interfield connections may both play a role in developing further postulates to extend the theory, especially if the analogy is a more detailed one than the Bateson example. In Mary Hesse's terms, the neutral analogy, or the areas of unexplored relationships, may be used in further development of the theory. Similarly, other information from the related field may be brought in to construct additional postulates.

Despite these commonalities in the roles played by analogies and interfield connections in theory construction, I would reply to the critic that there are also important differences that make the interfield connections better. It is much weaker to say that, since a *similar type of factor* has been found in other sciences, such a type may be operative here (appeal to analogy), than to say that the *actual factor* found in a related science behaves in a similar way to the postulated factors as a result of a *physical* relationship between them (interfield connections). Instances of this weaker plausibility provided by analogies are common in the history of science. In postulating that powers were coupling or repulsing, Bateson was appealing to analogies in the way Newton did when he proposed that there might be chemical and life forces similar to mechanical ones. Similarly, natural philosophers of the eighteenth century postulated different fluids to explain electricity, heat and nerve impulses. It is interesting to ask what justification exists for this appeal to similar types of explanatory factors. One speculation is that a weak unity of science thesis is operating: different phenomena should be explicable by similar types of factors; if one type, such as forces or fluids, has been successful in one field, try it in others.

On the other hand, one might postulate a stronger unity of science directive: instead of postulating analogous factors, see if one and the same factor may be the operative one. For example, instead of postulating a vital fluid analogous to an electrical fluid, check to see if actual electromagnetic forces are operating in living things. If Bateson had been appealing to detailed empirical knowledge about coupling and repulsing forces known to occur in cells, then he would have been using interfield connections rather than analogies. Since his analogies were not even specifically drawn from similar types of successful explanatory factors found elsewhere but were merely suggestive of some sort of 'power' or 'force' or 'vibration' which could 'couple' and 'repulse', I have called them 'vague analogies'.

Interfield connections postulate a relationship to empirically confirmed information in a related field in the ideal case, or make a prediction of the existence of such a relation in the weaker case. The interfield connection, in contrast to an analogy, is itself a scientific hypothesis that postulates a *physical relation*[2] between the entities or processes in the neighboring field and in the theory under construction. Thus, in contrast to analogies, one expects *empirical investigation* of the interfield connection, with the result that such a connection may be confirmed or disconfirmed. Also, since there is a physical relation, the payoff for theory construction may be greater: one has good reason to believe that unexplored properties will have relations to each other; there is no similar basis for expectations about neutral analogy being useful. For example, the physical relation postulated in the case of the theory of the gene was between hypothetical genes and observed chromosomes: the genes were claimed to be parts of chromosomes. Since parts usually do what their wholes do, then most of the properties of chromosomes provided potential information about genes. One is not merely importing knowledge from cytology as a heuristic device to construct the theory of the gene in such a way that that knowledge may be discarded once the theory is constructed. Instead, one is postulating a physical relation which may be confirmed or disconfirmed. Other kinds of physical relations besides part-whole obtain in other cases, *e.g.* causal, structure-function, or even identity, if one and the same factor is claimed to be operative (Darden and Maull, 1977, p. 49).

An important question about the use of analogies or interfield connections in theory construction is how the theorist chooses the ones to use. With respect to analogies, I have little to say in answer to that difficult question. Perhaps a quantitative measure of the amount of positive (and/or neutral) analogy would be useful. One would like one's analogue to have numerous properties similar to the subject. Furthermore, more areas of neutral analogy, properties that are not yet known to be similar or dissimilar, would provide more potential for further development.

With respect to interfield connections, I can provide more in the way of an answer to the question about choice of a related field. Harré suggested that a reductive hierarchy provides interscience explanations. Certainly the search for underlying structure has proved fruitful in the development of science. But detailed examination of actual interfield relations shows that they come in more varieties than the identities necessary for reductive accounts. Entities and properties are not always being identified with entities and properties at a lower level, a necessary condition for reduction. What the lower level is may not be easily decidable: is genetics with its observable, large scale characteristics and unobserved genes the lower level, or is cytology with its microscopically observable chromosomes the lower? Neither is a candidate to be reduced to the other. Science is more reticulate and less hierarchical, so one needs directives other than use the field at the next lower level. (For further discussion of the interfield analysis as an alternative to reduction, see Maull (1977).)

Thus, in addition to choosing fields which may provide information about underlying structure, other reasons exist for choosing another field for constructing interfield relations. In searching for fruitful interfield connections to use in constructing an explanation of data within a subject field, one may have background knowledge that another field is related if it has supplied ideas for other explanations within the subject field. More strongly, one may know that scientists in another field have been investigating the same phenomenon using differing techniques, as for example, cytologists used microscopic techniques to investigate questions about heredity while geneticists used breeding techniques. One would expect some relation to be established between the results in the two fields. Temporal relations may also indicate what fields are related. As Stephen Brush (1978b) has recently discussed, in order to explain a domain of geological phenomena certain interfield connections with astronomical theory about the formation of planets were postulated, since the astronomical theory dealt with phenomena temporally prior to the geological.[3]

I have proposed the hypothesis that interfield connections are better than analogies as a source of new ideas. Let me now draw on the differences between the two to provide evidence for that claim. First, the use of interfield connections produces a more plausible theory if the interfield connections have been empirically confirmed. One is not merely appealing to processes known to occur somewhere; one is appealing to known processes that one claims have physical relations to the postulated factors or processes. Secondly, with regard to extendability, if the related field does not already provide needed information, then further empirical work in that field can be done. Presumably with analogous relations, one already knows the properties of the analogue. But with interfield connections and the physical relationship they postulate, predictions for the related field can be made and tested in order to find additional information. Thus comes both the means of further theory construction and independent confirmation from another field for the constructed theory. Finally, using interfield connections in the construction of a theory provides a unification of that theory with other fields of science at the outset. In contrast to the account of unification of science by reduction of one completed theory by another completed theory (and the attendant difficulty of finding adequate identity relations without modifying the theories), this account shows that unification can be accomplished in the very way the theory is constructed, not 'after the fact'. The new theory constructed with interfield connections is already connected to another theory. Since interfield connections play these important roles of providing empirical evidence for the theory and of unifying it with other fields, they are not merely heuristic devices, useful as a source of ideas but then eliminable after theory construction. Instead, they are themselves important scientific hypotheses: things are connected in these ways.

My thesis should not be interpreted to be stronger than it is. I am not claiming that interfield links *must* be used in the construction of any adequate explanatory theory. Nor am I claiming that whenever they are used a successful (*i.e.*, subsequently confirmed) theory will necessarily be produced. In fact, I doubt that any necessary and sufficient schemas for constructing correct theories will be found. But I am claiming that in the case examined progress occurred with the use of interfield connections in theory construction and this analysis may apply to other past or future instances.

*Committee on the History and Philosophy of Science*
*The University of Maryland, College Park*

## NOTES

[1] 'New' is ambiguous here. It may refer to a concept entirely new to the history of thought, *e.g.*, the quantitative convertibility of a generalized energy in the late nineteenth century. Such developments are much rarer than the other sense of 'new', namely, different from the concepts associated with the domain and new to this field or this type of theory. In the second case, the concept itself may be common in other sciences, *e.g.* material units, but it simply is different from items of the domain.

[2] 'Physical relation' may not be the most general way of expressing the types of relations between fields which are scientific hypotheses. It fits this case very well since genes were claimed to be physically located on the chromosomes. But if this analysis is extended, *e.g.*, to relations between psychology and neurophysiology, then some term more general than 'physical' would be desirable.

[3] It has been suggested to me that in some parts of psychology and in the social sciences where related fields are not well developed, analogies may be the preferable source of new ideas. This seems plausible and needs to be investigated.

[4] This research was supported by National Science Foundation Grant SOC 77-23476. I would like to thank numerous people for helpful comments on earlier drafts: students and colleagues at The University of Maryland, participants at the Leonard Conference on Scientific Discovery, faculty at the Claremont Colleges, the University of Chicago, and the Johns Hopkins University.

## BIBLIOGRAPHY

Allen, G.: 1978, *Thomas Hunt Morgan*, Princeton Univ. Press, Princeton.

Bateson, B.: 1928, *William Bateson, Naturalist*, Cambridge Univ. Press, London.

Bateson, W.: 1902, *Mendel's Principles of Heredity – A Defense*, Cambridge Univ. Press, Cambridge.

Bateson, W.: 1913, *Problems of Genetics*, Yale Univ. Press, New Haven.

Bateson, W., R. C. Punnett, and E. R. Saunders: 1905, 'Further Experiments on Inheritance in Sweet Peas and Stocks: Preliminary Account', *Proceedings of the Royal Society* 77, reprinted in Punnett, 1928, v. 2, pp. 139–141.

Bateson, W., and R. C. Punnett: 1911, 'On Gametic Series Involving Reduplication of Certain Terms', *Journal of Genetics* 1, reprinted in Punnett, 1928, v. 2, pp. 206–215.

Blackwell, R. J.: 1969, *Discovery in the Physical Sciences*, Univ. of Notre Dame Press, Notre Dame, Indiana.

Boveri, T.: 1904, *Ergebnisse über die Konstitution der chromatischen Substanz des Zellkerns*, G. Fischer, Jena.

Bridges, C. B.: 1914, 'Direct Proof Through Non-disjunction that the Sex-linked Genes of Drosophila are Borne by the X-chromosome', *Science*, N. S., 40, 107–109.

Brush, S.: 1978a, 'Nettie M. Stevens and the Discovery of Sex Determination by Chromosomes', *Isis* 69, 163–172.

Brush, S.: 1978b, 'A Geologist Among Astronomers: The Rise and Fall of the Chamberlin-Moulton Cosmogony', *Journal for the History of Astronomy* 9, 1–41.

Carlson, E. A.: 1971, 'An Unacknowledged Founding of Molecular Biology: H. J. Muller's Contributions to Gene Theory, 1910–1936', *Journal of the History of Biology* 4, 149–170.

Coleman, W.: 1970, 'Bateson and Chromosomes: Conservative Thought in Science', *Centaurus* 15, 228–314.

Darden, L.: 1976, 'Reasoning in Scientific Change: Charles Darwin, Hugo de Vries, and the Discovery of Segregation', *Studies in the History and Philosophy of Science* 7, 127–169.

Darden, L.: 1977, 'William Bateson and the Promise of Mendelism', *Journal of the History of Biology* 10, 87–106.

Darden, L., and Maull, N.: 1977, 'Interfield Theories', *Philosophy of Science* 44, 43–64.

Darwin, C.: 1868, *The Variation of Animals and Plants under Domestication*, Orange Judd & Co., New York.

Gilbert, G. K.: 1896, 'The Origin of Hypotheses, Illustrated by The Discussion of a Topographic Problem', *Science*, N. S., 3, 1–13.

Hanson, N. R.: 1960, 'Is There a Logic of Scientific Discovery?', *The Australasian Journal of Philosophy* 38, 91–106.

Harré, R.: 1960, *An Introduction to the Logic of the Sciences*, Macmillan, London.

Harré, R.: 1970, *The Principles of Scientific Thinking*, Univ. of Chicago Press, Chicago.

Hempel, C. G.: 1966, *Philosophy of Natural Science*, Prentice-Hall, Englewood Cliffs, New Jersey.

Hesse, M.: 1966, *Models and Analogies in Science*, Univ. of Notre Dame Press, Notre Dame, Indiana.

Janssens, F. A.: 1909, 'La theorie de la chiasmatypie', *La Cellule* 25, 389–411.

Kuhn, T.: 1970, *The Structure of Scientific Revolutions*, 2nd ed., Univ. of Chicago Press, Chicago.

Lakatos, I.: 1970, 'Falsification and the Methodology of Scientific Research Programmes', in *Criticism and the Growth of Knowledge*, I. Lakatos and A. Musgrave (eds.), Cambridge Univ. Press, Cambridge, pp. 91–195.

Laudan, L.: 1977, *Progress and Its Problems*, Univ. of California Press, Berkeley.

Leatherdale, W. H.: 1974, *The Role of Analogy, Model and Metaphor in Science*, American Elsevier, New York.

Manier, E.: 1969, 'The Experimental Method in Biology, T. H. Morgan and the Theory of the Gene', *Synthese* 20, 185–205.

Maull, N.: 1977, 'Unifying Science without Reduction', *Studies in the History and Philosophy of Science* 8, 143–162.

Morgan, T. H.: 1910a, 'Chromosomes and Heredity', *The American Naturalist* 44, 449–496.

Morgan, T. H.: 1910b, 'Sex Limited Inheritance in Drosophila', *Science* 32, 120–122.

Morgan, T. H.: 1911, 'Random Segregation versus Coupling in Mendelian Inheritance', *Science* 34, 384.

Morgan, T. H.: 1926, *The Theory of the Gene*, Yale Univ. Press, New Haven.

Morgan, T. H., A. H. Sturtevant, H. J. Muller, and C. B. Bridges: 1915, *The Mechanism of Mendelian Heredity*, Henry Holt and Company, New York.

Popper, K.: 1960, *The Logic of Scientific Discovery*, Harper Torchbooks, New York.

Punnett, R. C. (ed.): 1928, *Scientific Papers of William Bateson*, Cambridge Univ. Press, Cambridge, 2 vols.

Roll-Hansen, N.: 1978, 'Drosophila Genetics: A Reductionist Research Program', *Journal of the History of Biology* 11, 159–210.

Shapere, D.: 1973, Unpublished MS. presented at IUHPS-LMPS Conference on Relations Between History and Philosophy of Science, Jyväskylä, Finland.

Shapere, D.: 1974a, 'Scientific Theories and Their Domains', *The Structure of Scientific Theories*, in F. Suppe (ed.), Univ. of Illinois Press, Urbana, pp. 518–565.

Shapere, D.: 1974b, 'On the Relations Between Compositional and Evolutionary Theories', in *Studies in the Philosophy of Biology*, F. J. Ayala and T. Dobzhansky (eds.), Univ. of California Press, Berkeley, pp. 187–204.

Stern, C., and E. Sherwood (eds.): 1966, *The Origin of Genetics, A Mendel Source Book*, W. H. Freeman, San Francisco.

Sturtevant, A. H.: 1913, 'The Linear Arrangement of Six Sex-linked Factors in Drosophila, as Shown by Their Mode of Association', *Journal of Experimental Zoology* 14, 43–59.

Sutton, W.: 1903, 'The Chromosomes in Heredity', *Biological Bulletin* 4, 231–251.

Vries, H. de: 1889, *Intracellular Pangenesis*, G. Fisher, Jena; translated by C. S. Gager, Open Court, Chicago, 1910.

KENNETH F. SCHAFFNER

# DISCOVERY IN THE BIOMEDICAL SCIENCES: LOGIC OR IRRATIONAL INTUITION?*

## I. INTRODUCTION

The thesis that the process of scientific discovery involves logically analyzable procedures, as opposed to intuitive leaps of genius, has generally not been a popular one in this century. Since the advent of logical empiricism in the early twentieth century, the logic of science has been generally understood to be a logic of *justification*. Scientific discovery has been considered to be of interest to historians, psychologists, and sociologists, but has usually been barred from the list of topics that demand logical analysis by philosophers.

This situation has not always been the case; historically, in point of fact, the logic of scientific discovery and the logic of scientific justification have more than once traded prince and pauper status in the philosophy of science. With some historical support, let us initially understand a logic of scientific discovery to be a set of well-characterized and applicable rules for *generating new knowledge*, and a logic of scientific justification to be a set of rules for *assessing* or *evaluating* the merits of proferred claims about the world in the light of their purportedly supportive evidence. (These rules may either be presented in the scientific literature *per se* or be given in a rational reconstruction, as in the work of philosophers of science.) If this is accepted as a viable characterization of two actual aspects of science, then it is not too difficult to understand that a believer in a logic of scientific discovery might well not see the utility of a separate logic of justification, for a logic of justification might appear to him only as a tool of an uncreative critic, and a *logic* of discovery itself could be said to provide sufficient justification for its own progeny. On the other hand, if it is thought that there is no such thing as a logic of discovery, and that all scientific conjectures, regardless of their source, must be subjected to rigorous critical evaluation, perhaps with the aid of principles collated and systematized as a logic of justification, then a logic of justification becomes of paramount importance in understanding science. Views representing these different extremes, as well as some less extreme positions, can be found both in the historical and in the contemporary literature of the philosophy of science.

In his *Rules for the Direction of the Mind*, René Descartes believed he was

171

*T. Nickles (ed.), Scientific Discovery: Case Studies*, 171–205.
*Copyright © 1980 by D. Reidel Publishing Company.*

proposing a method for arriving at new knowledge which possessed demonstrative force.[1] Francis Bacon in *Novum Organum* similarly expounded a
logic of scientific discovery which would have denied any distinction, except
perhaps a temporal one, between a logic of scientific discovery and a logic of
justification.[2] In the nineteenth century, John Stuart Mill recommended
various methods of experimental inquiry, which can be viewed as methods
for the discovery of new scientific knowledge, and he was severely criticized
by his contemporary, William Whewell, for so doing.[3] Whewell argued that
Mill's methods, which bore a close resemblance to Bacon's 'Prerogatives of
Instances', begged the question of the existence of a set of rules for scientific
discovery by taking "for granted the very thing which is most difficult to
discover, the reduction of the phenomena to formulae as are here [in Mill's
*Logic*] presented to us."[4] Whewell succinctly expressed his own view as
follows:

The conceptions by which facts are bound together are suggested by the sagacity of
discoverers. This sagacity cannot be taught. It commonly succeeds by guessing; and this
success seems to consist in framing several *tentative* hypotheses and selecting the right
one. But a supply of appropriate hypotheses cannot be constructed by rule, nor without
inventive talent.[5]

The debate has continued into the twentieth century and into our own
day. In 1917 F. C. S. Schiller introduced his well known distinction between
the logic of *proof* and the logic of scientific *discovery*, and recommended that
science "proclaim a logic of its own," different from traditional demonstrative
logic but a 'logic' nonetheless.[6] What is needed, Schiller wrote:

is not a logic which describes only the static relations of an unchanging system of knowledge, but one which is open to perceive motion, and willing to accept the dynamic
process of a knowledge that never ceases to grow and is never really stereotyped into a
system.[7]

Schiller wrote in the pragmatic tradition, but with the rise of logical
empiricism the existence of a significant logical component in the discovery
process of science was again denied by influential thinkers. For example,
Hans Reichenbach, who introduced the distinction, analogous to Schiller's,
between the 'context of discovery' and the 'context of justification', wrote:

The mystical interpretation of the hypothetico-deductive method as an irrational
guessing springs from a confusion of *context of discovery* and *context of justification*.
The act of discovery escapes logical analysis; there are no logical rules in terms of which
a "discovery machine" could be constructed that would take over the creative function
of the genius. But it is not the logician's task to account for scientific discoveries; all he

can do is to analyze the relation between given facts and a theory presented to him with the claim that it explains these facts. In other words, logic is concerned only with the context of justification. And the justification of a theory in terms of observational data is the subject of the theory of induction.[8]

Sir Karl Popper, who stands at polar positions from Reichenbach on a number of issues in the philosophy of science, contended in his important book, *The Logic of Scientific Discovery* — the translation of the German title *Logik der Forschung* is most misleading — that:

The initial stage, the act of conceiving or inventing a theory, seems to me neither to call for logical analysis nor to be susceptible of it. The question how it happens that a new idea occurs to a man — whether it is a musical theme, a dramatic conflict, or a scientific theory — may be of great interest to empirical psychology; but it is irrelevant to the logical analysis of scientific knowledge. This latter is concerned not with *questions of fact* . . . , but only with questions of *justification or validity* . . . .
Accordingly I shall distinguish sharply between the process of conceiving a new idea, and the methods and results of examining it logically.[9]

For Popper, as for Reichenbach, the logical analysis of science is restricted to an examination of the structures of already given scientific theories and of the testing and post-discovery evaluative procedures of scientists.

## II. HANSON ON A LOGIC OF SCIENTIFIC DISCOVERY

More recently, the writings of the late N. R. Hanson began to direct the attention of philosophers of science back toward the discovery process in science.[10] Hanson argued that the 'philosophical flavor' of the analyses of such notions as theory, hypothesis, law, evidence, observation, *etc.*, as developed by philosophers of science, was at great variance with the concepts as employed by creative scientists. The reason, Hanson claimed, was that the approach to an area that was still problematical, and which required 'theory-*finding*' thinking, analyzed science quite differently from that type of inquiry whose task was to "rearrange old facts and explanations into more elegant formal patterns."[11] As part of his approach, Hanson contended that what he termed the hypothetico-deductive or H-D, account of science was rather misleading. Hanson wrote:

Physicists do not start from hypotheses; they start from data. By the time a law has been fixed into an H-D system, really original physical thinking is over. The pedestrian process of deducing observation statements from hypotheses comes only after the physicist sees that the hypothesis will at least explain the initial data requiring explanation. This H-D account is helpful only when discussing the argument of a finished research report, or for understanding how the experimentalist or the engineer develops the theoretical physicist's hypotheses; the analysis leaves undiscussed the reasoning which often points to the first tentative proposals of laws.

. . . . . . . . . . . . . . . . . . . . . . . . . . . . . . . . . . . . . . . . . . . . . . . . . . .

... The initial suggestion of an hypothesis is very often a reasonable affair. It is not so often affected by intuition, insight, hunches, or other imponderables as biographers or scientists suggest. Disciples of the H-D account often dismiss the dawning of an hypothesis as being of psychological interest only, or else claim it to be the province solely of genius and not of logic. They are wrong. If establishing an hypothesis through its predictions has a logic, so has the conceiving of an hypothesis.[12]

Hanson later argued that the process of scientific discovery could at least partially be analyzed conceptually.[13] Discovery was Janus-faced: it had philosophically relevant aspects just as it had philosophically irrelevant aspects. Hanson maintained that 'verification' and 'confirmation' of scientific theories were also Janus-faced, and that one could analyze them from the point of view of scientists' *behavior*, but also from the point of view of the *ideas* of 'verification' and 'confirmation', a task which represented the research program of the logical empiricists such as Reichenbach and Carnap. For Hanson, it was irresponsible to turn over *all* analysis of discovery to the psychologists, historians, and sociologists, who would only analyze the behavioral aspect largely as fact-gatherers. Hanson wrote that "settling on the meaning of 'discovery' is too important to our understanding of science to be abandoned to scientific discoverers or to psychologists or to sociologists or to historians of science."[14] For Hanson, "the *idea* of discovery" was "conceptually too complex for any 'average' historical, psychological, or sociological analysis," and ought to be done by 'conceptual analysts', namely by philosophers of science.[15]

Hanson's analysis was presented in several of his writings in slightly different forms, ranging from an analysis of the logic of discovery to a typology of scientific discoveries. He tended to foucs, erroneously I shall argue, on the concept of *retroductive inference*. On this matter, Hanson contended he was largely following and developing the suggestions of Aristotle and C. S. Peirce in proposing that there is a kind of inference, termed 'retroductive inference', which captures the type of reasoning implicit in creative scientific discoveries. Like Peirce, Hanson stressed that retroductive reasoning was quite different from both inductive and deductive reasoning. In his most developed article on this logic, titled 'Is there a Logic of Scientific Discovery?', Hanson proposed the following schematic form for retroductive reasoning:

(1)     Some surprising, astonishing phenomena $p_1$, $p_2$, $p_3$ ... are encountered.

(2)     But $p_1$, $p_2$, $p_3$ ... would not be surprising were a hypothesis of $H$'s type to obtain. They would follow as a matter of course from something like $H$ and would be explained by it.

(3)     Therefore there is good reason for elaborating a hypothesis of type of $H$; for proposing it as a possible hypothesis from whose assumption $p_1$, $p_2$, $p_3$ ... might be explained.

In an important footnote to (1), Hanson added that

The astonishment may consist in the fact that p is at variance with accepted theories – for example, the discovery of discontinuous emission of radiation by hot black bodies, or the photoelectric effect, the Compton effect, and the continuous β-ray spectrum, or the orbital aberrations of Mercury, the refrangibility of white light, and the high velocities of Mars at 90 degrees. What is important here is *that* the phenomena are encountered as anomalous, not *why* they are so regarded.[16]

Other than for Hanson's emphasis that retroductive inference leads to a *type* of a hypothesis rather than to a detailed or specific hypothesis, this schema for retroductive reasoning or retroductive inference seems to be identical with Peirce's characterization of retroduction or 'abduction'. That a form of logical *inference* is intended is supportable by citing another footnote to (3) above, in which Hanson quotes Peirce approvingly:

This is a free development of remarks in Aristotle ... and Peirce .... Peirce amplifies: "It must be remembered that retroduction, although it is very little hampered by logical rules, nevertheless, is logical inference, asserting its conclusion only problematically, or conjecturally, it is true, but nevertheless having a perfectly definite logical form."[17]

It will be helpful to illustrate what Hanson refers to as retroductive inference by means of an example.

### III. THE NATURAL SELECTION THEORY OF ANTIBODY FORMATION

In March 1954, Niels Kaj Jerne, an immunologist working at the Danish State Serum Institute, was walking home through Copenhagen. Jerne had been conducting research on antibody 'avidity', – that property of an antibody which determines the rate at which it reacts with antigen.

In 1954, the extant theories of antibody generation were all species of an 'instructive' theory. Antibodies are crucially important macromolecules which vertebrates (including humans) manufacture to protect them against disease, and antigens are those aspects of invading bacteria, viruses, fungi, and toxins which stimulate (or generate) antibody synthesis. Antigens fit like highly specific *keys* into the antibody *locks*, thus initiating (unlocking) the immune response which destroys the antigens.[18] A component hypothesis of all *instructive* theories of antibody formation is that *antigens convey structural information* to the immune system, which then uses this information, like a template, on which to construct the complementary fitting antibody.[19]

Prior to his discovery of the natural selection theory of antibody formation, Jerne had been injecting rabbits with diptheria toxin and antitoxin, and had been examining the sera of horses for antibodies to bacteriophage at various stages in the development of the equine immune response. He had been struck by, and puzzled by, the existence of a large variety of different types of antibody — present at very low levels of concentration — in the sera of the horses.

Jerne recalls that walk through Copenhagen in March 1954 in the following terms, which is worth quoting *in extenso*:

> "Can the truth (*the capability to synthesize an antibody*) be learned? If so, it must be assumed not to pre-exist; to be learned, it must be acquired. We are thus confronted with the difficulty to which Socrates calls attention in *Meno* (Socrates, 375 B.C.), namely that it makes as little sense to search for what one does not know as to search for what one knows; what one knows one cannot search for, since one knows it already, and what one does not know one cannot search for, since one does not even know what to search for. Socrates resolves this difficulty by postulating that learning is nothing but recollection. The truth (*the capability to synthesize an antibody*) cannot be brought in, but was already inherent."

The above paragraph is a translation of the first lines of Søren Kierkegaard's 'Philosophical Bits or a Bit of Philosophy' (Kierkegaard, 1844). By replacing the word "truth" by the italicized words, the statement can be made to present the logical basis of the selective theories of antibody formation. Or, in the parlance of Molecular Biology: synthetic potentialities cannot be imposed upon nucleic acid, but must pre-exist.

I do not know whether reverberations of Kierkegaard contributed to the idea of a selective mechanism of antibody formation that occurred to me one evening in March 1954, as I was walking home in Copenhagen from the Danish State Serum Institute to Amaliegade. The train of thought went like this: the only property that all antigens share is that they can attach to the combining site of an appropriate antibody molecule; this attachment must, therefore, be a crucial step in the sequences of events by which the introduction of an antigen into an animal leads to antibody formation; a million structurally different antibody-combining sites would suffice to explain serological specificity; if all $10^{17}$ gamma-globulin molecules per ml of blood are antibodies, they must include a vast number of different combining sites, because otherwise normal serum would show a high titer against all usual antigens; three mechanisms must be assumed:

(1) a random mechanism for ensuring the limited synthesis of antibody molecules possessing all possible combining sites, in the absence of antigen, (2) a purging mechanism for repressing the synthesis of such antibody molecules that happen to fit to auto-antigens, and (3) a selective mechanism for promoting the synthesis of those antibody molecules that make the best fit to any antigen entering the animal. The framework of the theory was complete before I had crossed Knippelsbridge. I decided to let it mature and to preserve it for a first discussion with Max Delbrück on our freighter trip to the U.S.A., planned for that summer.[20]

The 'selective mechanism' cited in (3) above was somewhat more elaborated by Jerne in his first publication on the theory in 1955.[21] There he suggested that the antibody or globulin molecules that make the best fit to an antigen entering the animal

may then be engulfed by a phagocytic cell. When the globulin molecules thus brought into a cell have been dissociated from the surface of the antigen, the antigen has accomplished its role and can be eliminated. [Jerne added that] the introduction of the selected molecules into a cell or the transfer of these molecules into another cell is the signal for the synthesis or reproduction of molecules identical to those introduced, i.e., of specific antibodies.[22]

We can reformulate Jerne's inference to (1)–(3) above in the retroductive pattern as follows:

(1) (a)  The presence of a large variety of types of antibody at low levels of concentration detectable by combining with various antigens is a puzzling phenomena. (This is puzzling in the light of the then current instructive theories of antibody formation.)

(b)  The only property that all these antigens share is that they combine with antibody molecules, therefore this must be an essential property in antibody generation.

(2)  But (a) and (b) in (1) above would not be surprising were a selective type of theory to obtain. Such a theory would have to have as component hypotheses: (i) "a random mechanism for ensuring the limited synthesis of antibody molecules possessing all possible combining sites, in the absence of antigen," (ii) "a purging mechanism for repressing the synthesis of such antibody molecules that happen to fit to auto-antigens, and" (iii) "a selective mechanism for promoting the synthesis of those antibody molecules that make the best fit to any antigen entering the animal."

(3)      Therefore there is a good reason for elaborating a selective type of
         theory, on the basis of which 1(a) and 1(b) would be explained.

## IV. CRITICISM OF THE RETRODUCTIVE MODEL

The above example illustrates the retroductive mode of inference, but it also
raises some serious questions about its completeness and how it significantly
differs from what Hanson castigated as the H-D approach. Hanson's analysis
has been critized in recent years both directly (by P. Achinstein[23]) and
indirectly (by G. Harman[24]). These authors make some valid points against
the simple retroductive model outlined and illustrated above; I believe, however, that they do not go far enough in their criticism, and also that they are
not sufficiently sensitive to an important distinction in the logic of scientific
discovery between (1) a logic of *generation* and (2) a logic of *preliminary
evaluation*. In the following pages I shall attempt to develop these points.

Achinstein has argued that Hanson (and Peirce) neglect the

*background* of theory which the scientist often has to begin with, and which may provide
at least part of and in some cases the entire basis for an inference to a law.

That this is not completely accurate is seen from the quotation given earlier
from Hanson.[25] It is, however, worth stressing that not only the perception
of an anomaly, but also the inference to a new theory, may be based on a
background. It must be added through, that it is not only background *theory*,
but also various experimental results and (low level) generalizations which
may not yet be systematized into any theory, or be part of any *one* specific
theory, which serve as parts of a premise set from which an inference to a
new theory or hypothesis is made. The property of combination with antibody which antigens possess is an example of such a generalization. The
variety of types of antibody in horse sera is an example of an experimental
result in the premise set.

Another difficulty which is raised by the simple retroductive model is that
it licenses peculiar inferences. Both Harman and Achinstein have maintained
that an inference to a new theory or hypothesis is legitimate only if the new
theory is a *better explanation* of the anomalous facts, $p_1, p_2, p_3 \ldots$, in
Hanson's notation, than any other *competing* theory or hypothesis. (Harman
terms such an explanation the best explanation.) Otherwise *any* sort of
theory could be inferred from $p_1, p_2, p_3 \ldots$. As Harman notes:

In general, there will be several hypotheses which might explain the evidence, so one
must be able to reject all such alternative hypotheses before one is warranted in making

the inference. Thus one infers from the premise that a given hypothesis would provide a "better" explanation for the evidence than would any hypothesis, to the conclusion that the given hypothesis is true.[26]

I believe that this conclusion — to the truth of the hypothesis — is too strong, and, following Hanson, would want rather to argue to the plausibility or probability of the hypothesis. With that exception, however, I think Harman is making an important but incomplete point.

In my view, the notion of *inference* in *reasoning to* an hypothesis is obscure in Hanson and in his critics. Hanson, in his far ranging writings on retroduction, never distinguished clearly between (1) a *logic of generation*, by which a new hypothesis is first articulated, and (2) a logic of preliminary evaluation, in terms of which an hypothesis is assessed for its plausibility. Neither is the distinction clear in Achinstein's and Harman's accounts of retroductive logic nor in most recent commentators on the logic of discovery — the only exceptions all appear to be researchers in the artificial intelligence (AI) area. Herbert Simon is one such exception, and he has in the past ten years developed a theory of scientific discovery in great detail, which draws on his own and others' AI research. It would take us far beyond the scope of this paper to discuss his important work in the detailed form which it requires, and the reader must be referred elsewhere.[27] The distinction between generation and preliminary evaluation is not equivalent to an inductive-deductive distinction, nor to the still again different demonstrative-nondemonstrative distinction.

## V. THE LOGIC OF GENERATION

Let us understand the task of a logic of generation as the production of new hypotheses — its function is to articulate heretofore unasserted conclusions.[28] These heretofore unasserted conclusions may range from the strikingly original, through the moderately novel, as when standard ideas are put together in new ways to solve a problem, to the trivial. A logic of generation need not *per se* distinguish among and assess the quality or originality of its progeny.

This 'new' knowledge can fall either into the demonstrative area, such as a new conjecture or new theorem in mathematics, or into the nondemonstrative area involving an empirical discovery. I take demonstrative reasoning to involve reasoning to *necessarily* true conclusions from true premises, and non-demonstrative reasoning to cover all other forms of reasoning. The point about the demonstrative-nondemonstrative distinction is to stress that it is not synonomous with the generative-evaluative distinction; rather, it cuts across it.

The logic of *evaluation* involves *assessing* proffered claims of support or of entailment. It can appear in weak or strong forms and can be employed in nondemonstrative and demonstrative areas. I have summarized in an incomplete way some examples of these types of reasoning under these different rubrics in table 2.1 below. The distinction between 'type' and 'mode' shown in the table is for convenience only.

TABLE 2.1.
Types and modes of scientific inference

| Model of Inference \ Type of Inference | Nondemonstrative Inference (induction, analogy, and retroduction) | Demonstrative Inference (essentially deductive) |
|---|---|---|
| Generative: | Weak: Retroduction, and/or analogical reasoning.<br>Strong: Traditional naive Baconian induction and enumerative induction; 'evocation' in the logic of diagnosis. | Weak: Mathematical heuristics.<br>Strong: Well-established algorithms for generating solutions. |
| Evaluative: | Weak: Retroduction, logic of comparative theory. evaluation; Bayesian logic of non-statistical hypothesis testing.<br>Strong: Eliminative induction; statistical hypothesis testing (Neyman-Pearson, Fisherean, and Bayesian methods). | Weak: Proofs by approximation.<br>Strong: Standard checking of logic/mathematical theorems. (Certain inductive logics also belong here, such as Carnap's.) |

The reader will note that I have listed retroduction both in the weak generative, nondemonstrative category and in the weak evaluative, nondemonstrative class. I did this because of the ambiguity I believe we find in Hanson, Achinstein, and Harman, among others, as into which class retroductive inference falls.

## VI. THE DOUBLE ASPECT OF THE LOGIC OF DISCOVERY

The thesis that will be urged in this paper is that *both* aspects of discovery — generation and weak evaluation — are important parts of a logic of discovery. It is, on this view, crucial to make the generative-evaluative distinction for two reasons. The first is that a *fully* developed logic of discovery requires both a logic of generation and a logic of weak evaluation. The second reason is that, in the opinion of this author, there are significant aspects of the 'discovery process' which are constituted by the logic of preliminary or weak evaluation which need articulation. It is part of the thesis to be urged here that both aspects of the discovery process are, at least partially, amenable to logical and philosophical analysis and that such analysis is needed to fully understand the complexities of scientific inquiry. In the remaining sections of this essay, I shall elaborate on both points. We begin with a discussion of some current inquiry into a logic of generation.

Contemporary investigation into a *generative* logic for hypotheses and theories is in its infancy. Much of it comes from artificial intelligence (AI) research in problem solving by computers.[29] The logic of generation is currently restricted to essentially non-novel solutions, though as will be noted below, there are areas in which nonoriginal hypotheses, when taken in combinations, are nonetheless difficult for even experienced scientists to generate.

I will comment briefly on one example of generative logic developed in the chemistry of molecular structures, and more extensively on an example drawn from the logic of clinical diagnosis in internal medicine. These examples will enable us to see both the distinction between generation and weak evaluation more clearly, and will also indicate more specifically the current limitations of a generative logic of scientific discovery.

J. Lederberg and his former colleagues at Stanford University, E. A. Feigenbaum and B. G. Buchanan, have for a number of years been developing a computer program which discovers the chemical structure of molecules on the basis of the molecules' mass spectrum and, if available, the nuclear magnetic reasonance spectrum.[30] The program, known as DENDRAL, contains the following basic elements, according to its originators:

Some of the essential features of the DENDRAL program include:
1. Conceptualizing organic chemistry in terms of topological graph theory, that is, a general theory of ways of combining atoms.
2. Embodying this approach in an exhaustive hypothesis generator. This is a program which is capable, in principle, of "imagining" every conceivable molecular structure.
3. Organizing the generator so that it avoids duplication and irrelevancy, and moves from structure to structure in an orderly and predictable way. The key concept is that induction becomes a process of efficient selection from the domain of all possible structures. Heuristic search and evaluation are used to implement this efficient selection.[31]

We note already in item 3 above that some principle(s) of preliminary evaluation is required in the logic of discovery. DENDRAL's finer structure can be conceptualized in terms of the above distinctions between generation, weak evaluation, and strong evaluation, the latter which DENDRAL's originators refer to as 'validation'. They have written that:

The Structure Generator incorporates:
1. an algorithm that allows it to proceed systematically from one possible candidate to the next, that is, a legal move generator that defines the space;
2. general criteria for instability of organic molecules that allow it to avoid working on chemically irrelevant structures;
3. procedures for treating subgraphs as if they were atoms, allowing particularly important combinations of atoms to be treated as a unit in the combinatorial work of the generator. Because of the structure of molecular graphs, this task environment lends itself to partial solutions using the techniques described below.

The structure Generator program knows nothing of the theory of mass spectrometry. Given a chemical formula, it will generate all the isomers (structural variants) that are chemically plausible a priori. These are the candidates that are input to the 'test' part of the generate-and-test procedure.[32]

Note that what I would term the preliminary or weak evaluation is contained in item 2.

In describing the 'validation' or strong evaluation aspect, which is part of their 'Predictor program', Feigenbaum, Buchanan and Lederberg wrote:

The Predictor program is the 'expert' on the general theory of mass spectrometry. It answers this question for the system: Though the candidate may be chemically plausible on a priori grounds, is it a good candidate to explain the given mass spectrum? In other words, does its predicted spectrum fit the data?

The Predictor incorporates a general theory of the fragmentation and recombination processes that can take place in a mass spectrometer, insofar as these are known to our chemist collaborators. The Predictor program is continually under development as the theory of mass spectrometry develops.[33]

A similar set of distinctions appears in AI research on the logic of clinical diagnosis. Our example here is taken from the investigations of J. D. Myers and H. Pople at the University of Pittsburgh.[34] They have programmed into a computer characteristics of about 75% of all diseases known to contemporary internal medicine. This data base is then used as background for a discovery program originally known as DIALOG (*diagnostic logic*) and now a INTERNIST-I. DIALOG or INTERNIST-I can take 'new' cases, such as the Massachusetts General Hospital patients presented weekly in the pages of the *New England Journal of Medicine*, and provide a diagnosis of the patient's illness which will often contain diseases joined together in complicating unique ways.

The INTERNIST-I program's logic is illustrative in that it incorporates a number of the distinctions made above and in addition has found it necessary to develop the phase of preliminary evaluation by further breaking it down into several further subphases. INTERNIST-I contains a generative phase, during which signs and symptoms of a disease *evoke* disease models. These models, which are potential explanations of the signs and symptoms, are then processed in a weighing phase of preliminary evaluation, to be described in more detail below. The program further partitions the disease models by employing a concept of dominance, which cuts the set of evoked models down to a shortlist of highly plausible, alternative candidates termed a 'considered' list. This list is then tested in an interrogative phase, during which the computer requests additional data in an attempt to reach a conclusive diagnosis. The program here proceeds through a broad *ruleout* phase if more than five plausible disease models are being considered, or in a narrower *discriminate* mode when the 'considered' list contains two to four models. Here questions are asked of the physician, adding more information to the computer's data, which might count heavily for one disease model and strongly against another. If there is sufficient information to warrant detailed testing of only *one* model, an activity which I would term 'strong evaluation', the program enters into a *pursuing* phase,[35] which may terminate in a *conclusion*, the final diagnosis. (In any of its phases, the program may indicate which data it is temporarily *disregarding*. This is data not explained by the top-ranked model.) If the detailed testing does not yield a strongly supported conclusion which excludes significant alternative diseases, the program reverts to its discriminate interrogative phase for further investigation. It would take us beyond the scope of this paper to provide the full and complex details of the INTERNIST-I program. (The program is constantly evolving and already has been developed into an INTERNIST-II program, which considers parallel diagnosis with considerably augmented heuristic searching power.) Some additional detail and a specific example largely based on the earlier research with INTERNIST-I will, however, be introduced to illustrate the points at which the logic of generation and the logic of preliminary evaluation function as the two main components of a logic of discovery in the biomedical sciences.

Before I commence with a specific example of INTERNIST-I, it might be useful to point out a fundamental weakness of the INTERNIST-I program as a model of scientific discovery. The data base of INTERNIST-I is developed by first constructing a 'disease tree'. This is, in the words of the authors of the program "a classification scheme for *all possible* diagnoses." We thus see that a *new* disease, akin to a novel or original hypothesis, is not generatable by INTERNIST-I. (This limitation was also part of the DENDRAL program, since in that program an hypothesis generator had to be programmed to generate a priori 'every conceivable molecular structure'.) I will return to this point about novelty again in a succeeding section.

## VII. AN EXAMPLE OF THE LOGICS OF GENERATION AND EVALUATION IN CLINICAL DIAGNOSIS

The disease tree introduced above is constructed by selecting "successive general areas of internal medicine."[36] Subcategories based on similar pathogenetic mechanisms are then specified within each general area. Further subdivision continues, terminating at the level of individual diagnosis. Under each individual diagnosis, the appropriate manifestations, such as history, signs, symptoms, and results of laboratory tests, are entered. These manifestations are each ranked by *type* and by *import*. 'Type' enables a request for additional data on a patient to be assessed for expense and danger; the order of preference is to inquire first of the patient's history data, then through symptoms, signs, and the various levels of laboratory data. The 'import' assessment indicates how likely a manifestation is to be a 'red herring', or how easily it can be disregarded.[37]

Each possible manifestation, say $M_a$, is assigned an 'evoking strength' on a scale of 0 to 5. This represents the logic of generation phase, since the disease model, say $D_i$, is generated by the manifestation which is read into the computer for a case under consideration. The number 0 is assigned as the evoking strength "if the manifestation is too nonspecific to yield any diagnostic conclusion," whereas a 5 is assigned if the manifestation is "pathognomonic" for a specific disease.[38]

Each specific disease, $D_j$, is also assigned a complementary manifestation frequency, ranked on a scale of 1 to 5. This represents the "frequency with which patients with a proven diagnosis $D_j$ will display $M_b$ as a manifestation of that disease." [39] As 1 indicates that $M_b$ occurs rarely in patients with $D_j$, whereas a 5 indicates that $M_b$ is an almost necessary condition for $D_j$, occurring in excess of 95% of patients with a proven diagnosis of that diease.

The program also 'condenses' the data. This process is important. It permits the program to 'generalize', and to construct a patient's differential diagnosis on an appropriately general level, by enabling it to rule out general classes of disease before detailed pursuit of an individual diagnosis. (The condensing or generalization process proceeds by determining "for each node of the disease hierarchy [or tree], the [mathematical] intersection of the manifestation list of its subnodes . . . ," which results in "a list of the manifestations common to all the subnodes of a given node"; it is thus the manifestation list for that higher node. By this method, jaundice becomes a "manifestation of hepatocellular injury (and of cholestasis and others) . . . " [40]

A specific case begins with an entry of the history, signs, symptoms, and available laboratory test results. As an example which represents an actual clinical case, we can consider the following data entry. [41] Successive data based on history, symptoms, signs, and laboratory tests are entered, with the computer requesting additional data through its 'please continue' phrase until the doctor indicates by a 'no' that the preliminary data is complete:

*Symptoms*: (TIMEIN)-(TIMER INITIALIZED)-*(DOCTOR)-SYMPTOMS?-*AGE-25-TO-50-PLEASE CONTINUE-*SEX-FEMALE-PLEASE CONTINUE-*FEVER-PLEASE CONTINUE-*JAUNDICE-PLEASE CONTINUE-*NAUSEA-VOMITING-PLEASE CONTINUE-*ANOREXIA-PLEASE CONTINUE-*URINE-DARK-HX-PLEASE CONTINUE-*LIVER-ENLARGED-PLEASE CONTINUE-*EOSINOPHILIA-PLEASE CONTINUE-*BILIRUBIN-CONJUGATED-BLOOD-INCREASED-PLEASE CONTINUE-*SGOT-GTR-THAN-400-UNITS-PLEASE CONTINUE-*SGPT-GTR-THAN-600-UNITS-PLEASE CONTINUE-*PROTHROMBIN-TIME-INCREASED-PLEASE CONTINUE-*SKIN-RASH-MACULOPAPULAR-ALLERGY-HX-PLEASE CONTINUE-*NO

Each manifestation, *e.g.*, jaundice, evokes a disease model $D_i$ which is capable in the weakest possible sense of explaining that manifestation. Each evoked model consists of four components: (1) a list of all manifestations which have been observed but which cannot be explained by $D_i$ (this list is termed the shelf of this model); (2) a list of all observed manifestations together with the evoking strengths that are consistent with $D_i$; (3) a list of all manifestations which normally would be found with $D_i$ but which in this case have not been; and (4) a list of all other manifestations consistent with $D_i$ but about which there is no relevant information available as yet. [42] If a model had been previously evoked, the new evocation updates the previous evocation. This procedure represents the generative phase of the discovery process.

The program then begins preliminary evaluation. It assigns *weights* to each $D_i$ on the basis of the following criteria:

(a) counting in favor of each model is a factor proportional to the combined evoking strengths of all observed manifestations explained by that model.

(b) counting against a model are two factors: data not explained by the model, and data expected but found negative. Data not explained are weighed in proportion to their importance: Data not found are weighed in proportion to their frequency of occurrence in the considered disease.

(c) in certain cases, a 'bonus' value is awarded; those models that are linked either causally with disease nodes that have already been confirmed in the diagnosis have their weights increased by an amount proportional to the degree of association between these linked nodes.[43]

This set of evoked disease models, as ordered on the basis of this preliminary weighing, is then further evaluated.

The program evaluates the top ranked model $D_T$, together with a subset of alternatives. The subset is chosen by a partitioning heuristic which divides the evoked set of disease models into a 'major' portion (the best alternative candidates) and a 'minor' portion (which usually contains numerically more disease models). The heuristic is employed since it "has the effect of grouping with the top ranked model those diagnoses that may reasonably be considered *mutually exclusive alternatives* to it." [44] (More specifically, the major subset is partially constructed by a process that uses a special concept of 'dominance'. A disease model $D_A$ is said to *dominate* a disease model $D_B$ if the net shelf of $D_A$ is a subset of the net shelf of $D_B$. In alternative language, $D_A$ dominates $D_B$ if manifestations *un*explained by $D_A$ are but a subset of those unexplained by $D_B$, i.e., less exactly or more crudely, $D_A$ explains more of the findings than does $D_B$. The major portion of the first cut contains (1) disease models that are either dominated by $D_T$ *or* which dominate $D_T$,[45] *and* (2) which are 'within a fixed range of $D_T$' as measured by the criteria given above on pages 184–185.

After this first cut, the program proceeds with a more elaborate preliminary evaluation. In point of fact, at this point the evaluation stages move increasingly toward what can be construed as strong evaluation. If there are still a number (five or more) of plausible candidates on the major portion of the considered list, the program proceeds in the *ruleout* mode. In this phase, it asks questions that have a high degree of probability of eliminating one or more of the considered models, *e.g.*, by requesting information about manifestations which have a very high degree of occurrence in the disease being evaluated.

In the example we are considering, the ruleout mode is represented by the first series of moves after the initial data entry (history, signs, symptoms, and laboratory results) described above on page 185. Specifically, those moves are represented in our example by the following sequence of questions and answers:[46]

(RULEOUT:  DRUG-HYPERSENSITIVITY-HEPATOCELLULAR-REACTION  HEPA TITIS-B HEPATITIS-A INFECTIOUS-MONONUCLEOSIS CHRONIC-ACTIVE-HEPA TITIS LEPTOSPIROSIS)

WEIGHT-LOSS-GTR-THAN-TEN-PERCENT ?*NO

RIGOR(S) ?*NO

(RULEOUT:  DRUG-HYPERSENSITIVITY-HEPATOCELLULAR-REACTION  HEPA TITIS-B HEPATITIS-A TOXIC-HEPATOCELLULAR-DISEASE INFECTIOUS-MONO NUCLEOSIS DRUG-HYPERSENSITIVITY-CHOLESTATIC-REACTION)
PLEASE ENTER FINDINGS OF PAIN-ABDOMINAL
*NO

ABDOMEN-PAIN-GENERALIZED ?*NO

ABDOMEN-PAIN-RIGHT-UPPER-QUADRANT ?*NO

(RULEOUT:  DRUG-HYPERSENSITIVITY-HEPATOCELLULAR-REACTION  HEPA TITIS-B HEPATITIS-A INFECTIOUS-MONONUCLEOSIS DRUG-HYPERSENSITIVI TY-CHOLESTATIC-REACTION TOXIC-HEPATOCELLULAR-DISEASE)

ARTHRITIS ?*NO

SKIN-URTICARIA ?*NO

PLEASE ENTER FINDINGS OF ABDOMEN-PALPATION
*NO

ABDOMEN-TENDERNESS-GENERALIZED ?*NO

ABDOMEN-TENDERNESS-RIGHT-UPPER-QUADRANT ?*YES

LYMPHADENOPATHY ?*NO

SPLENOMEGALY ?*NO

(RULEOUT:  DRUG-HYPERSENSITIVITY-HEPATOCELLULAR-REACTION  HEPA TITIS-A TOXIC-HEPATOCELLULAR-DISEASE HEPATITIS-B DRUG-HYPERSENSI TIVITY-CHOLESTATIC-REACTION)
EXPANDING

SKIN-SPIDER-ANGIOMATA ?*NO

(RULEOUT: DRUG-HYPERSENSITIVITY-HEPATOCELLULAR-REACTION ACUTE-
TOXIC-HEPATOCELLULAR-INJURY HEPATITIS-A HEPATITIS-B DRUG-HYPER
SENSITIVITY-CHOLESTATIC-REACTION ACUTE-ALCHOHOLIC-HEPATITIS)
PLEASE ENTER FINDINGS OF URINALYSIS-ROUTINE
*BILLIRUBIN-URINE-INCREASED

*NO

UROBILINOGEN-URINE-INCREASED ?*NO

PROTEINURIA ?*NO

PLEASE ENTER FINDINGS OF LIVER-FUNCTION-TESTS
*NO

ALKALINE-PHOSPHATASE-80-TO-169-IU ?*NO

ALKALINE-PHOSPHATASE-GTR-THAN-160-IU ?*NO

PLEASE ENTER FINDINGS OF WBC-AND-DIFFERENTIAL
*NO

WBC-LESS-THAN-4000 ?*NO

WBC-1400-TO-30000 ?*NO

The program continues in the ruleout mode until either the discriminate
mode is entered, or until the type of data requested has reached the level of
laboratory specificity (recall the type order is history, symptoms, signs,
laboratory data). At this point, because of the cost/danger of obtaining new
laboratory data, the program shifts to the narrow interphase mode.

Passing through this narrow interphase mode, the mode of inquiry artificially shifts to the *discriminate* phase, which usually only is used when the
considered list is reduced to two to four models.

In the discriminate mode, evaluation is comparative between the two top
models, with questions asked which are very likely to count heavily for one
model while counting heavily against the other. In our example, the printout
of the program for the narrow/discriminate phase read as follows: [47]

(DISREGARDING: EOSINOPHILIA)
(NARROW: TOXIC-HEPATOCELLULAR-DISEASE DRUG-HYPERSENSITIVITY-
HEPATOCELLULAR-REACTION)

ANEMIA-HEMOLYTIC ?*YES

PLEASE ENTER FINDINGS OF DRUG-ADMINISTRATION-HX
*HALOTHANE-ADMINISTRATION-HX

PLEASE CONTINUE
*NO

DRUG-ADMINISTRATION-RECENT-HX ?*YES

PLEASE ENTER FINDINGS OF LIPIDS-BLOOD
*NO

CHOLESTEROL-BLOOD-DECREASED ?*YES

(DISCRIMINATE:     DRUG-HYPERSENSITIVITY-HEPATOCELLULAR-REACTION
HEPATITIS-B)

DRUG-ADDICTION-HX ?*NO

LDH-BLOOD-INCREASED ?*YES

PLEASE ENTER FINDINGS OF HEPATITIS-CONTACT-HX
*NO

HEPATITIS-CONTACT-ABOUT-180-DAYS-BEFORE-ILLNESS ?*NO

When the considered list is reduced to only one highly plausible model, the program enters what I would term its strongest evaluative phase termed 'confirm' or *pursuing* by Pople, Myers, and Miller. The program proceeds in this mode, asking additional questions if needed, which are 'clinchers': manifestations which have a strong evoking strength for the specific model are asked. The program continues in the pursuing phase until (1) either the spread between the two top models is sufficiently *increased* that it meets *criterion*,[48] at which point the program *concludes*, or (2) until the spread is *reduced* sufficiently so that the program reenters the discriminate mode. In our example, the pursuing phase is very rapid, one step:

(PURSUING: DRUG-HYPERSENSITIVITY-HEPATOCELLULAR-REACTION)
(CONCLUDE: DRUG-HYPERSENSITIVITY-HEPATOCELLULAR-REACTION)
(DIAGNOSIS: DRUG-HYPERSENSITIVITY-HEPATOCELLULAR-REACTION)
(CPU TIME UTILIZED: 61.569000 SECONDS)

In more complex examples, the pursuing phase is quite long.[49]

The point of this rather detailed example from the logic of diagnosis is to underscore the difference between the logic of generation and the logic of preliminary evaluation. Another feature of the example is that it indicates the necessity in a fully developed logic of discovery for a set of generators which will articulate the hypothesis that is to explain the data, and which can be evaluated in a preliminary way as part of the discovery process and in a more developed way as part of the logic of test and of justification.

I argued above that Hanson, in his extensive writings, never made these distinctions clear. He also never provided 'a way up', to use his terminology, a means of generating a hypothesis in the first place. As such, without a logic of generation, *the retroductive mode of inference is little more than the H-D model itself.*

What little more it is seems to lie in Hanson's and others' conceptualization of a distinction akin to the distinction urged here between preliminary evaluation and fully developed justification. Hanson did press for a distinction between:

(1)      reasons for accepting a hypothesis $H$, from
(2)      reasons for suggesting $H$ in the first place.[50]

In our view, however, Hanson equivocated on the word 'suggesting', sometimes using it in the *generative* sense, and sometimes in the *preliminary evaluative* sense. Hanson also wished to make a sharp distinction between reasons for (1), which he identified with reasons we might have for thinking $H$ *true*, and reasons for (2), which he identified with those reasons which would make $H$ *plausible*. Hanson distinguished his position from one he attributed to Herbert Feigl as follows:

> One might urge, as does Professor Feigl, that the difference [between (1) and (2)] is just one of refinement, degree, and intensity. Feigl, argues that considerations which settle whether $H$ consitutes a plausible conjecture are of the *same type* as those which settle whether $H$ is true. But since the initial proposal of a hypothesis is a groping affair, involving guesswork amongst sparse data, there is a distinction to be drawn; but this, Feigl urges, concerns two ends of a spectrum ranging all the way from inadequate and badly selected data to that which is abundant, well diversified, and buttressed by a battery of established theories.[51]

On this issue I tend to agree with Feigl against Hanson. The *criteria* of acceptance in the preliminary act of discovery stage are identical with those which will be preferred in the context of justification. In the context of justification, however, the substantive evidence will be systematically organized, often further developed, and will in general be much more powerful.

These criteria are familiar ones: empirical adequacy, coherence, and the like. Hanson argued that simplicity considerations were involved in the logic of discovery, but in my view he did not go far enough. Harman also has proposed similar criteria, as when he noted in his discussion of inference to the best explanation:

> There is, of course, a problem about how one is to judge that one hypothesis is sufficiently better than another hypothesis. Presumably such a judgment will be based on considerations such as which hypothesis is simpler, which is more plausible, which explains more, which is less *ad hoc*, and so forth.[52]

Harman, however, did not elaborate on these criteria.

The logic of discovery, then, comprises the two aspects of a logic of generation and a logic of preliminary evaluation. Actual discoveries in the biomedical sciences involve both these aspects. Unfortunately, recollections of the generative phase is often couched in 'illuminationist' terminology in which irrational aspects are heavily stressed − which gives support to Popper's and others' thesis of the illogicality of discovery. Application of the account of a logic of discovery sketched above can, I believe, accommodate *some* irrational elements for two reasons. First, it is not essential for the thesis sketched above that *all* aspects of the generative aspect of the discovery process be fully articulated − I believe we can make some progress in understanding the discovery process even if parts of the scenario are murky and contain gaps. Secondly, even in those cases in which a solution comes in a 'eureka' experience without − it is often contended − conscious preparation, there is still the preliminary evaluative aspect of a logic of discovery into which we should inquire.

It may be useful at this point to examine a specific case which contains an irrational, 'eureka' element and which also advances a truly novel hypothesis. This case study is Burnet's discovery of the clonal selection theory of acquired immunity, and is a natural development from our earlier example of Jerne's natural selection theory of antibody formation.

## VIII  THE CLONAL SELECTION THEORY

Burnet's theory was initially proposed in 1957[53] and further elaborated by him in 1958 in the Abraham Flexner lectures at Vanderbuilt, on 'Clonal Selection as Exemplified in Some Medically Significant Topics', published in 1959 as *The Clonal Selection Theory of Acquired Immunity*.[54] The theory initially encountered resistance and was even 'falsified' several times in its early life. Such falsifications subsequently turned out to be experimental artifacts, and the power of the theory was gradually appreciated, with the theory becoming generally accepted by 1967.[55] Currently the clonal selection theory serves as *the general theory* of the immune response, and all finer structured theories involving detailed cellular and molecular components are required to be consistent with it.

The discovery of the clonal selection theory illustrates some important logical and epistemic features of scientific discovery in general. First let me present Burnet's own account of the discovery.[56] I shall then comment on the philosophically interesting aspects of the example.

Burnet recalls the episode in the following terms:

In 1957, I had been deeply interested in immunity for many years and had published a good deal about the possible interpretations of antibody. They had necessarily been in terms of the concepts that had been current at the time in biochemistry and general biology. The development, first of ideas, and then of an experimental approach in the field of immunological tolerance and already impressed me with the crucial nature of those phenomena for immunological theory.

Then in 1955, I had seen Niels Jerne's paper on a natural selection theory of antibody production.

Niels Jerne is a Dane, a few years younger than I am, whom I met for the first time at Geneva in 1956. Since then I have come to know him well and I am sure that he is the most intelligent immunologist alive. I did not always think so because, in 1954 [actually in 1956], I had published what I now know was a rather bad, over-ambitious book with an interpretation of antibody production which was already beginning to look unconvincing by the time I finalized the proofs. Jerne did not like the interpretation and told me that I was merely creating metaphysical entities which explained nothing. This, of course, did not predispose me to adopt his quite different theory.[57]

At this point Burnet describes Jerne's natural selection theory which I have presented above. Burnet critically focused on the uptake mechanism introducing the antigen-natural antibody complex into the macrophage which would then process the antibody and initiate massive production of identical antibodies:

The crux of the theory was that this complex of antigen and natural antibody was taken up by a phagocytic cell, inside which the antigen was destroyed and the natural antibody used as a pattern to make more antibody of the same type. This had many virtues in explaining tolerance and some other phenomena but even in 1955 it did not fit at all with the way proteins were known to be synthesized and there was still the awkward failure to account in any satisfactory fashion for the diversity of natural antibodies. So we were mutually unimpressed by the other's ideas.

I came back to Australia pondering heavily on why Jerne's theory was so attractive, though obviously wrong. I said earlier that at that time I was then working with two phenomena which suggested (a) that auto-antibodies might be produced by *proliferating* semi-malignant cells, and (b) that the Simonsen spots on the membrane CAM of the chick embryo resulted in part from the *proliferation* of lymphocytes reacting immunologically. Rather suddenly 'the penny dropped'. If one replaced Jerne's natural antibodies by the cells which produced them and applied a selective process in a Darwinian sense to the antibody-producing cells, the whole picture fell into shape. I wrote it out in a short paper of two pages which I published in *The Australian Journal of Science*. It still, I believe, gives a a brief and clear account of the clonal selection theory basically acceptable in 1967.[58]

I shall return to Burnet's comments concerning his concurrent work on auto-antibodies and the Simonsen spots to provide some additional background in a moment. Let me however quote Burnet's concise formulation of the basic postulates of his clonal selection theory.

This formation represents to Burnet the '*essence*' of his theory. It does articulate in a somewhat more explicit, modern, and concise manner what was asserted in 1957.[59]

(1) Antibody is produced by cells, to a pattern which is laid down by the genetic mechanism in the nucleus of the cell.

(2) Antigen has only one function, to stimulate cells capable of producing the kind of antibody which will react with it, to proliferate and liberate their characteristic antibody.

(3) Except under quite abnormal conditions one cell produces only one type of antibody.

(4) All descendants of an antibody-producing cell produce the same type of antibody.

(5) There is a genetic mechanism [Burnet prefers somatic mutation] capable of generating in random fashion a wide but not infinite range of patterns, so that there will be at least some cells that can react with any foreign material which enters the body.[60]

It will be useful to dwell for a moment on the background theories, generalizations, and experiments, which were involved in the generation and preliminary evaluations of the clonal selection theory.

Note that Burnet identifies as relevant background: his own earlier 1954 [or 1956] *theory*, Jerne's *theory*, and Jerne's *methodological* criticism imputing "metaphysical entities which explained nothing" to Burnet's earlier theory [this is probably the Burnet (and Fenner) self-marker hypothesis],[61] as well as his (Burnet's) criticism of Jerne's theory that "even in 1955 it [Jerne's theory] did not fit at all with the way proteins were known to be synthesized . . . ." (Apparently Burnet is referring have to what he noted in his 1957 essay announcing the clonal selection theory, namely, that "its major objection is the absence of any precedent for, and the intrinsic unlikelihood of the suggestion, that a molecule of partially denatured antibody could stimulate a cell, into which it had been taken, to produce a series of replicas of the [undenatured] molecule."[62])

These considerations serve to provide the puzzles or problems with which inquiry leading to a new hypothesis or theory begins. However it must be noted that in this case several key elements came from outside the specific area of focus — namely from the work with the Simonsen spots on the CAM, and from Burnet's speculations concerning cancer and auto-antibodies.

CAM is an abbreviation for chorioallantoic membrane — that delicate sheath which surrounds the developing embryos of birds and which normally lies against the inside of the protective eggshell. Early in his career — in the early 1930s — Burnet had developed a means of exposing the CAM without destroying it, and was able to use it as a fertile medium upon which to grow disease viruses. This was in itself an important contribution to research on the causes and cures of virus diseases.[63]

Over twenty years later — in early 1957 — M. Simonsen discovered the graft-versus-host (GVH) reaction, in which transplanted cells immunologically attack the host into which they have been transferred.[64] The GVH reaction, often termed the Simonsen phenomena, was investigated by Burnet through smearing blood cells from adult fowls on the CAM of chick eggs. Burnet found that *single* white blood cells were stimulated to divide and attack the host CAM, producing white spots on the membrane. The "recognition that single cells could produce immune responses on the CAM"[65] was for Burnet an extremely important experimental background discovery lending to the clonal selection theory.

Burnet also noted above that he had considered that "auto-antibodies might be produced by *proliferating* semi-malignant cells,"[66] and that this speculation had also played an important role in the discovery of the clonal selection theory. This may well be the case, though Burnet's articles on cancer, which were completed in 1956 and published in April 1957 in the *British Medical Journal*, do not explicitly discuss anto-antibodies.[67] What we *do* find in Burnet's speculations on cancer, however, is the thesis that cancer is probably a consequence of a *series* of somatic mutations in cells, mutations which successively free the altered cells from those levels of biological controls — Burnet speculated these were about six in number — which maintain the cells in their appropriate locus and form, and at their usual rate of replication. Burnet explicitly noted that

there is every reason to believe that mutation is at least as frequent in somatic cells as in the germ cells. The essential difference between a mutation occurring in a germ cell and . . . a somatic cell depends simply on . . . the cells descendent . . . . A somatic mutation can influence only those body cells which directly descend from the mutant cell.[68]

We see here the essence of the clonal branching aspect, based on random somatic mutation, which later that year was incorporated into the clonal selection theory of acquired immunity. Clonal branching, and the Simonsen phenomena, thus appear to have functioned as relevant analogies in the generation of the clonal selection theory.

Burnet's clonal selection theory serves as an illustration of the manner in which earlier and contemporaneous reserach markedly affects scientific discovery. It demonstrates the role of earlier theories and the function of criticism of those theories. Criticism on various levels, ranging from experimental through theoretical to methodological points of view, provides puzzles or problems that initiate the discovery process. We also see what can be called the irrational and creative element described by Burnet when he noted that "Rather suddenly 'the penny dropped'."

It would clearly be Utopian to expect that at this point in the development of a logic of generation that Burnet's discovery of the clonal selection theory could ever be fully rationally constructed, much less have been anticipated by such an applied logic. (Similar but future novel discoveries might be so reconstructable, if the right data are gathered). It may be useful nonetheless to speculate on the manner in which such a rational reconstruction, using the terminology developed thus far in this essay, might be employed to understand what was involved in the discovery of this theory.

The analogies of the Simonsen phenomena (individuals cells produce antibody) and the clonal branching aspect of cells rendered malignant by somatic mutations were mechanisms in Burnet's repertoire of *possible* models. These are conjecturally evoked by the puzzles of natural antibodies and of self-tolerance that were highlighted by Jerne's theory. The linkage between the puzzles $p_1$, $p_2$, $p_3$ — accepted facts or generalizations which are not adequately accounted for by extant hypotheses — and new hypotheses evoked to account for them are, however, extremely weak. If we were to speculatively attempt to develop this approach further, we might imagine a computer programmed with all known basic biochemical and genetic mechanisms and a combining operator which would associate combinations of these mechanisms to constitute 'new' models. We might also envisage a subprogram which could operate on known mechanisms and construct analogical forms of them. The actual process of combination or analogizing which occurred in Burnet's mind is not available for inspection and reconstructive imitation, though he may have given future investigations enough clues to at least partially mimic the private generative process involved in this case.

Turning to the possibly more public process of preliminary evaluation, we can also surmise what occurred immediately after 'the penny dropped', namely a review of the problems with Jerne's and his own earlier theory. Burnet has noted that he "wrote [the theory] out in a short paper . . . "[69] It is *likely* that that paper closely reflects the state of mind of Burnet in the early stages of evaluation of the clonal selection theory. Inspection of the paper, which appeared in the October 27, 1957, issue of *The Australian Journal of Science*, shows striking parallels with the factors preceding and concomitant with the generation of the hypothesis, with the crucially important exception that in the paper the hypothesis is now present to give a focus to the related constraints and to *solve* in a preliminary and tentative way the problems that initiated the inquiry. Let us now examine that paper.

In the 1957 paper, Burnet begins with a brief review of the extant theories of antibody production – the direct template theory, his and Fenner's own indirect template theory with self-markers, and Jerne's 1955 theory. He points out early in the paper that Jerne's theory offers a better explanation of self-tolerance than the self-marker theory, but that one of its basic flaws is that there is no

precedent for . . . a molecule of partially denatured antibody 'stimulating' a cell, into which it had been taken, to produce a series of replicas of the molecule."

Burnet then adds that he believes the

advantages of Jerne's theory can be retained and its difficulties overcome if the re-cognition of the foreign pattern is abscribed to clones of lymphocytic cells and not to circulating natural antibodies.

A statement of the clonal selection theory follows essentially as outlined above on page 193. Burnet closes his paper by repeating the advantages the clonal selection theory has (1) over his own earlier theory – namely its explanation of existing natural antibodies [recall this was the major empirical finding impelling Jerne to his theory] and a simpler interpretation of self-tolerance – and (2) over Jerne's theory, namely the unlikelihood of the par-tially denatured antibody stimulating replicas of itself. At this point Burnet also added a *new* advantage of the clonal selection theory over Jerne's theory: the clonal selection theory could account for the fact that some aspects of the immune response – namely certain types of sensitization and homograft immunity – seem to be mediated directly by cells, not involving liberation of classic antibody. The clonal selection theory could permit cells to be sensitized directly, whereas Jerne's theory seemed to demand the production of anti-body for such a process.

Burnet's paper is very much a preliminary communication. Detailed deduction of consequences from the elaborated theory did not occur until later, though even in the preliminary communication we see the marshalling of *additional* evidence beyond what was used in the very early stages of hypothesis evaluation. The theory was not fully developed until the following year when Burnet gave the Flexner lectures, and was not extensively tested and criticized until several years later. As already mentioned, it was not generally accepted until about 1967.

It seems to me that Burnet's original paper, when analyzed in the light of his autobiographical remarks quoted earlier, constitutes an exercise in the logic of preliminary evaluation of the clonal selection theory. It does, however, seem to be akin to the 'discriminate' mode hypothesis evaluation discussed in connection with the logic of diagnosis and is thus somewhat removed from the surmised earlier stage of evaluation commented on in the autobiographical note.

## IX. OBJECTIONS TO A LOGIC OF SCIENTIFIC DISCOVERY

The distinguished philosopher of science, C. G. Hempel, has offered a series of clear and powerful arguments against a logic of scientific discovery of a strong or 'Baconian' form.[70] Some of his arguments will apply to the two-aspect account of a logic of scientific discovery as a logic of generation and a logic of preliminary evaluation urged earlier. Let us turn to Hempel's precisely articulated views on this subject.

Hempel presents essentially three arguments against the possibility of a logic of scientific discovery. First, a logic of discovery would presumably begin from *facts*, but there are an *infinite* number of facts, and some sorting into relevant and irrelevant facts must be presumed. However, secondly, such a sorting cannot be based purely on a *problem* with which scientific inquiry might be said to begin, for a 'problem' is too vague and general to provide criteria for the selection of relevant facts. The only candidate, Hempel argues, which qualifies as a sorting device for relevant facts, is a *hypothesis*, and on his view relevance is a relation between a hypothesis and its consequences: a fact is relevant if and only if it or its negation is a deductive consequence of a hypothesis.[71] Thus, Hempel argues, we must begin our inquiry with hypotheses which are 'happy guesses', freely invented to account for facts.

Hempel also offers another, third argument against a logic of scientific discovery.

Scientific hypotheses and theories are usually couched in terms that do not occur at all in the description of the empirical findings on which they rest, and which they serve to explain . . . . Induction Rules . . . . [a strong logic of discovery] would therefore have to provide a mechanical routine for constructing, on the basis of given data, a hypothesis or theory stated in terms of some quite novel concepts which are nowhere used in the description of the data themselves. Surely, no general mechanical rule of procedure can be expected to achieve this.[72]

My response to these trenchant objections will proceed by anticipating some aspects of the nature of theory construction and the dynamics of research programs in the biomedical sciences. These issues are pursued in greater depth elsewhere and can only be sketched here.[73]

## X. PROBLEMS, PROVISIONAL FIELDS, AND THE LOGIC OF SCIENTIFIC DISCOVERY

It should be obvious from the historical examples cited earlier that scientific inquiry as performed by trained, competent researchers always begins *in medias res*. An individual working on the forefront of biomedical science has been schooled for many years in previous theories, experimental results, techniques of analysis, and also has a usually rough and ready set of standards by which he judges the worth of scientific contributions.

For such an individual, a 'problem', though it does not generate a hypothesis, does normally fall into and thus identify both a *subdiscipline* and what I will term a 'provisional field'. I now think that this term 'provisional field' may raise more problems than it solves — see Maull's 'Comment' following and my 'Reply' — and I now prefer the term 'set of provisional explicanda' to refer to this aggregate.) The subdiscipline, in its ideational aspects, comprises a body of knowledge, including extant received theories, significant generalizations not necessarily contained in or yet explained by a theory, and experimental results. The sub-discipline is often inter-field, to use Darden and Maull's terminology.[74] Surrounding the subdiscipline, to employ a spatial metaphor, lie clusters of often entrenched theories which provide relevant background for the subdiscipline. Around the subdiscipline of antibody diversity theorizing, and interacting strongly with it for example, are located theories of genetics, a theory (or theories) of protein synthesis, and evolutionary theory, which serve both as sources of potentially fruitful ideas in the subdiscipline and as constraints: a new theory in the subdiscipline is required to be consistent with these background theories (or at least with the *entrenched* background theories).

Of a still more abstract nature, but nonetheless important, are a set of rather vague, methodological principles employed by scientific investigators. These seem to be sufficiently general that they are roughly constant throughout all disciplines and subdisciplines in biomedical science, and may well be common to all scientific theorizing in general. Examples of these are a distrust of *ad hoc* hypotheses, a correlative dislike of 'metaphysical' speculation in science, and a respect for 'simplicity' and 'elegance'. Also falling within this methodological realm are principles which reflect back on the contents of the subdiscipline or to the background theories. As an example one might consider the methodological rule that any new theory *ought* to account for a number of important experimental results which a previously accepted theory in the subdiscipline explained. Another example, already alluded to, is the *consistency requirement* between any new theory propounded in the sub-discipline and entrenched background theories.[75] This is a very powerful type of constraint: it resulted in Jerne's natural selection theory being quickly eliminated from consideration by immunologists in the late 1950s, and, applied in another way, was a major reason for the acceptance of the clonal selection theory over the competing instructive theory in the 1960s.[76]

What I referred to above as a 'provisional field' (more accurately a set of provisional *explicanda*), is a part of the subdiscipline and is closely associated with a problem'. It represents an extrinsic control over any solution to a problem which can be solved by a new hypothesis or theory. The provisional field is that complex of experimental results and generalizations which any new adequate theory is expected to account for, whether this be by identifying a generalization as a new 'ideal of natural order', to use Toulmin's phrase, or by explaining the generalization or experiment. It is akin in important respects to certain features of D. Shapere's concept of a 'domain', to which the concept is indebted.[77] Though it is difficult to specify the general features of such provisional fields at a high level of aggregation, their specific contents usually are achieved by consensus of leading researchers in a subdiscipline.

In connection with the Jerne and especially the Burnet examples above, we saw the provisional field of a theory of antibody generation sketched in outline in terms of the presence of natural antibodies in serum, mnemistic immune response, the existence of self-tolerance, the lack of antigen in antibody producing cells, the purely cellular aspects of one part of the immune response, and the like.

The provisional field and the series of methodological standards of a subdiscipline are thus evoked by a problem.

I would thus maintain, in reply to the possible objection of Hempel raised in the previous section, that a problem is sufficient to generate the set of antecendent constraints which function as a minimum logic of scientific discovery as components in the preliminary evaluation of a new hypothesis or theory. Furthermore, I would contend that Hempel's view, which analyzes the relevance of a 'fact' on the basis of its (or its contrary's) deducibility from an hypothesis, is too strong to serve as an account of how facts govern hypotheses or theories in the preliminary evaluative stage. The argument against Hempel's strong view is, I believe, a simple and straightforward one. On the view urged by Hempel, a fact $O$ can only confirm/falsify an hypothesis

$H$ if either $H$ (in conjunction with appropriate initial conditions) entails $O$ (confirmation) or entails some $O'$ which is inconsistent with the observed $O$ (falsification). But there are many facts ($O$'s) in the subdiscipline which are required to be explained by any new adequate $H$, and which often will not stand in either of the deductively characterized relations to $H$. That is, some significant fact in the subdiscipline, $O_S$, may have a deductive *non*-relation, $- (H \to O_S)$ & $[(H \to O)$ & $(O \neq O_S \neq -O)]$, where this *non*-relation is sufficient to reject the $H$. If this is the case, then the concept of relevance *cannot* be fully given by a deductive account.

Accordingly, I would maintain that we have answers to Hempel's first two objections to a logic of discovery cited above. There are *not* an infinite number of relevant facts; there are a limited number which are selected in advance by a problem which is sufficiently precise to demarcate a provisional field within a subdiscipline.

Hempel's third objection is more difficult to answer. At best, in the light of our analysis of the logic of diagnosis and in terms of the Burnet clonal selection theory example, we can speculate that a generative logic may be devisable which can evoke novel mechanisms *via* a programmable procedure. How this might have been done in connection with the Burnet example was outlined above on page 195. Further developments of the artificial intelligence work look very promising, [78] but at this point the only considered decision that can be reached is that the issue of a logic of discovery as a logic of generation is an unsettled one. On the other hand, if the thesis of the double aspect of the logic of discovery is defensible, even if the likelihood of such a strong logic of discovery is unsettled, the weaker thesis is independently supportable. I have argued above that a logic of preliminary evaluation constitutes an important and often misconstrued, if not actually denied, component of scientific inquiry. I would maintain, in conclusion, that denial of the significance of this part of scientific research would leave philosophy of science poorer and science itself less intelligible.

*Department of History and Philosophy of Science*
*University of Pittsburgh*

### NOTES AND REFERENCES

* I would like to thank Richard Burian, F. Macfarlane Burnet, J. Lederberg, Thomas Nickles, H. E. Pople, Jr., Herbert A. Simon, and William C. Wimsatt for reading an earlier draft of this paper and for helpful comments. I also acknowledge with gratitude support from the U.S. National Science Foundation.

[1] René Descartes, *Rules for the Direction of the Mind*, in René Descartes, *Philosophical Essays*, tr. L. J. Lafleur, Bobbs-Merrill, Indianapolis, 1960. Originally published in 1701.

[2] Francis Bacon, *The New Organon*, Bobbs-Merrill, Indianapolis, 1960. Originally published in 1620.

[3] John Stuart Mill, *A System of Logic*, Longmans, Green and Co., London, 1959. Originally published in 1843.

[4] William Whewell, 'Mr. Mill's Logic', in R. Butts (ed.), *William Whewell's Theory of Scientific Method*, Univ. of Pittsburgh Press, Pittsburgh, 1968, pp. 265–308. Originally published by Whewell in 1849.

[5] Whewell, 'Mr. Mill's Logic', p. 286.

[6] F. C. S. Schiller, 'Scientific Discovery and Logical Proof', in C. Singer (ed.), *Studies in the History and the Methods of the Sciences*, Vol. I, Clarendon Press, Oxford, 1917, pp. 235–289.

[7] Schiller, 'Scientific Discovery . . . ', p. 273.

[8] H. Reichenbach, *The Rise of Scientific Philosophy*, Univ. of California Press, Berkeley, 1958, p. 231. Also see H. Reichenbach, *Experience and Prediction* Univ. of Chicago Press, Chicago, 1938, pp. 7 and 382. For a defense of Reichenbach's distinction in the light of more recent developments in the philosophy of science, see W. C. Salmon's 'Bayes's Theorem and the History of Science', in R. Stuewer (ed.), *Minnesota Studies in the Philosophy of Science*, Vol. V, Univ. of Minnesota Press, Minneapolis, 1970, pp. 68–86.

[9] Karl R. Popper, *The Logic of Scientific Discovery*, Basic Books, New York, 1959, pp. 31–32.

[10] N. R. Hanson, *Patterns of Discovery*, Cambridge Univ. Press, Cambridge, 1958. N. R. Hanson, 'Is There a Logic of Scientific Discovery?', in H. Feigl and G. Maxwell (eds.), *Current Issues in the Philosophy of Science*, Holt, Rinehart and Winston, New York, 1961, pp. 20–42; N. R. Hanson, 'Retroductive Inference', in B. Baumrin (ed.), *Philosophy of Science: The Delaware Seminar*, John Wiley, New York, 1963, pp. 21–37; N. R. Hanson, 'An Anatomy of Discovery', *Journal of Philosophy* 64 (1967), 321–352.

[11] N. R. Hanson, *Patterns of Discovery*, pp. 2–3, my emphasis.

[12] N. R. Hanson, *Patterns of Discovery*, pp. 70–71.

[13] N. R. Hanson, 'An Anatomy of Discovery'.

[14] N. R. Hanson, 'An Anatomy of Discovery', p. 323.

[15] N. R. Hanson, 'An Anatomy of Discovery', p. 324.

[16] N. R. Hanson, 'Is There a Logic . . . ?', p. 33.

[17] N. R. Hanson, 'Is There a Logic . . . ?', p. 33, n. 23.

[18] This simile comparing antigens and antibodies to keys and locks is, I believe, a variant of Paul Ehrlich's original simile, which likened toxins and antitoxins to locks and keys.

[19] Instructive theories in immunology were also paralleled by instructive theories in genetics. See K. F. Schaffner, 'Logic of Discovery and Justification in Regulatory Genetics', *Studies in History and Philosophy of Science* 4 (1974), pp. 349–385, for a discussion and references.

[20] N. K. Jerne, 'The Natural Selection Theory of Antibody Formation; Ten Years Later', in J. Cairns, G. S. Stent, and J. D. Watson, (eds.) *Phage and The Origins of Molecular Biology*, Laboratory of Quantitative Biology, Cold Spring Harbor, N. Y., 1969, pp. 301–313.

[21] N. K. Jerne, 'The Natural-Selection Theory of Antibody Formation', *Proceedings of the U.S. National Academy of Sciences* **41** (1955), 849–857.

[22] N. K. Jerne, 'The Natural-Selection Theory . . . ', pp. 849–850.

[23] P. Achinstein, 'Inference to Scientific Laws', in R. Stuewer (ed.), *Minnesota Studies in the Philosophy of Science*, Vol. V, Univ. of Minnesota Press, Minneapolis, 1970, pp. 87–104; P. Achinstein, *Law and Explanation*, Oxford Univ. Press, London, 1971, esp. Ch. 6.

[24] G. H. Harman, 'The Inference to the Best Explanation', *Philosophical Review* **64** (1965), 88–95; G. H. Harman, 'Enumerative Induction as Inference to the Best Explanation', *Journal of Philosophy* **65** (1968), 529–533.

[25] *Cp*. P. Bowman, 'Comment on Achinstein', in *Minnesota Studies . . .* , Vol. V, pp. 107–109.

[26] G. H. Harman, 'The Inference to the Best Explanation', p. 89.

[27] In Chapter 2 of my forthcoming book, *Discovery and Explanation in the Biomedical Sciences*, I discuss Simon's analysis of discovery in some detail. For a collection of Simon's papers dealing with scientific discovery, see Herbert A. Simon, *Models of Discovery*, Reidel, Dordrecht, Holland, 1977, esp. Section 5.

[28] To be more accurate we should say heretofore unasserted conclusions of which the investigator (or computer – see below) is unaware (or is not in the retrievable memory).

[29] See N. J. Nilsson, *Problem Solving Methods in Artificial Intelligence*, McGraw-Hill, New York, 1971, for a good introduction to this area.

[30] See for example E. A. Feigenbaum, B. G. Buchanan, and J. Lederberg, 'On Generality and Problem Solving: A Case Study Using the DENDRAL Program', *Machine Intelligence* **6** (1971), 165–190.

[31] Feigenbaum, Buchanan, and Lederberg, p. 168.

[32] Feigenbaum, Buchanan, and Lederberg, p. 169.

[33] Feigenbaum, Buchanan, and Lederberg, p. 170.

[34] H. E. Pople, Jr., J. D. Myers, and R. A. Miller, 'The DIALOG Model of Diagnostic Logic and Its Use in Internal Medicine', June 1974, mimeo, privately circulated; H. E. Pople, Jr., J. D. Myers, and R. A. Miller, 'DIALOG: A Model of Diagnostic Logic for Internal Medicine', *Advance Papers of the Fourth International Joint Conference on Artificial Intelligence*, Vol. 2, Artificial Intelligence Laboratory, Cambridge, 1975, pp. 848–855; H. E. Pople, Jr., 'The Formation of Composite Hypotheses in Diagnostic Problem Solving: An Exercise in Synthetic Reasoning', *Advance Papers of the Fifth International Joint Conference on Artificial Intelligence*, Artificial Intelligence Laboratory, Cambridge, 1977.

[35] An independent, but similar, phase of scientific inquiry has been considered by L. Laudan and A. Grünbaum, who have termed such a phase in the analysis of budding research traditions a 'logic of pursuit' or a 'context of pursuit'. For detailed comments on the latter, see L. Laudan, *Progress and Its Problems*, Univ. of California Press, Berkeley, 1977, esp. pp. 108–114.

[36] Pople, Myers, and Miller, 'DIALOG . . . ', p. 850.

[37] Pople, Myers, and Miller, 'DIALOG . . . ', p. 850.

[38] Popler, Myers, and Miller, 'DIALOG . . . ', p. 850.

[39] Pople, Myers, and Miller, 'DIALOG . . . ', p. 850.

[40] Pople, Myers, and Miller, 'DIALOG . . . ', p. 850.

[41] This case is taken from the unpublished essay by Pople, Myers, and Miller, 'DIA-

LOG Model of Diagnostic Logic . . . . ' For a more complex, published case involving multiple diagnoses, see Pople, Myers, and Miller, 'DIALOGUE . . . ', pp. 852–854.

[42] Pople, Myers, Miller, 'DIALOG . . . ', pp. 850–851.

[43] Pople, Myers, Miller, 'DIALOG . . . ', p. 851.

[44] Pople, Myers, Miller, 'DIALOG . . . ', (my emphasis).

[45] Pople, Myers, Miller, 'DIALOG . . . '.

[46] See Note 41 for source of case.

[47] See Note 41 for source of case.

[48] The rationale for criterion used here to reach conclusion is complex, but essentially involves *doubling* the spread in terms of the sum of the weights assigned to the disease(s) on the basis of its manifestations.

[49] See Pople, Myers, and Miller, 'DIALOG . . . ', p. 853.

[50] N. R. Hanson, 'Is There a Logic . . . ?', p. 22.

[51] N. R. Hanson, 'Is There a Logic . . . ?', p. 22.

[52] G. H. Harman, 'The Inference to the Best Explanation', p. 89.

[53] F. M. Burnet, 'A modification of Jerne's Theory of Antibody Production Using the Concept of Clonal Selection', *The Australian Journal of Science* 20 (1957), 67–69.

[54] F. M. Burnet, *The Clonal Selection Theory of Acquired Immunity*, Vanderbilt Univ. Press, Nashville, 1959.

[55] See the 1967 volume of the Cold Spring Harbor meeting on 'Antibodies', with opening remarks by Burnet and a concluding summary by Jerne: in *Cold Spring Harbor Labotatory of Quantitative Biology* 32 (1967).

[56] The quotations and general account of Burnet's discovery of the Clonal Selection Theory follow his autobiography's account in *Changing Patterns*, Heinemann, Melbourne, 1968, see esp. Ch. 15. I have had to question only one minor point concerning that account on the basis of the primary literature: see below, esp. Note 67. A parallel account which adds essentially only one further point of information can be found in Burnet's introductory address, cited in Note 55 above. The additional point is a reference to work by Mackay and Gajdusek (published in 1958) noting that a case of Waldenstrom's macroglobulenemia had an extremely high titer in Gajdusek's AICF test. This also, Burnet remarks in that address, impressed "the cellular aspect of immunity on me" (p. 2).

[57] F. M. Burnet, *Changing Patterns*, pp. 203–204.

[58] F. M. Burnet, *Changing Patterns*, pp. 204–205.

[59] Burnet's 1957 formulation of the Clonal Selection Theory is as follows: "The plasma γ-globulins comprise a wide variety of individually patterned molecules and probably several types of physically distinct structure. Amongst them are molecules with reactive sites which can correspond probably with varying degrees of precision to all, or virtually all, the antigenic determinants that occur in biological material other than that characteristic of the body itself. Each type of pattern is a specific product of a clone of mesenchymal cells and it is the essence of the hypothesis that each cell automatically has available on its surface representative reactive sites equivalent to those of the globulin they produce. For the sake of ease of exposition these cells will be referred to as lymphocytes, it being understood that other mesenchymal types may also be involved. Under appropriate conditions, cells of most clones can either liberate soluble antibody or give rise to descendant cells which can.

It is assumed that when an antigen enters the blood or tissue fluids it will attach to the surface of any lymphocyte carrying reactive sites which correspond to one of its antigenic determinants. The capacity of a circulating lymphocyte to pass to tissue sites and there to initiate proliferation is now relatively well established (*cf.* Gowens, 1957; Simonsen, 1957). It is postulated that when antigen-natural antibody contact takes place on the surface of a lymphocyte the cell is activated to settle in an appropriate tissue, spleen, lymphnode or local inflammatory accumulation, and there undergo proliferation to produce a variety of descendants. In this way preferential proliferation will be initiated of all those clones whose reactive sites correspond to the antigenic determinants on the antigen used. The descendants will include plasmacytoid forms and lymphocytes which fulfill the same functions as the parental forms. The net result will be a change in the composition of the globulin molecule population to give an excess of molecules capable of reacting with the antigen, in other words the serum will not take on the qualities of specific antibody. The increase in the number of circulating lymphcytes of the clones concerned will also ensure that the response to a subsequent entry of the same antigen will be extensive and rapid, i.e., a secondary type immunological response will occur.

Such a point of view is basically an attempt to apply the concept of population genetics to the clones of mesenchymal cell within the body." (From F. M. Burnet 'A Modification of Jerne's Theory . . . ', p. 68).

[60] F. M. Burnet, *Changing Patterns*, p. 213).

[61] For a discussion of the 'self-marker' hypothesis and difficulties with it, see F. M. Burnet, *The Clonal Selection Theory*, pp. 51–56.

[62] F. M. Burnet, 'A Modification of Jerne's Theory . . . ', p. 67.

[63] See F. M. Burnet, *Changing Patterns*, pp. 91–92, for an account of his early work with the chorioallantoic membrane.

[64] See M. Simonsen, 'Graft versus host reactions', *Progress in Allergy* 6 (1962), 349–467.

[65] F. M. Burnet, *Changing Patterns*, p. 199.

[66] F. M. Burnet, *Changing Patterns*, p. 204.

[67] See F. M. Burnet, 'Cancer – A Biological Approach', *British Medical Journal* 1 (1957), 779–786 and 841–847.

[68] F. M. Burnet, 'Cancer', p. 843.

[69] F. M. Burnet, 'A Modification of Jerne's Theory . . . '. The quotations immediately following are from this essay.

[70] C. G. Hempel, *Philosophy of Natural Science*, Prentice-Hall, Englewood Cliffs, New Jersey, esp. pp. 11–18.

[71] C. G. Hempel, *Philosophy of Natural Science*, p. 12.

[72] C. G. Hempel, *Philosophy of Natural Science*, p. 12.

[73] See my forthcoming book cited in Note 27 and also my 'Theory Structure in the Biomedical Sciences', *The Journal of Medicine and Philosophy*, March, 1980.

[74] See Darden, this volume, for a discussion and references.

[75] For a discussion of some of these methodological principles utilized in theory choice, see K. F. Schaffner 'Outlines of a Logic of Comparative Theory Evaluation with Special Attention to Pre- and Post-Relativistic Electrodynamics', in R. Stuewer (ed.), *Minnesota Studies in the Philosophy of Science*, Vol. V, Univ. of Minnesota Press, Minneapolis, 1970, pp. 311–364.

[76] For a discussion of the acceptance of clonal selection theory, see my forthcoming book, cited in Note 27 above.

[77] D. Shapere, 'Scientific Theories and Their Domains', in F. Suppe (ed.), *The Structure of Scientific Theories*, Univ. of Illinois Press, Urbana, 1974, pp. 518–565.

[78] For a discussion of recent AI developments in the area of scientific discovery, see B. Buchanan's 'Step towards Mechanizing Discovery', forthcoming in K. F. Schaffner, A. Grünbaum, and L. Laudan (eds.), *Logic of Discovery and Diagnosis in Medicine*, Univ. of California Press, Berkeley. For an AI approach to analogy, see P. H. Winston's article in *Artificial Intelligence* **10** (1978), 147–172.

NANCY L. MAULL

# COMMENT ON SCHAFFNER

Schaffner makes a persuasive case for his distinction between the logic of preliminary evaluation and the logic of justification. Less convincing, I find, are his admittedly tentative claims for a logic of hypothesis generation. He bases these claims on an analysis of problem-solving machines. Indeed, he hopefully puts forward his machine examples as illustrating a limited generative logic at work. He further speculates, anticipating the inevitable objection, that a computer may even be able to construct 'new' solutions. I would like to raise two questions about the usefulness of these appeals to diagnostic machines.

1. The Lederberg machine, according to Schaffner, 'discovers' molecular structure on the basis of mass spectrometry and (if available) NMR. In fact, the computer is able to select a structure only by 'imagining' every conceivable structure and by making an 'efficient' choice. Similarly, the program for clinical diagnosis contains "a classification scheme for *all possible* diagnoses" (p. 184), among which the machine 'chooses'. In spite of complexities however, the machines remain *classifying devices*. But what can taxonomic decision-making tell us about scientific discovery? Schaffner — in the good company of Simon and Hanson — seems to find these taxonomic models illuminating. It is not clear why.

Simon himself is quite explicit about the core presuppositions of taxonomy-as-discovery. He claims that the process of discovering laws is a process of "recoding in a parsimonious fashion, sets of empirical data" (1973, p. 475). Similarly Hanson (1965), using a cue from Gestalt Theory, called the process 'pattern recognition'. The idea that discovery is a pattern recognition (or an instance of classification) is attractive, it seems to me, chiefly because several undeniably important discoveries do look (at least on the surface) like pattern recognition. As Hanson tells us, for example, Kepler found his laws by pattern recognition. The data for the orbit of Mars simply had to fit some closed curve. We now know, in fact, that a computer could have solved this problem as well as Kepler. And, of course, that does nothing to lessen Kepler's achievement.

But how much weight can a *general* theory of discovery put on such examples? Surely it is a mistake to think that all (or even most) scientific

207

*T. Nickles (ed.), Scientific Discovery: Case Studies, 207–209.*
*Copyright © 1980 by D. Reidel Publishing Company.*

discoveries arise as solutions to problems of classifying this or that entity or process according to some pre-given scheme. (Schaffner's own examples – the discoveries of Burnet and Jerne – defy such analysis.) Could it be that behind the suggestion that discovery can be understood as correct classification lies an old and honorable error? I am thinking, of course, of Aristotle's identification of systematic classification with theoretical explanation. To put it another way, there seems to be a failure here to distinguish certain domain problems – what the computers seem to be solving – from theoretical problems. (For the distinction between domain problems and theoretical problems, see Shapere, 1974.)

Reformulated, my question is this: Is Schaffner, by his use of the computer examples, suggesting only a difference in degree (rather than one of kind) between classification and explanation?

2. The computer, one notes, starts with a problem. Its initial or 'logic of generation' phase, according to Schaffner, then consists in evoking or 'calling up' one or more hypothetical solutions for preliminary evaluation. Since his machines are preprogrammed with all possible solutions, it is tempting to compare Schaffner's proposed logic of *discovery* with the Platonic doctrine of *anamnesis* – the paradoxical theory that acquiring new knowledge is really a process of *uncovering* what we already know. But that is not the parallel which I want to pursue here.

Instead, I want to call attention to a rather striking lacuna in Schaffner's account of the 'generation phase': this is the process by which the problem itself is generated and first recognized *as* a problem. Why does Schaffner slight this process? One reason perhaps is that computers do not generate their own problems. So much the worse for machines, most of us would say. And indeed, what justification can Schaffner give for initiating the discovery process with a full-fledged problem rather than with the gradual generation of the problem? In fact, it seems plausible to claim that without an account of the generation of a problem, we often have a highly foreshortened view of particular cases of discovery. Surely one good reason why Kepler's discovery is not 'merely' a case of pattern recognition or data classification is that he, and no one else, posed that particular problem about the orbit of Mars. Similarly, but in a case that not even Hanson would be tempted to label classificatory, Newton 'discovered' the problem of providing an explanation of Kepler's laws where no such problem had been recognized before. Conversely, not *every* fact inadequately accounted for by currently accepted hypotheses poses a problem. In sum, Schaffner's computer models presuppose problems. Computers never 'have' problems. Of course, Schaffner himself accepts the idea that, in his words, it is "puzzles or problems that initiate the discovery process" (p. 195). And in response to Hempel's argument that problems are too vague to provide criteria for the selection of relevant facts, Schaffner replies that "The provisional field and the series of methodological standards of a subdiscipline are . . . evoked by a problem" (p. 199). This very claim, however is the occasion for my final criticism. Schaffner can only say the *problems generate fields* because he has slighted the process whereby problems themselves are generated. If he had attended to these processes, however, he would have seen that the very generation of problems presupposes an organized and patterned field in which problems can be recognized *as* problems. (For the concept of a scientific field, see Darden and Maull, 1977.)

*Department of Philosophy*
*Yale University*

## BIBLIOGRAPHY

Darden, L. and N. Maull: 1977, 'Interfield Theories', *Philosophy of Science* 44, 43–64.
Hanson, N. R.: 1965, *Patterns of Discovery*, Cambridge Univ. Press, Cambridge.
Shapere, D.: 1974, 'Scientific Theories and Their Domains', in F. Suppe (ed.), *The Structure of Scientific Theories*, Univ. of Illinois Press, Urbana, pp. 518–565.
Simon, H. L.: 1973, 'Does Scientific Discovery Have a Logic?', *Philosophy of Science* 40, 471–480.

KENNETH F. SCHAFFNER

# REPLY TO MAULL

Professor Nancy Maull, in her trenchant comments above, raises two difficulties for the account of a logic of discovery which I have sketched. Let me
reply briefly to each of these difficulties.

1. Maull asks whether the computer examples (more accurately these
are artificial intelligence or AI examples) are not too limited to illuminate a
*general* theory of discovery. I believe that in part she misses the point of those
examples, which was to illustrate in detail the need for a logic of *generation* if
one was to have a logic of discovery in the strong sense, and also to show how
various other fine-structured stages of inquiry functioned in the discovery-
justification continuum. I stressed the problem of generating novel hypotheses
in my paper and indicated that at present computer programs are deficient in
this regard, but I also suggested in a fairly speculative manner how one might
go about formulating a logic of generation which could accommodate the
Burnet example. Clearly, an enormous number of knowledge representation
problems would first have to be solved and a program written which would
perhaps function to combine mechanisms from one field with those of
another, or subtly vary existing mechanisms using an analogy subprogram.
Whether these approaches would be very different in kind from current
approaches, or whether they would only represent a difference in degree,
is an open question which requires further elaboration of the degree-kind
distinction among other things. I do not think the question can be usefully
answered by appealing to Aristotelian (?) distinctions between explanation
and classification, which are likely to beg the very question they are supposed
to answer in this area.

2. Maull also raises another difficulty, which she develops into two criticisms. First, she wants to call attention to a rather striking lacuna "in the
generation phase", namely the process by which the problem itself is generated
and first recognized *as* a problem. Maull conjectures that "one reason perhaps
is that computers don't generate their own problems."

I think this last conjecture is beside the point, though the general criticism
is well taken. However, it is *too* general a criticism, for it seems to me that
what constitutes a problem is not one which is idiosyncratic to the context
of discovery (and to attempts to represent logical features of the discovery

211

*T. Nickles (ed.), Scientific Discovery: Case Studies, 211–212.*
*Copyright © 1980 by D. Reidel Publishing Company.*

process). Problems exist also in the context of justification; in point of fact, my colleague Larry Laudan recently published a book-length analysis oriented toward problem solving in general, and toward elucidating various subtypes of problems in particular. This is not to say that it would not be useful to develop a general analysis of problem generation, but I do not see that it specifically affects either discovery in general or my account in particular. Recall that Hanson began his account from problems; I still believe it is a useful point from which to begin an analysis of scientific discovery.

Maull's second part of her second criticism is that "Schaffner can only say the *problems generate fields* because he has slighted the process whereby problems themselves are generated" and that "the very generation of problems presupposes an organized and patterned field in which problems can be recognized *as* problems." This objection is, I think, based on an understandable misinterpretation. Initially I used the term 'provisional field' in the paper for want of a better term, and although I defined it and distinguished it from subdisciplines, I think it still has the connotations of the latter and led Maull to draw the conclusion she drew. In the final version, printed above, I have added a sentence saying that I now prefer the term 'set of provisional *explicanda*' (*SPE*) in lieu of 'provisional field'. *Why* something is seen as falling into this set *is* a topic for further analysis: here I can only suggest that it is a function of the explanatory success of earlier theories and certain empirically based clusterings of phenomena which associates the events and generalizations to be explained. Theories are rejected if they do not explain an adequate number of the *SPE*. Repeated failure to account for something which it is felt should be explained by an adequate theory in that subject area can lead to disciplinary reassignment, as when optical phenomena were reassigned to the electromagnetic domain from the mechanical. Similar reassignments occur when certain diseases are reassigned from the genetic realm to the environmental or *vice versa*. A scientist, however, begins from a set of provisional explicanda which is generally held to be the touchstone of a good theory, and sees as a *general* problem the need to provide 'explanations' of that *SPE* − a process which usually leads to specific subproblems.

WILLIAM C. WIMSATT

# REDUCTIONISTIC RESEARCH STRATEGIES AND THEIR
# BIASES IN THE UNITS OF SELECTION CONTROVERSY*

## 1. MOTIVATING REMARKS ON GENETIC DETERMINISM

> "A hen is but an egg's way of making another egg."
> Samuel Butler

Butler's satiric comment encapsulates the reductionistic spirit that made Darwinism objectionable to many in his own day, but has fared ever better as a prophetic characterization of the explanatory tenor of modern evolutionary biology. It preceded August Weismann's doctrine of the continuity of the germ plasm advanced in his inaugural lecture in 1883 by some five years. As Weismann's views became one of the anchor points of the modern 'neo-Darwinian' theory of evolution, they led to many modern recapitulations and elaborations of Butler's epigram. Thus Richard Dawkins writes (1976, p. 21):

Was there to be any end to the gradual improvement in the techniques and artifices used by the replicators to insure their own continuance in the world? . . . They did not die out, for they are past masters of the survival arts. But do not look for them floating loose in the sea . . . . Now they swarm in huge colonies, safe inside gigantic lumbering robots, sealed off from the outside world, communicating with it by torturous indirect routes, manipulating it by indirect control. They are in you and me; they created us, body and mind; and their preservation is the ultimate rationale for our existence. . . . Now they go by the name of genes, and we are their survival machines.

Thus Dawkins announces his intention to " . . . argue that the fundamental unit of selection, and therefore of self-interest, is not the species, nor the group, nor even, strictly, the individual. It is the gene, the unit of heredity" (1976, p. 12). His purple prose gives ample food for worries that the account of evolved structure and behavior in general, and social behavior in particular, will do at best 'simple justice' to its complexities and smack of genetic determinism. But his conclusions are well anchored in the dominant interpretation of the modern 'genetical theory of natural selection', as R. A. Fisher called his theory (Fisher, 1930), and are espoused by many major students of evolutionary biology, as exemplified in the works of G. C. Williams (1966), J. Maynard Smith (1975), and E. O. Wilson (1975). Dawkins has in fact been a

213

T. Nickles (ed.), Scientific Discovery: Case Studies, 213–259.

clear and ingenious expositor and elaborator of this view (1976, 1978), despite his often colorful language.

The quote from Dawkins is a direct reflection of the genetic determinism espoused by many and perhaps by most evolutionary biologists today. I take this view to involve two theses, one ontological and the other dynamic about the nature of evolutionary processes:

*T1*:        (Ontological thesis): Genes are the only significant units (or individuals) required for the analysis of evolutionary processes.

*T2*:        (Dynamical thesis): Processes at the genetic level determine (and are the primary and ultimate) explanations for processes at all higher levels.

Genetic determinism has its origins in a misconstrual of the nature of reductionism and of reductive explanation promulgated by most philosophers and by many biologists. A correct view makes it plausible, even inevitable:

(1)        that there should be a variety of significant units of selection at various level of organization, thus denying *T1*;

(2)        that understanding evolutionary processes requires the invocation and analysis of causal mechanisms and nomic regularities concerning their behavior at each of these levels of organization for the explanation of phenomena at a variety of levels, including that of the individual gene and a number of higher levels, thus denying *T2*;

(3)        that sociobiology, properly conceived, should be viewed as the incorporation of an evolutionary perspective into the analysis of processes at these levels (and for most aspects of human social evolution, invoking cultural, rather than biological evolution) rather than the replacement of sociology, psychology, anthropology and the other social sciences by an extended, genetically based ethology of the selfish gene; and

(4)        that the apparent success and power of genetic reductionist theories derives from distortions produced by cognitive biases arising from the uncritical application of a variety of reductionistic problem solving heuristics and research strategies.

In what follows I will outline the standard philosophical account of reduction and how it relates to genetic determinism (Section 2); discuss its inadequacy in handling problems of computational complexity (Section 3); show in particular how a claim of *in principle* reduction made by G. C.

Williams fails for the simplest extension to a more complex system than that of two alleles at one locus (Section 4); and then discuss at length (Sections 5 through 8) how the problem solving heuristics actually used by reductionistically inclined scientists result in systematic distortions biasing the case against recognizing the need for invoking various higher-level units of selection. A study of these biases suggests recommendations for methodological procedures which should at least partially mitigate their effects, and will in any case serve as a warning to those who must use these heuristic procedures.

## 2. THE PHILOSOPHERS' VIEW OF REDUCTION

The view I will be discussing in generically based on the model of Nagel (1961) and has been elaborated in somewhat different representative directions by Schaffner (1967, 1969, 1974, 1976), Ruse (1974, 1976), and Causey (1972a, 1972b, 1976). It has been widely criticized by a number of authors, including Hull (1973, 1974, 1976), Nickles (1973, 1976), Dresden (1974), Darden and Maull (1976), Maull (1977), Bantz (this volume), Bogaard (1979), and myself (1974, 1976a, 1976b, 1978). My (1978) is an extensive review of the literature, but probably the best self-contained systematic critical analysis of these issues is to be found in McCauley (1979, Chapters 4 and 5). Schaffner's own most recent work (1979), while not explicitly discussing reduction, seems to me to be a powerful and productive move away from his earlier position, in this direction. I mention these sources to indicate where extensive discussion of relevant issues can be found. Most of this discussion will be presupposed here.

The traditional view of reduction holds that it is the

(i)    *in principle*
(ii)   *deducibility* of upper-level entities, properties, theories, and laws
(iii)  in terms of the properties, laws, and relations *of any degree of complexity* of entities at the lower level.

I have emphasized in this characterization the three clauses which have caused the greatest problems for this traditional view. This view has two corollaries:

C1:    Upper level entities are thus shown to be 'nothing more than' collections of lower level entities (and their relations).
C2:    Upper level laws and causal relations are illusory or are shown to be 'nothing more than' a shorthand for and to be determined by lower level laws and causal relations.

These corollaries have a direct relation to the ontological and dynamical theses of genetic determinism, for *C1* and *C2* respectively provide the reasons for holding *T1* and *T2* to be true, if the relation of the various higher-level units of selection and phenomena to those at the genetic level is one of reduction, as it has been traditionally construed. *T1* is true *because* then the various higher-level units of selection are 'nothing more than' collections of genes (and their relations) in a fashion demonstrated by the deductive or definitional relations between terms, and similarly for *T2* and *C2*.

A major problem for applying the Nagelian model of reduction, one recognized by many of its defenders (see, *e.g.*, Schaffner, 1974), is that it appears not to fit the practice of reductionistically inclined scientists. As various writers have observed, if the Nagel model of reduction as a kind of deduction or its extensions is accepted as an adequate model of reduction, there may not *be* any cases of reduction in science (see Schaffner, 1974; Hull, 1973, 1976). A standard retreat in the face of this problem has been from claiming deducibility in practice, to defending a claim of deducibility or analyzability *in principle*. For this reason, I have added the 'in principle' qualifier explicitly in my characterization of reduction, since at least this modification is required to describe the actual practice of scientists.

Further problems have arisen with the characterization of the analysis of the upper level in lower level terms as a kind of deduction. Dresden (1974), Bogaard (1979), Bantz (this volume), Sklar (1973, 1976), and others have pointed to the role of approximations, which prevent this 'derivation' from being characterized as a deduction. I have emphasized (1976*b*, pp. 685–689) how the interpretation and elaboration of the implicit *ceteris paribus* clause in purported deductions makes any supposed 'translation' context-dependent in a way that undercuts the usefulness of the deductive model and falsifies corollaries *C1* and *C2*. Furthermore, not only are there problems with filling out the *ceteris paribus* clause, but, rather anomalously on this model, scientists appear to have no interest in trying to do so.

I have argued elsewhere that the 'in principle' deducibility, analyzability, or translatability is best seen not as the primary structure, focus, or thesis of reductionism but as a derivative corollary of the use of identificatory hypotheses in reductive explanation. If an upper level entity, phenomenon, *etc.* is *identified* with a lower-level complex of entities, properties, and relations, then Leibniz's law tells us that any property of one is a property of the other. Thus *Leibniz's Law* tells us that if the purported identity holds, any upper level thing must *in principle* be analyzable in lower-level terms. These 'in principle' claims thus become an important heuristic method for moving from an identity claim to specific hypotheses about heretofore unmatched properties at the upper and lower levels (see Wimsatt, 1976*a*, pp. 225–237, and 1976*b*, pp. 697–701). This heuristic use of identity claims and *in principle* arguments provides further support for the view advocated here: that *the power, limitations, and character of reductionistic approaches in science is better analyzed in terms of the reductionistic research strategies one is led to adopt than in terms of idealized deductive accounts and ontological theses derived from them.* For more on this use of identity claims, see my (1976*a*; 1976*b*, Section 3; and 1978). It will not be further elaborated here.

The use of simpler models and approximations in reductionistic modelling produces a gap between promise and performance that has interesting consequences. The metaphysical position that the reductionist defends holds that a reductionistic analysis of upper-level phenomena must exist in terms of lower level entities, properties, and relations *of some degree of complexity* – preferably in terms of monadic properties; but if not these, then at least in terms of *some* (possibly complex and relational) properties of the lower-level entities. This is another formulation of the claim of *in principle* deducibility of reducibility that I have argued is a corollary of the use of identity claims and Leibniz's Law.

The holist, as anti-reductionist, is taken normally as denying this metaphysical claim, and thus to be holding the equally metaphysical (and to most people, radically implausible) claim that no analysis of whatever complexity in lower-level terms could be adequate. But, despite appearances, the in principle claim of the reductionist is seldom in dispute. *In the cases I know in population biology, in neurophysiology, and in the history of genetics, the issue between scientists who are reductionists and holists is not over the in principle possibility of an analysis in lower level terms but on the complexity and scope of the properties and analyses required.* The more holistically inclined scientists usually argue that higher-order relational properties of the lower-level entities are required, and the reductionists argue that a given simple, lower level model (often one using only monadic properties) is adequate. To the extent that this is true, the portrayal of the dispute between reductionist and holist as over the *in principle* claim (a portrayal favored by most philosophers, and by many scientists) is seriously in error and turns a usually serious, comprehensible, and important empirical dispute into a usually one-sided and poorly motivated metaphysical one.

This reading of the dispute might seem to have the apparent disadvantage of dissolving it, for both holists and reductionists now appear to be species of reductionist – "complex" reductionists or "simple" reductionists respectively.

But this species of complex reductionist is still recognizably a holist. (See Wimsatt, 1978.) Complex relations relate several to many lower-level entities, and require the recognition of these complexes as entities. Further, if the relationships overlap in their relata, these higher-level entities become tied together in terms of still higher level systems in ways suggested by the discussion of descriptive and interactional complexity in (Wimsatt, 1974). The further presence of many-one mappings between lower and higher-level state descriptions, required even by the existence of recognizable stable higher-level phenomena generates a kind of autonomy and independence of the dynamics and the explanations at the higher level from detailed lower level specifications and laws. (See Wimsatt, 1979, Sections 4 and 5, for the discussions of 'sufficient parameters' and 'robustness'; Wimsatt, 1976a, pp. 248–251 on 'explanatory primacy'; and Wimsatt, 1976b, section 6, pp. 689–692, and the appendix, pp. 701–704, for the discussion of the 'screening off' relation of Salmon.)

The net effect of these considerations is that the holist can get the significance and autonomy of upper level entities, laws, and phenomena which he desires while accepting a kind of in principle (but 'complex') reductionism. The arguments in what follows for the significance of higher-level units of selection are to be interpreted as espousing this kind of holism.

### 3. THE PROBLEM OF COMPUTATIONAL COMPLEXITY AND THE USE OF REDUCTIONISTIC RESEARCH HEURISTICS

I claimed in the last section that the claims of 'in principle' deducibility or translatability are best seen as corollaries of Leibniz's Law, and thus as consequences of the use of compositional identities in reductive explanations. However, there is much to be learned by looking at the standard explication of these claims. Most of these seem to suppose that a claim of *in principle* translatability is to be explicated in terms of effective computability or, mirroring Laplace's definition of a deterministic system, as a translation which could be produced by a sufficiently powerful computer which was given a total state description of the micro-level of the appropriate system, together with all of the micro-level laws which applied to that system. Richard Boyd (1972) has given a brilliant criticism of this possibility in general. I wish here to make only some more pragmatic criticisms (see also my 1978).

First of all, it is unclear how a practicing scientist could make use of results concerning the effective computability of any system he is studying, since all of these results would presuppose a total knowledge of the system, which he does not possess, and a theory of that system organized in a fashion unlike any of the theories which he knows. It is all right to *talk* about writing the Schrödinger wave equation for a particular organism, but in fact physical and quantum chemists don't even do it for chemical bonding in simple molecules (see Bantz, this volume; Bogaard, 1979; and Dresden, 1974). Instead they use simple heuristic approximations even for these far simpler cases.

But suppose that one could. Does this mean that we would, or even could study systems in this way? It does not. This can be seen by studying the game of chess. Chess is a totally deterministic game. At each stage, the possible moves and their outcomes are exhaustively specifiable − and indeed relatively straightforwardly specifiable. This means that if we specify in advance how many moves we wish to allow in the game, we can in principle write down a branching tree, beginning with the initial state of the chessboard and ending with branches of nodes corresponding to all possible games of that many moves or less. (Some of the games may already have terminated in fewer moves.)

There are twenty possible opening moves (two for each of eight pawns and two knights). Suppose that this number of alternatives continues throughout the game on the average, as a geometrical mean. (This is almost certainly an underestimate.) And suppose that we wish to consider games of 100 moves (fifty pairs of moves). Then we are considering on the order of $20^{100}$ possible games. This is a large number, but clearly it is a task which is effectively computable. But that is not much consolation. Consider the size of the task: $20^{100} = 2^{100} \times 10^{100}$. Since $2^{10} \approx 10^3$, then $20^{100} \approx 10^{130}$. Now for some other relevant numbers. There are about $10^{79}$ elementary particles in the universe. There have been about $10^{19}$ seconds since the big bang. And the shortest known time for a physical event is on the order of $10^{-24}$ second, the time it takes for light to traverse the diameter of an atomic nucleus. Putting this all together, we arrive at an upper estimate of the number of events in the universe to date of about $10^{122}$. Then this task, a trivial one for a universal Turing machine, is nonetheless *not* doable by the most universal computer we could imagine − the universe as a computer! It would fall 8 orders of magnitude short of having had enough *actual* states (as opposed to possible ones) to represent all of these games, and we have not even raised the questions of how these games could be mapped into the states in a usable manner and how rapidly the different parts of a computer spanning $10^{10}$ light years will be able to communicate with one another! Clearly, this effectively computable (and therefore, to many logicians and mathematicians, *in principle possible*) task is, *physically speaking, in principle impossible*.

This kind of observation led Herbert Simon and others since to look for other kinds of models than the exhaustive, brute force algorithmic approach for human problem solving. First, for decision making (Simon, 1957), then for proving theorems (Newell, Shaw, and Simon, 1958), and subsequently for other problems (Newell and Simon, 1961; Simon, 1966a, 1966b, 1969, 1973), Simon espoused a 'principle of bounded rationality' (Simon, 1957, pp. 198–199) which asserts that we are generally faced with problems of such complexity that we cannot solve them exactly, and therefore, if we are to get any solutions at all we must do so by introducing various simplifying and approximative techniques. Thus was born the idea of a heuristic.[1] As I use that notion here, I take a heuristic procedure to have three important properties:

(1)     By contrast with an algorithmic procedure, the correct application of a heuristic procedure does not guarantee a solution; and if it produces a solution, does not guarantee that the solution is correct. Thus valid deduction from true premises is not a heuristic procedure. Most or all inductive procedures are, however (see Shimony, 1970).

(2)     The expected time, effort, and computational complexity of producing a solution using a heuristic procedure is appreciably less than that expected using an algorithmic procedure. This is indeed the reason why heuristics are used. They are a 'cost-effective' way of producing a solution, and often the only physically possible way.

(3)     The failures and errors produced using a heuristic are not random, but systematic. I conjecture that *any heuristic, once we understand how it works, can be made to fail.* That is, given this knowledge of the heuristic procedure, we can construct classes of problems for which it will always fail to produce an answer, or for which it will always produce the *wrong* answer. This property of systematic production of wrong answers will be called the *bias(es)* of the heuristic.

Not only can we work forward from an understanding of a heuristic to predict its biases, but we can also work backwards, hypothetically, from the observation of systematic biases as data to conjecture as to the heuristic which produced them; and if we can get independent evidence as to the nature of the heuristics, we can propose a well-founded theory of the structure of our heuristic reasoning in these areas. This was elegantly done for the first time by Tversky and Kahneman (1974), in their analysis of fallacies of probabilistic reasoning and the cognitive heuristics which produce them. To my mind, Simon's work and that of Tversky and Kahneman have opened up a whole new set of questions, a new area of investigation of pragmatic inference in science, which should revolutionize our discipline in the next decade, and increasing numbers of workers are moving in this direction. (See, for example, the papers of Schaffner, Darden, and Bantz in this volume.)

The notion of a heuristic has far greater implications than can be explored in this paper. In addition to its centrality in human problem solving, it is a pivotal concept in evolutionary biology and in evolutionary epistemology. It is a central concept in evolutionary biology because any biological adaptation meets the conditions given for a heuristic procedure. First, it is a commonplace among evolutionary biologists that adaptations, even when functioning properly, do not guarantee survival and production of offspring. Secondly, they are, however, cost-effective ways of contributing to this end. Finally, any adaptation has systematically specifiable conditions, derivable through an understanding of the adaptation, under which its employment will actually *decrease* the fitness of the organism employing it, by causing the organism to do what is, under those conditions, the wrong thing for its survival and reproduction. (This, of course, seldom happens in the organism's 'normal' environment, or the adaptation would become maladaptive and be selected against.) This fact is indeed systematically exploited in the functional analysis of organic adaptations. It is a truism of functional inference that learning the conditions under which a system malfunctions, and how it malfunctions under those conditions, is a powerful tool for determining how it functions normally and the conditions under which it was designed to function. (See, *e.g.*, Gregory, 1967; Lorenz, 1965; Valenstein, 1973; and Glassman, 1978, for illuminating discussion of the problems, techniques, and fallacies of functional inference under a variety of circumstances.)

The notion of a heuristic is central to evolutionary epistemology, because Campbell's notion of a 'vicarious selector' (1974, 1977), which is central to his conception of a hierarchy of adaptive and selective processes spanning subcognitive, cognitive, and social levels, is that of a heuristic procedure. A vicarious selector for Campbell is a (1) substitute (2) less costly selection procedure acting to optimize some index which is only contingently connected with the index optimized by the selection process it is substituting for. This contingent connection allows for the possibility — indeed the inevitability — of systematic error when the conditions for the contingent concilience of the substitute and primary indices are not met. An important ramification of

Campbell's idea of a vicarious selector is the possibility that one heuristic may substitute for another (rather than for an algorithmic procedure) under restricted sets of conditions, and that this process may be repeated, producing a nested hierarchy of heuristics. I believe that this is an appropriate model for describing the nested or sequential structure of many approximation techniques, limiting operations, and the families of progressively more realistic models found widely in 'progressive research programs', as exemplified in the development of 19th century kinetic theory, early 20th century genetics, and in several areas of modern population genetics and evolutionary ecology. (On this last, see, *e.g.*, Roughgarden, 1979.)

The ultimate end of this paper is to discuss some of the heuristic used in reductionistic modelling and to show how their systematic biases have given illegitimate support to a reductionistic vision of evolutionary processes culminating in genetic determinism. But before I do this, it is necessary to show that, how, and why a brute force, quasi-algorithmic, reductionistic approach cannot work in evolutionary biology and population genetics, for just such an approach has been suggested by G. C. Williams.

### 4. WILLIAMS'S 'IN PRINCIPLE' REDUCTIONISM AND THE CASE OF TWO LOCI

A problem-solving heuristic which Simon (1966) has called 'factoring into subproblems' appears in a variety of guises in reductionistic modelling. Simon illustrates the heuristic and its advantages using the problem of finding the right combination for a combination lock. Imagine a bicycle lock with ten wheels of ten positions each. If there is only one combination which will work, one would expect to look through about half of the possible $10^{10}$ combinations on the average before finding it. On the other hand, suppose that the lock is a cheap or defective one for which one can tell individually for each wheel when it is in the right position. Then an average of 5 tries on each wheel, for a total of 50 tries would be expected to find the right combination. The advantage that accrues from being able to break the problem down into subproblems, being able to find out parts of the combination, rather than having to solve the whole problem at once, is given by the ratio of the number of alternatives which must be inspected. This is, in this case, $(5 \times 10^9)/(5 \times 10) = 10^8$.

Similar advantages accrue for similar combinatorial reasons if problems of evolutionary dynamics can be treated in terms of the frequencies of individual alleles, with no epistatic interactions and no probabilistic associations between alleles at different loci due to linkage or assortative mating rather than in terms of the gametic or zygotic genotype frequencies required if these assumptions do not hold. Here the simplification occurs in the number of dimensions in the phase space required to adequately describe and predict evolutionary changes, and, correlatively, in the number of state variables in the equations required to describe and predict the dynamics of evolutionary change. Table 1, derived and extended from Table 56 of Lewontin (1974, p. 283), summarizes the dimensionality of the problem under different simplifying assumptions. It is worth noting that if *no* simplifying assumptions are

TABLE 1

Sufficient dimensionality required for the prediction of evolution of a single locus with *a* alleles where there are *n* segregating loci in the system

| Level of Description: | | Zygotic Classes | Gametic Classes | Allele Frequencies | Allele Frequencies |
|---|---|---|---|---|---|
| Dimensionality: | | $\dfrac{a^n(a^n+1)}{2}-1$ | $a^n-1$ | $n(a-1)$ | $(a-1)$ |
| Assumptions: | | none | 1 | 1, 2 | 1, 2, 3 |
| *n*: | *a*: | | | | |
| 2 | 2 | 9 | 3 | 2 | 1 |
| 3 | 2 | 35 | 7 | 3 | 1 |
| 3 | 3 | 377 | 26 | 6 | 2 |
| 5 | 2 | 527 | 31 | 5 | 1 |
| 10 | 2 | 524799 | 1023 | 10 | 1 |
| 32 | 2 | $9.22 \times 10^{18}$ | $4.29 \times 10^9$ | 32 | 1 |

*Assumptions*:
  (1)  random union of gametes (no sex linkage, no assortative mating)
  (2)  random statistical association of genes at different loci (linkage equilibrium).
  (3)  no epistatic interaction (inter-locus effects are totally additive).

(Table is adapted and extended from Table 56 of Lewontin, 1974, p. 283.)

made, even the simplest multi-locus case of two alleles at each of two loci is analytically intractable. This should not be surprising: the problem of dimensionality nine (there are nine possible genotypes, with independently specifiable fitness parameters) is already more complicated than the three-body problem of classical mechanics. Like the three-body problem, it has been solved for a variety of special case (see Roughgarden, 1979, Chapter 8, pp. 111–133) but has not been solved in general.

In the light of this, G. C. Williams makes a claim which gives substantial hope, for it appears to promise that the problem can be treated as one of the lowest dimensionality. His view is that since the operation of any higher-level selection processes can be mathematically expressed as resulting from the operation of selection coefficients acting independently at each locus to change the frequency of individual alleles or genes, there is no need to postulate the existence of any higher-level units of selection or selection forces. This view will look most familiar to philosophers, since it bears the strongest resemblance to traditional philosophical accounts of theory reduction. Williams expresses it as follows:

> Obviously it is unrealistic to believe that a gene actually exists in its own world with no complications other than abstract selection coefficients and mutation rates. The unity of the genotype and the functional subordination of the individual genes to each other and to their surroundings would seem, at first sight, to invalidate the one-locus model of natural selection. Actually these considerations do not bear on the basic postulates of the theory. No matter how functionally dependent a gene may be, and no matter how complicated its interactions with other genes and environmental factors, it must always be true that a given gene substitution will have an arithmetic mean effect on fitness in any population. One allele can always be regarded as having a certain selection coefficient relative to another at the same locus at any given point in time. Such coefficients are numbers that can be treated algebraically, and conclusions inferred from one locus can be iterated over all loci. Adaptation can thus be attributed to the effect of selection acting independently at each locus. (Williams, 1966, pp. 56–57)

Williams goes on, in the next two pages, to illustrate how this algebraic manipulation can be accomplished in a simplified genetic environment of two alleles at each of two loci, and we are to imagine the extrapolation to cases of many alleles at many loci. Complicated it would be, but *in principle*, of course (we are told), it could be done, "by iterating over all loci."

This claim might appear to involve another variant of the 'factoring into subproblems' heuristic which Simon has studied and written upon at length, and which he has called 'the hypothesis of near decomposability' (see Simon, 1969, pp. 99ff, in Ando *et al.*, 1963, and further references given there; and also Wimsatt, 1974, for further discussion).

The hypothesis of near decomposability involves the assumption that a complex system can be decomposed into a set of subsystems such that all strong interactions are contained within subsystems' boundaries, and interactions between variables or entities in different subsystems are appreciably weaker than those relating variables or entities in the same subsystem. In this case, an approximation to the behavior of the system, in the short run, can be gotten by ignoring the intersystemic interactions and analyzing each subsystem as if it were isolated, studying only internal variables in their common approach to equilibrium. Its behavior in the long run can be approximated by ignoring the intra-systemic interactions of the subsystems (assuming that there is intra-systemic equilibrium), representing each subsystem by a single index, and considering the equilibration of the various subsystems with one another as a system involving the interaction only of these lumped index variables.

There are thus two different approximations involved in studying the short-run and the long run behavior of the system. Each substantially reduces the complexity of the problem, if the assumptions allowing the approximation are justified.

Indeed the hypothesis of near decomposability *is* used in this way in a number of multi-locus models — in particular when it is assumed: (1) that the system starts at or near linkage equilibrium. (This is the condition when all genotypes occur at frequencies given by the products of the frequencies of their constituent genes, a condition equivalent to the assumption of a multi-locus Hardy-Weinberg equilibrium.) (2) That selection between genotypes is relatively weak (a condition that guarantees that the population never deviates far from the multi-locus Hardy-Weinberg equilibrium). Indeed, these assumptions (as well as that of random association of gametes, implying no assortative mating) are made in the original model (Lewontin and White, 1960) for which the two-locus fitness surfaces, which provide below a counterexample to Williams's claim, originally were derived. Under these conditions, recombination can be neglected as a significant contributor to genotype frequencies, and the dynamics can be treated as if they are affected by segregation and selection only. This is equivalent to using the 'long range' approximation in studying the behavior of a nearly decomposable system, since if the system is far from linkage-equilibrium, recombination may be a far greater contributor to genotype frequency of some genotypes than either segregation or selection, and thus behaves like an intra-systemic 'strong' interaction which goes relatively rapidly to equilibrium.[2] The observation that under some conditions there can be permanent and substantial linkage disequilibrium (see Lewontin, 1974; Roughgarden, 1979; and also Maynard Smith, 1978, Chapter 5) is equivalent then to saying that the system cannot be treated as nearly decomposable.

In fact, however, Williams's claim in the above quote appears to be far stronger than a near decomposability claim and is not made on the basis of these assumptions about linkage equilibrium and random assortment of gametes produced by random mating. He claims that the problem can be solved one locus at a time and then extended to a global solution by "iterating over all loci." His claim is thus not that the genetic system is nearly decomposable, but that it is simply decomposable, like the simpler of Simon's two locks. Without all of the qualifications in Lewontin's table, this claim is simply incorrect, and can be shown to be so for the simplest case involving more than one locus — that of two alleles at each of two loci. The reasons for the failure of Williams's claim can best be seen after a discussion of this case.

A claim that evolutionary processes can be analyzed in the manner Williams suggests, as being of the lowest possible dimensionality, involves at least the claim that a deterministic theory of the change of gene frequencies at a given locus can be constructed using only the frequencies of the alternative alleles of that locus. In the simplest case of two alleles at one locus, this involves saying that it is a function only of the frequency of a single gene since if $q$ is the frequency of gene $a$, then $1-q$ must be the frequency of the other gene, $A$ because there are not other genes at that locus. (It is a function also of the fitnesses, $W_{11}$, $W_{12}$ and $W_{22}$ of the genotypes $AA$, $Aa$, and $aa$, but these are assumed to be constant parameters of the system in this discussion.)

Consider Figure 1 as a graph of gene frequency from different initial points (.05 for the bottom curve, .95 for the top curve) as it changes in successive generations. If this were the graph of an actual case (Lewontin describes it as of a "hypothetical laboratory population") it would falsify Williams's claim. Why? Consider the topmost curve. At all points between the initial high value of gene frequency of .95 and the minimum value (of about.7, reached in generation 4) a population which is decreasing in gene frequency at that value (between generations 0 and 4) is later increasing in gene frequency at that value (in generations 5 and later). *But if gene frequency can either increase or decrease from a given value, then gene frequency* (of that gene or its allele) *alone is not an adequate basis for a deterministic theory of evolutionary change.*

Williams gets into trouble at this point because his claim is neither a theory of evolution in terms of gene frequencies, nor even a schematic description of the form of such a theory. His statement that " . . . it must always be true that a given gene substitution will have an arithmetic mean effect on fitness in any population" (1966, p. 56) suggests the following procedure for evaluating this effect on fitness. Imagine a gigantic (non-interventive) DNA sequencer that, given a population, will determine all of the genes in that population and their frequencies. Perform this genetic census at two points in time — or

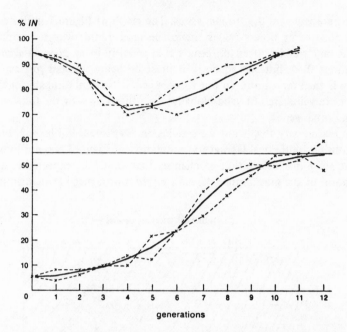

Fig. 1. The frequency of an inversion, IN, in hypothetical laboratory populations. The heavy lines represent the average behavior of replicates, while the *X*'s represent individual data points. (Reprinted from Lewontin, 1974, Figure 23, p. 274, with permission of Columbia University Press.)

perhaps in each generation. For each gene, its frequency in the interval will either increase (in which case it is being selected for), decrease (it is selected against), or remain constant (it is neutral).

This data can then be used in one of two ways. It all can be used to describe the evolutionary trajectory of the population in its phase space. But then this is not a *theory* of evolutionary change but a description. The fitnesses, $W_{11}$, $W_{12}$, and $W_{22}$ inferred from this are merely biological redescriptions of what is happening in successive generations[3] and may undergo arbitrary changes as the 'curve-fitting' parameters that they are. Or the changes observed in one generation may be used to estimate fitness values which are then used to predict future changes. This is more of a process of trend extrapolation using an assumed model rather than a theory itself, but it is at least not totally tautological. To have a predictive tool or theory then, Williams must intend his remarks to describe a process of trend extrapolation.

But here is where the trouble arises. The graph of Figure 1 indicates that local estimates of fitness values *cannot* be used in this way to extrapolate evolutionary trends. After all, gene *a* is apparently being selected *against* in generations 0–4, but subsequently it must be being selected *for*, as it frequency is then *increasing*. To put it more generally, local estimates of fitness or selective value are not valid globally, for other values of the frequency of that and other genes.

The reason why this is not the case becomes apparent in Figure 2. Indeed, Lewontin's hypothetical laboratory population of Figure 1 was not hypothetical at all, but a description of changes that would be expected in a field population of the grasshopper, *Moraba scurra*, whose mean Darwinian fitness

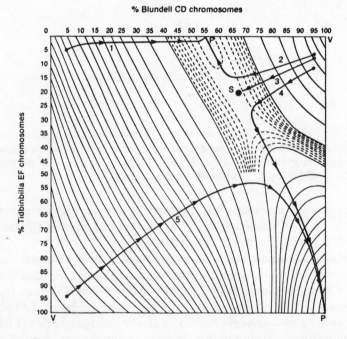

Fig. 2. Projected changes in the frequency of two polymorphic inversion systems in *Moraba scurra* from different initial compositions, based of fitness estimates from nature. The trajectories, shown by arrow-marked lines, are calculated by the solution of differential equations of gene frequency change. Lines crossing the trajectories are contours of equal mean population fitness, $W$. (Reprinted from Lewontin, 1974, Figure 24, p. 280, with permission of Columbia University Press.)

$\bar{W}$, is given as a function of the frequency of two alleles of each of two loci in the adaptive topography of Figure 2. An adaptive topography is a plot of contours of equal mean population fitness, $\bar{W}$, as a function of the gene frequencies (in this case, at two loci). Since in many simpler models (particularly where the genotypic fitnesses are constant) a population will tend to evolve in directions of increasing $\bar{W}$, the adaptive topography gives a visual means of making qualitative predictions about the direction and relative rates of local evolutionary change.

Indeed, the lower curve of Figure 1 is just trajectory 1 of Figure 2, and the upper (problematic) curve of Figure 1 is trajectory 4 of Figure 2. And the results that appeared as indeterministic in terms of the trajectory of gene frequencies of a single locus are seen to be deterministic once the frequencies of each of two loci are specified. Thus, the initial points of trajectories 1 and 4 should not have been specified as .05 and .95, the frequency of a gene at the first locus, as is implicit in Figure 1, but as (.05, .05) and as (.95, .12), the frequencies of the genes at both loci as in Figure 2. Note also that trajectory 2, with initial point (.95, .07) − in which the frequency at the first locus is the same as for trajectory 4, but that at the second locus is different − also shows a violation of the deterministic assumption if only the first locus frequency is looked at when the two trajectories are compared. Trajectory 2 is not by itself evidence of a violation, as is trajectory 4, however.

It is quite clear from this adaptive topography that what will happen in evolution is a function of the joint values of gene frequency at two loci, and no set of measurements or extrapolations looking at frequencies of just one locus at a time can provide an adequate basis for prediction. This is true in this case because of epistatic interactions between loci, which is a sufficient condition for having to go to a phase space of greater dimensionality for prediction.

Williams's proposal, then, fails in this case, in 3 ways:

(1)     It does not result in a deterministic theory of evolutionary change in terms of the gene frequencies of individual loci which can be "iterated over all loci" to produce a global solution.

(2)     It fails to do so because epistatic interactions among loci prevent local estimates of fitness at single loci from being projectable or extrapolatable *if gene frequencies at other loci are free to change simultaneously*. (The appearance of projectability usually arises when estimates are only done locally under conditions in which there *is* no change or no significant change at other loci. But this

cannot be assumed in general.) Williams in effect errs by assuming that single locus fitnesses are independent of context, when in fact they are functions of the context of other loci. Illegitimate assumptions of context-independence are a frequent error in reductionistic analyses. See Wimsatt, 1976, p. 688 and 1980 for further discussion.

(3)     In fact, Lewontin's data on *Moraba scurra* represents not gene frequency changes at single loci, but the frequencies of chromosomal inversions involving *many* loci. For reasons which I will not detail here, inversions can often act as units of selection, and Lewontin has devoted much of the earlier portions of his book to arguing that Williams's aim of measuring the fitness effects of single gene substitutions is bedevilled with a host of practical and theoretical problems. So Lewontin's one chromosome example of Figure 1 and the two chromosome counterexample of Figure 2 are already at a higher level of organization than that supposed by Williams's single locus genetic reductionism.

What goes for two loci or chromosomes, goes as well for many. In this light, Williams's remarks suggesting genetic reductionism are better seen as having more import as a kind of genetic bookkeeping than as promising a reductionistic theory of evolutionary change in terms of gene frequencies. The latter is a tempting mirage which vanishes upon closer inspection of the complexities and heuristics of the actual theory.

## 5. A GENERAL CLASS OF REDUCTIONISTIC RESEARCH HEURISTICS AND THEIR BIASES

The kind of mistake implicit in the reductionistic argument discussed in the preceding section seems so straightforward, yet is so pervasive and is made by so many leading practitioners of the discipline that it cries out for a deeper explanation. Any engineer knows that systematic failures in a mechanism indicate a design problem, which he then tries to locate, and eliminate. This procedure is so important (and so often ignored in the traditional education of engineers) that at least one company holds seminars for the engineers of its various divisions to teach approaches and methods for doing it, as described in Moss (1979). The heuristics in our reasoning processes have similar possibilities for systematic error, and we should similarly try to analyze these failures to get an understanding of where they are likely to occur and, where possible, eliminate or moderate their effects through redesign. If redesign (through teaching different or modified heuristics) is impossible or impractical, we can at least, through an understanding of the causes of failure, be warned when they are likely to lead us into error, so that our troubleshooting efforts may be concentrated there.

I will describe here a general class of heuristics and their biases which have a common origin in the nature of reductionistic analyses, thus providing the appropriate warnings. After defining the notion of unit of selection, I will describe some of the results of Wade's review of the models of group selection (1978) which nicely illustrates several of these biases with systematic failures in the literature. I will then discuss a design modification in our heuristic which should serve to eliminate or at least moderate the effects of several of these biases.

(1) Any analysis of a system presupposes a division of the world, however tentatively, into the system being studied and its environment. This division may be made on grounds of interest, which will in turn often be determined by judgments of the scope of one's field (a molecular geneticist is unlikely, at least initially, to consider social forces as part of the subject matter of his discipline), other jurisdictional criteria, and probably most frequently, intuitive judgments about the natural chunks and boundaries in his area.[4] Judgments of what can be manipulated relatively independently of 'outside' forces are likely to enter into any of these, and this in turn implies judgments of near decomposability or near isolatability in the individuation of systems.

(2) A reductionist adds to this a further consideration: by his description as a reductionist, he is interested in understanding the behavior of his system in terms of the interaction of its parts. This means that his *interest* at least (though not necessarily his scope of investigation) will be focussed on the entities and interrelations between them *internal* to the system he is studying.

(3) The third and last constraint is to recognize the practical impossibility of generating an exhaustive, quasi-algorithmic, or exact analysis of the behavior of the system in its environment. This is an application of Simon's 'principle of bounded rationality' discussed earlier. So the reductionist must start simplifying. In general, simplifying assumptions will have to be made everywhere, but given his interest in studying relations *internal* to the system, he will tend to order his list of economic priorities so as to simplify first and more extremely in his description observation, control, and analysis of the environment than in the system he is studying. After all, simplifications internal to the system face the danger of simplifying out of existence the very phenomena and mechanisms he wishes to study.

This fact alone, derived just from these three very general assumptions, is sufficient to generate and explain a wealth of heuristics and their attendant biases arising in the reductionistic analysis of systems. These heuristics and biases can be classified roughly as biases of conceptualization, biases of model building and theory construction, and as biases of observation and experimental design, though any rigid classification would fail because of the interdependence and intercalation of these activities in the course of a scientific investigation.

I will here describe them only relatively cursorily, leaving their further elaboration for other occasions. (I have already discussed item 2 in 1976*a*, pp. 244–245, and in 1978. Extensive discussions of items 1, 3, and 4 are forthcoming in Wimsatt, 1980. Wade's work, and the work I have done so far in population biology relates most strongly to items 4, 5, and 6, though 1, 7, 8, and 9 are also implicated.) This is possible because even a statement of the heuristic naturally suggests the pervasiveness of their use and multiplicity of their possible effects, and it is in any case necessitated by space limitations.

These heuristics and/or biases are as follows:

A. *Conceptualization*:

1.    *Descriptive Localization*. Describe a relational property as if it were monadic, or a lower order relational property; thus, *e.g.*, fitness as a property of phenotype (or even of genes) rather than phenotype-environmental relation.

2.    *Meaning Reductionism*. Assume lower level redescriptions to change meanings of scientific terms; higher level redescriptions not. Result: philosophers (who view themselves as concerned with meaning relations) are inclined to a reductionistic bias.

3.    *Interface Determinism*. Assume that all that counts in analyzing the nature and behavior of a system is what comes or goes across the system-environment interface. This has two versions: (a) Black-box behaviorism: all that matters about a system is how it responds to given inputs. (b) Black-world perspectivalism: all that matters about the environment is what comes in across the system boundaries and how it responds to system inputs. Either can introduce reductionistic biases when conjoined with the assumption of white box analysis . . . that the order of study is from a system, with its input-output relations, to its subsystems, with theirs, and so on. The analysis of functional properties in particular, is rendered incoherent and impossible by these assumptions.

B. *Model Building and Theory Construction*:

4.    *Modelling Localization.* Look for an intrasystemic mechanism to explain a systemic property, rather than an intersystemic one. Structural properties are regarded as more important than functional ones, and mechanisms as more important than context.

5.    *Simplification.* In reductionistic model building, simplify environment before simplifying system. This strategy often legislates higher level systems out of existence or leaves no way of describing systemic phenomena appropriately.

6.    *Generalization.* When starting out to improve a simple model of system, environment: focus on generalizing or elaborating the internal structure at the cost of ignoring generalizations or elaborations of the external structure.

   *Corollary.* If the model doesn't work, it must be because of simplifications in description of internal structure, not because of simplified descriptions of external structure.

C. *Observation and Experimental Design*:

7.    *Observation.* Reductionist will tend not to monitor environmental variables, and thus will often tend not to record data necessary to detect interactional or larger scale patterns.

8.    *Control.* Reductionist will tend to keep environmental variables constant, and will thus often tend to miss dependencies of system variables on them. ('*Ceteris paribus*' is viewed as a qualifier on environmental variables.)

9.    *Testing.* Make sure that a theory works out only locally (or only in the laboratory) rather than testing it in appropriate natural environments, or doing appropriate robustness analyses to suggest what are important environmental variables and/or parameter ranges.

These heuristics and their biases can be particularly powerful for two reasons: (1) There is, on the face of it, no way to correct for their effects — at best not in the most obvious way by producing the exact and general analysis of the behavior of the system in its environment to use as a check against the models produced. That may be all right in theory, but it won't work out in practice, as I have tried to suggest in Section 3. Nonetheless, there are at least two possible corrective measures which will be discussed later. The first is robustness analysis — a term and procedure first suggested by Richard Levins in his (1966). The second, which I will call 'multi-level reductionistic analysis' involves using these heuristics simultaneously at more than one level of organization — a procedure which allows discovery of errors and their correction in at least some circumstances, and which in fact implicitly followed as a species of 'means-end analysis' (see Simon, 1966) in the construction of interlevel theories involving compositional identities (see Wimsatt, 1976*a*, pp. 230–237, and 1976*b*, Section 8).

(2) Secondly, it should be clear that *these heuristics are mutually supporting*, not only in their effective use in structuring and in solving problems, but also *in reinforcing, multiplying, and above all, in hiding the effects of their respective biases*. This effect of bias amplification is very serious, and one of the biggest reasons why the effects of these biases are so hard to detect and why the proponents of extreme reductionistic positions can be so resistant to recognizing potential counterexamples to their position. Whatever can be said for theories or paradigms as self-confirming entities (and much that has been said is too excessive, and would render progressive science impossible), as much and perhaps more can be said similarly for heuristics. Indeed, I suspect that most of the blame and criticism of theories in this regard is more accurately laid at the doorstep of the heuristics used by those applying these theories and extending these paradigms.

Consider how this could work. Heuristics 1 and 4, applied in the early stage of an investigation, give apparent conceptual and theoretical reasons for locating a phenomenon of interest (say, that an organism has a given fitness in a given environment) as having causes primarily or wholly within the system under study. In the process of model building, the environment may be simply described (*e.g.*, as totally constant in space and in time, a frequent assumption in population genetics) that relevant variables (and the possibility of their variation) are ignored (Bias 5), further leading to and being reinforced by tendencies not to observe variation in relevant environmental variables (Bias 7) and to make efforts to assure (or, too often, merely to assume) that they are constant as controls in the experimental analysis of the system (Bias 8). Any failures in the model are then assumed to be caused by a failure to model intrasystemic interactions in sufficient detail (Bias 6), leading to another cycle, beginning with biases 1 and 4 applied to the properties and phenomena which were anomalous for the first model. This may result in further simplifications in the environment to offset the loss in analytical tractability arising from the increased internal complexity now assumed, or it may result in focussing in on a particular subsystem to be modelled in further detail, with much of the rest of the system now becoming part of the systematically simplified, ignored, and controlled environment. If even a part of this scenario or one like it is correct, we should not be surprised if quite remarkable failures went undetected for appreciable lengths of time. At present this remains just a hypothetical scenario, probably only one of many possible scenarios for producing this result. It would be very difficult to establish that the whole scenario, or one like it, was played out in any given case, in part because of the practice of not describing chains of hypothetical reasoning or discovery in scientific papers. Moreover, the practitioners are usually not themselves aware of the microstructure and background presuppositions of their reasoning processes, a fact which has bedevilled attempts to use protocols in which experimental subjects try to describe their reasoning processes as a basis for constructing theories of problem-solving behavior even for much simpler tasks (see Newell and Simon, 1972). Nonetheless, the scientific literature does contain suggestive evidence of several of these heuristics in operation, and it could be hoped that future research would turn up more. A remarkable example of cumulative and systematic biases was unearthed by the work of Michael Wade on the models of group selection, which will be discussed after some preliminary discussion of the notion of a unit of selection in the next section.

### 6. DARWIN'S PRINCIPLES AND THE DEFINITION OF A UNIT OF SELECTION

Charles Darwin's argument in *The Origin of Species* is adumbrated[5] by R. C. Lewontin (1970, p. 1) as a scheme involving three essential principles:

1. Different individuals in a population have different morphologies, physiologies, and behaviors (*phenotypic variation*).
2. Different phentoypes have different rates of survival and reproduction in different environments (*differential fitness*).
3. There is a correlation between parents and offspring in the contribution of each to future generations (*fitness is heritable*).

Where (and while)[6] these three principles hold, evolutionary change will occur. Lewontin argues not only that these requirements are necessary for evolution to occur, but also that they are sufficient. They also embody that is generally regarded as Darwin's major contribution over prior evolutionists in that they specify a mechanism, natural selection, which produces this change.

Mechanism or not, these principles specify very little about the units which must meet these conditions. Although they are specified in terms of phenotypes and their properties (a form appropriate to Darwin's original theory, and one to which modern evolutionists still pay lip service), Lewontin immediately applies them to genes (the units of the neo-Darwinian theory, under the impetus of Weismannism). Lewontin exploits the fact that these requirements say little about the units which must meet them, to argue that selection can operate — simultaneously and in different directions — on a variety of units (the unspecified individuals) at a number of levels of organization. In his view, he discusses selection processes at the micro- and macro-molecular levels, and as operating on cell organelles, cells (in the immune system, in developmental processes, and, he could have added, in cancer), gametes, individual organisms, varieties of kin groups, populations, species, and even ecological communities.

These principles give necessary conditions for an entity to act as a unit of selection, as well as necessary and sufficient conditions for evolution to occur. The three conditions must all be met by the same entity, in a way that can be summarized by saying that entities of that kind must show *heritable variance in fitness.*[7]

These conditions fail to be sufficient for the entity to be a unit of selection, however, for they guarantee only that the entity in question is either a unit of selection *or is composed of units of selection.* A further condition, which is sufficient, is given in the following definition:

A *unit of selection* is any entity for which there is heritable *context-independent* variance in fitness among entities at that level which does not appear as heritable context-independent variance in fitness (and thus, for which the variance in fitness is *context-dependent*) at any lower level of organization.

Much of population genetic theory involves the notion of additive variance in fitness. It is this quantity which, in Fisher's fundamental theorem of natural selection (Fisher, 1930) determines the rate of evolution. To say that variance in fitness is totally additive is to say that the fitness increase in a genotype is a linear function of the number of genes of a given type present in it. But this entails that the contribution to fitness of a given gene whose effect on fitness is totally additive is independent of the genetic background in which it occurs, which is to say that the variance in fitness is context-independent. Additivity is thus a special case of context-independence. It is assumed for reasons of analytical tractability, but the properties which flow from this assumption derive from its relation to context-independence.

One very important result follows when this assumption holds at a given level of organization. *If variance in fitness is totally additive at a given level of organization over a given range of conditions on the environment and the system, then, under those conditions there are no higher-level units of selection*! This is true because fitness of any higher level unit is then a totally aggregative or mass effect of the fitnesses of the individual entities at that level of organization. With no context-dependence of fitness, the *organization* of these units into higher-level units does not matter. There are no epistatic interactions to tie complexes of these entities together as units of selection. The higher level unit is totally reducible in its effects to the action of various lower level units, acting in a context-independent manner.[8]

It may be that this assumption (a product of bias 4 or 5 applied at the level of the gene to increase the analytical tractability of the model) is one of the major reasons contributing to the plausibility of Williams's reductionistic vision. It is clear that once this assumption is made, it becomes plausible to attribute adaptation (and thus fitness) "to the effect of selection acting independently at each locus" (Williams, 1966, p. 57) and leads naturally to regarding fitness as a property of genes (a case of bias 1). It is also true that many or most population geneticists believe and argue (as James Crow has, in personal conversation) that most variance in fitness is additive – presumably, at the level of the contributions of individual genes. This is an empirical claim and represents a view not shared by all population geneticists. Sewell Wright has systematically argued throughout his professional life and his magisterial four-volume treatise that the opposite is true, that epistatic interactions are all pervasive and important (personal conversation; see *e.g.*, Wright, 1968, Chapter 5, especially pp. 71–105). Michael Wade's current research indicates the importance of epistatic interactions at the *individual* level (that is, between individuals in populations) in group selection (personal conversation; see Wade and McCauley, 1979, and McCauley and Wade, 1979). What is clearly true is that biases 7, 8, and 9 would in general contribute substantially to failures to detect nonadditive variance if it exists because of artificially induced constancies in or ignorance of environmental conditions capable of producing nonadditive components of variance in fitness.

To summarize then, if variance in fitness at a given level is totally additive, the entities of that level are composed of units of selection, and there are no higher level units of selection. If the additive variance in fitness at that level is totally analyzable as additive variance in fitness at lower levels, then the entities at that level are composed of units of selection at these lower levels, rather than being units of selection themselves. To put it in terms of Salmon's (1970) analysis of statistical explanation, the higher level units of selection as causal factors are then 'screened off' by the lower level units of selection. In their causal effects, they are then 'nothing more than' collections of the lower level entities, and any independent causal efficacy is illusory. This is a necessary and sufficient condition for the truth of Williams's genetic reductionism.

But in general, we would expect this partitioning of variance in fitness into additive and nonadditive components at different levels to show a number of levels − genes, gene complexes, chromosomes, individuals, even groups − at which additive variance at that level appears only as nonadditive variance at lower levels. There are units of selection at each level at which this occurs, and if it does, genetic reductionism and determinism are false.

## 7. WADE'S REVIEW OF THE MODELS OF GROUP SELECTION

For groups to act as units of selection, they must show heritable context-independent (or additive) variance in fitness[9] which is not merely a summative redescription of additive components of variance in fitness to be found at lower levels of organization. Groups must meet Darwin's three principles as well as the additional constraint of a context-independent component of fitness to so qualify.

Organisms meet Darwin's principles by having phenotypic differences (variability) which are heritable, and which have a differential effect on their survival and/or reproduction. Similarly, there may be differences of group structure and interaction (variability of group phenotype) which are transmitted to offspring groups or migrant propagules (group heritability), and which affect the rate of survival and/or reproduction of groups. The heritability of group variability may be either genetically or phenotypically transmitted. The latter results in models for cultural transmission and evolution. The former results in models for the genetic transmission of group traits and biological evolutionary models of group selection. Wade's results concern these models, in which what is inherited, to a greater or lesser degree, is the set of gene frequencies of a group's gene pool. Thus, if migration from a group occurs at random with respect to the genotypes of the individual migrants, the fact that the migrants are drawn from a given group will confer a kind of heritability or correlation between the gene frequencies of the parent population and the gene frequencies in the migrant propagule. (Indeed, this may be true to some extent even if migration from the group is not at random with respect to genotype.) This, together with a differential rate of production of migrant propagules by groups of different genetic compositions or differential rates of survival of such groups or both should make group selection a reality.

The various mathematical models of group selection surveyed by Wade all admit of the possibility of group selection. But almost all of them predict that group selection should be a significant evolutionary factor only very rarely, because they predict that group selection should have significant effects only under very special circumstances — for extreme values of parameters of the models which should be found in nature only rarely. Wade undertook an experimental test of the relative efficacy of individual and group selection — acting in concert or in opposition in laboratory populations of the flour beetle, *Tribolium*. This work (reported in Wade, 1976 and 1977) produced surprising results: group selection appeared to be a significant force in these experiments, one capable of overwhelming individual selection in the opposite direction for a wide range of parameter values, apparently contradicting the results of all of the then extant mathematical models of group selection. This led Wade to a closer analysis of these models, with results reported in Wade (1978), and described here.

All of the models surveyed made simplifying assumptions, most of them different. Five assumptions however were widely held in common: of the twelve models surveyed, each made at least three of these assumptions, and five of the models made all five assumptions.[10] Crucially, for present purposes, the five assumptions are biologically unrealistic and incorrect, and each independently has a strong negative effect on the possibility or efficacy of group selection. It is important to note that these models were advanced by a variety of different biologists, some sympathetic to and some skeptical of group selection as a significant evolutionary force. *Why then did all of them make assumptions strongly inimical to it?* Such a 'coincidence', highly improbable at best, cries out for explanation. I will attempt to offer some explanations after I have presented and discussed some of the assumptions. These assumptions are given in Wade (1978, p. 103):

(1)    It is assumed that the frequency of a single allele within a population can produce a significant change in the probability of survival of that population, or in the genetic contribution which the population makes to the next generation.

(2)    All populations contribute migrants to a common pool, called the "migrant pool" (Levins, 1970), from which colonists are drawn at random to fill vacant habitats.

(3)    The number of migrants contributed to the migrant pool by a population is often assumed to be independent of the size of the population. Thus, the frequency of an allele in the migrant pool

can be represented by the mean allele frequency of all contributing populations.

(4)     It is assumed, often implicitly, that the variance between populations (which is a prerequisite for the operation of group selection) is created primarily by genetic drift within the populations and, to a lesser extent, by sampling from the migrant pool.

(5)     Group and individual selection are assumed to be operating in opposite directions with respect to the allele in question. In short, the allele is favored by selection between groups but disfavored by selection within groups.

The first assumption replicates Williams's reductionistic approach by assuming that the trait under selection at the group and individual levels is a single-locus trait, rather than one with a polygenic basis. However, Wade's experiments rule out this possibility for the trait under selection, population size. In these experiments, a number of replicate populations given the same selection treatment (selection for increasing size at both group and individual levels; selection for decreasing size at both levels; or selection for increasing size at one level and decreasing size at the other) were assayed to determine a number of demographic parameters of the population. These demographic parameters included fecundity, fertility, body size, developmental time, and cannibalism rates for various developmental stages on other developmental stages, all factors having a known and modelable effect on population size, and on the number of different organisms of different age classes as a function of time. (The demographic effects of changes in these variables are visualized in the comparison of the number of individuals of different life stages in high productivity and low productivity populations as a function of time in Figure 4 of McCauley and Wade, 1979.) The results of this assay showed that replicate populations drawn from the same genetic stock, and given the same selection treatment, to which they responded in the same way at the level of the macroscopic trait, population size, achieved this response by different combinations of changes in the underlying demographic parameters affecting population size![11] This means that even if each of these underlying demographic parameters is affected by a single-locus trait (which seems unlikely), *population size cannot be a single locus trait because similar values of it are produced by simultaneous independent variations in the underlying demographic variables.* Thus none of the single-locus models (nine out of the twelve) can adequately describe the selection processes in this experiment. In virtue of the increased dynamical complexity and richness of two-locus systems as compared with one-locus systems (see above, Section 4, and Roughgarden, 1979, Chapter 5) and the further poorly understood complexities of systems involving more than two loci (see Lewontin, 1974, Chapter 6 and Maynard Smith, 1978, Chapter 5), this is a very significant failure. Little if anything learned from the single-locus case is generalizable to cases involving two or more loci.

The second assumption, the analytical device of a 'migrant pool', is, if anything, more serious. In this assumption, all migrants contributed by any population are thrown into a common pool (see Figure 3) from which all new

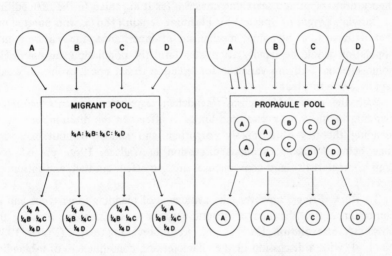

Fig. 3. Diagram illustrating the differences between a migrant pool and a propagule pool. (Reprinted from Wade, 1978, Figure 1, p. 109, with permission of the *Quarterly Review of Biology*.)

founding populations are drawn. This assumption may not be unrealistic in some cases in the five intrademic selection models (it is equivalent to assuming panmixia or random mating within populations), but is surely radically unrealistic for any of the seven traditional models, where it is equivalent to assuming panmixia throughout the entire range of a species. If there is any systematic genetic differentiation throughout the range of a species, and if the averge one-generation migration distance of an individual of that species is appreciably less than the dimensions of its range, then there will be mating which is assortative (or nonrandom) merely because of the limited distance a migrant can travel, thus falsifying this assumption. In general, parent populations will contribute migrants to founding populations in their immediate vicinity, and this assumption will be seriously incorrect.

The seriousness of this simplifying assumption can be better seen by exploring its analogies with the theory of blending inheritance at the individual level. In 1868, Darwin proposed his 'provisional hypothesis of pangenesis' in which large numbers of gemmules secreted by the various cells of an organism were combined in sexual reproduction in such a way that the characters of the offspring (produced by equal contributions of gemmules from both parents) were an intermediate blend of the characters of the parents. This theory drew rapid and searching criticism (as it appeared in the fifth edition of Darwin's *Origin of Species*) by Fleeming Jenkin (1867), who pointed out that with such a 'blending' mechanism of inheritance, the variation in a population would be rapidly attenuated until the population was essentially homogeneous. With no variation for selection to act upon, evolution would rapidly come to a halt.

With the rise of Mendelism (Mendelian segregation prevents significant blending in the 1-locus case, and limits its effects in the multi-locus case [12]) blending theories were rapidly forgotten, and almost without exception (one being Wallace's excellent discussion in Wallace, 1968, pp. 61–65), their characteristics and consequences are forgotten, ignored, and misunderstood.[13]

To R. A. Fisher, however, the avoidance of blending inheritance and its consequences for the loss of genetic variance was a *sine qua non* for the possibility of evolution. He began his ground-breaking treatise (Fisher, 1930, pp. 1–4) with a discussion of the character and consequences of a blending mode of inheritance. To Fisher, Mendel had clearly made the world safe for Darwinism, and he turns somersaults in a marvelously Whiggish rewriting of history to try to show that Darwin didn't really take blending inheritance too seriously.

Whatever the merit of that attempt, he makes the consequences of blending inheritance very clear. Fisher's fundamental theorem of natural selection says that a measure of the rate at which gene frequencies change is directly proportional to the additive genetic variance in fitness. No additive genetic variance in fitness means no gene frequency change, and no evolution. In those beginning pages, he derives (or actually, claims the derivability of) a formula for the rate of loss of variance resulting from mating under blending inheritance. In each generation, the variance is attenuated by a factor of $\frac{1}{2}(1 + r)$, where $r$ is the correlation between parental genotypes. Fisher expected this correlation to be small ($r$ ranges between $-1$ and $+1$, and is 0 in a randomly mating population). Thus, he argued that the existing genetic variance in a population would be approximately halved in successive generations. With no production of new variation, evolution would go as far in a given generation as it would in all successive generations, in accordance with the series $1 + \frac{1}{2} + \frac{1}{4} + \ldots$ (see Wallace, 1968).

Blending inheritance at the individual level is analogous to the assumption of the panmixia of migrants, or of a mating pool at the group level. Wade and I (Wade and Wimsatt, in preparation) have rederived and extended Fisher's results in a manner applicable to the case of group selection. Blending inheritance at the individual level involves two parents, each making equal contributions. The group case may involve contributions of migrants from $n$ parental populations, in possibly unequal contributions, $w_1, w_2, \ldots, w_n$, which sum to 1. In a deterministic model which neglects sampling error in drawing migrants from parental populations, and in which the draw does not affect the gene frequencies in the parental population (requiring, in effect, infinite population size of both parental and migrant propagule populations with the migrants constituting a negligible proportion of the parent populations), a close analogy with Fisher's result can be derived. In this case, if the correlation between any two parental populations is constant at $r$, then the factor by which variance is attenuated in successive generations, $\alpha$, is given by equation 1:

(1)      $\alpha = r + (1 - r) \Sigma w_i^2$

This reduces to Fisher's derived factor of $\frac{1}{2}(1 + r)$ if there are two parents making equal contributions. With $n$ parents making equal contributions, and no correlation between them, $(r = 0)$, $\alpha = 1/n$. This is a 'worst case' assumption, but it is exactly what is assumed in the 'migrant pool' models! The assumption of the migrant pool with any significant number of populations guarantees that the variance goes to 0 essentially immediately.

Suppose that variance is being created anew in each generation by some mechanism or mechanisms, at a rate $V_0$. Then the equilibrium pool of variance (which will determine the rate of evolution) is that amount of variance for which variance is being lost at the same rate at which it is being created, which happens when:

(2)      $V = [1/(1 - \alpha)] \, V_0$

Under the migrant pool assumption, this is simply $[n/(n - 1)] \, V_0$, which is essentially $V_0$ for $n$ large. In effect, there is no pool of variance, but selection can only act on the variance created anew in each generation.

It can be seen by inspection of equations (1) and (2) what conditions will allow the accumulation of an appreciable amount of variance. This happens when $\alpha$ is close to 1, which is true if either there is high correlation between parent populations ($r$ close to 1), or if one population predominates in contributing to a given migrant propagule ($\Sigma w_i^2$ will be close to 1 if and only if $w_i$ is close to 1 for some $i$). In the migrant propagule model of Wade (1978), the latter condition holds, since all of a migrant population comes from a single parent population, and $\alpha = 1$ (in this deterministic model, which doesn't allow for sampling error, and thus does not exactly apply to Wade's case).[14] In more realistic models for the diffusion of migrants in a cline, the former condition would tend to hold. So Wade's models, and realistic population models which don't make the migrant pool assumption, would be expected to allow much more rapid evolution at the group level. Plausible values of the parameters $r$ (or $r_{ij}$'s in a more general treatment) and the $w_i$'s suggest rates of evolutionary change at the group level which are easily 10 to 100 times greater than expected under comparable circumstances with the migrant pool assumption.[15]

Why the seriousness of the migrant pool assumption should have been overlooked, and the significance of how it was finally detected and analyzed, relate to the use of heuristics, and will be discussed in the closing section.

The third assumption, that parent populations contribute the same number of migrants to the migrant pool, independent of their size, in each generation, was also made for reasons of analytical tractability given by Wade (1978). Its effects however are far more immediately obvious, and thus it is similarly more anomalous that the assumption would ever have been made. In models where it is made, differences in reproductive output of different populations is due entirely to the populations' surviving for a different number of generations, since all alike have the same reproductive output in generations in which they are surviving.

Such an assumption would never have been made in models of selection at the individual level. Darwin was acutely aware of the importance of differential reproduction as a selective force, one overshadowing in its potential intensity the effects of differential survivorship, and other evolutionists since have generally retained this awareness. It is doubly mysterious as an assumption at the group level, since presumably one of the primary ways in which increased group fitness could be manifested would be through increased production of individuals, leading to increased population size and increased migration of individuals once maximum population size is attained. One would thus expect a strong correlation between population size and migration rate, rather than the constancy assumed in these models. The association between size and reproductive rate is even more direct than it is for most cases of selection at the individual level. This assumption also has an enormous effect under some conditions, particularly those in which the average population survives for several or more (population) generations in which the averge output of migrant propagules is appreciably more than one per generation. This can be seen by looking at the coefficients in Table 2, which gives the ratio in per generation reproductive rates required of two replicators, one

TABLE 2

Ratio of intrinsic growth rates, $r$, of shorter-lived replicator ($r_b$) to longer-lived replicator ($r_a$) that shorter-lived replicator needs to offset shorter lifetime.**

| lifetime of replicator (generation) | Intrinsic growth rate (per generation) of a replicator with a lifetime of 1000 generations | | | | | |
|---|---|---|---|---|---|---|
| | 1.001 | 1.01 | 1.1 | 2 | 10 | 100 |
| 1000 | 1 | 1 | 1 | 1 | 1 | 1 |
| 100 | 1.0062 | 1.0032 | * | * | * | * |
| 10 | 1.071 | 1.067 | 1.033 | 1.001 | * | * |
| 5 | 1.148 | 1.143 | 1.101 | 1.0062 | * | * |
| 4 | 1.188 | 1.183 | 1.139 | 1.015 | * | * |
| 3 | 1.259 | 1.254 | 1.205 | 1.040 | 1.0003 | * |
| 2 | 1.413 | 1.403 | 1.352 | 1.118 | 1.005 | * |
| 1 | 1.998 | 1.990 | 1.909 | 1.500 | 1.100 | 1.010 |
| .9 | 2.158 | 2.149 | 2.062 | 1.611 | 1.141 | 1.017 |
| .75 | 2.518 | 2.507 | 2.405 | 1.863 | 1.244 | 1.042 |
| .50 | 3.997 | 3.980 | 3.816 | 2.914 | 1.733 | 1.210 |
| .25 | 15.988 | 15.921 | 15.260 | 11.485 | 5.958 | 3.001 |
| .10 | 1023.25 | 1018.92 | 976.46 | 728.44 | 345.95 | 133.18 |
| .01 | $1.27 \times 10^{30}$ | $1.26 \times 10^{30}$ | $1.21 \times 10^{30}$ | $8.97 \times 10^{29}$ | $4.04 \times 10^{29}$ | $1.30 \times 10^{29}$ |

*  Added increase is less than 1 in $10^4$, Fisher's rough lower limit for selective differences to be significant.
** selection coefficient is $r-1$.
$r$ short ($r_b$) calculated from $r_{1000}$ ($r_a$) from the relationship $r_b = \{[(r_a)^a - 1]^{b/a} + 1\}^{1/b}$ and from the approximation $r_b = [(r_a)^b + 1]^{a/b}$ when $(r_a)^a$ is greater than $1 \times 10^{100}$.

with very long lifetime (1000 generations) and with different assumed reproductive rates given across the horizontal dimension, the shorter with a lifetime given in the vertical column and with a per generation replication rate such that the shorter lived replicator will have the same long-range replication rate as the longer lived one. Thus the shorter lived replicator must increase its per generation reproductive rate above that of the longer lived replicator (indicated by the amount that the ratio of reproductive rates exceeds 1) in order to offset the effect of its shorter lifetime. Comparisons in which the longer lived replicator is not so long lived (say ten generations) can be gotten by dividing the appropriate coefficient of the shorter lived replicator (say 1.0062 for comparison with a five-generation replicator if the ten-generation replicator is having two offspring each generation) by its appropriate coefficient (1.0001, in this case), to get the ratio (1.0061, in this case) of per generation reproductive rates for the five- and ten-generation replicators.

What should be obvious from looking at this table is that unless one is comparing replicators which survive for a relatively short time *and* have very low reproductive rates, almost negligible differences in reproductive rate more than offset even quite substantial differences in lifetime. This is so because in this range of parameter values, changes in lifetime have a roughly linear effect on the outcome, whereas changes in reproductive rate have a roughly exponential effect. Thus a small increase in reproductive rate offsets a large decrease in lifetime.[16] Given this fact, it is a bizarre and dangerous simplification in a model to assume that per generation reproductive rate (a variable to which the net reproductive rate, which determines the intensity of selection, is very sensitive) is constant across populations, and that differences in selection are due entirely to differences in the lifetimes of different populations (a variable to which it is quite insensitive). Presumably, group selection would often, even usually, act to optimize both the reproductive rate and the survival time of populations, but these considerations, suggest that the former variable bears watching much more closely than the latter.

I will make no lengthy comments on the last two simplifications discovered and discussed by Wade. Clearly the fourth rules out spatial heterogeneity as a significant selective force at either the individual or group levels, an assumption which also exerts a bias against higher-level units of selection which will be discussed on another occasion.

The fifth simplification, the assumption that group and individual selection are operating to change gene frequencies in opposite directions, has two interesting features.

The first is that it almost certainly has its origin not in any arguments about what would be true in nature, but in the joint action of a consideration of testability and a simplifying assumption. The consideration of testability is that, because of the complexity of interaction of fitness components and the difficulty of determining the relevant parameter values (see Lewontin, 1974), it would be helpful in determining the efficacy of group selection if we could find a trait whose presence clearly signalled the operation of group selection, because it would be selected *against* at the individual level and thus could *only* owe its presence to group selection. This does *not* mean that group selection would usually or generally tend to be opposed to individual selection in nature. Nonetheless, it was probably responsible for the concentration of analytical models on circumstances designed to investigate this condition. In the context of models of two alleles at one locus, the natural way to implement this condition is to assume that the effect of selection at one level was to increase a given gene frequency and that of selection at the other level was to decrease it. In the context of such simplified models, the move from traits to genes or genotypes is easy — all too easy as recent sociobiology has shown — but this way is fraught with error, as a longer discussion on another occasion (see Note 7) will show.

It is worth noting only how implausible the assumption that individual and group selection are opposed becomes once multi-locus models are considered. The effects of selection may generally be described as a vector in which each component is the change in one of the stable variables (*e.g.*, frequencies of genes, gametic, or zygotic genotypes) describing the population. *Only in a phase space of one dimension, such as that characterizing the model of two alleles at one locus, are change vectors constrained to life in the same or in opposite directions. In spaces of any higher dimensionality, the probability that they will be identically or oppositely directed is of measure zero, and the resultant of the two vectors may similarly lie in any direction whatsoever.* Any residual plausibility of this assumption is clearly an artifact of being guided only by the simplest possible model of evolutionary change.

But then if group selection no longer has to overcome forces of individual selection to which it must be opposed, it matters little what or how strong are the selective forces acting at the individual level in evaluating the possibility that group selection can be efficacious. This is particularly detrimental to many of Williams's (1966) arguments. In a multi-dimensional space, even relatively weak selective forces at the group level, when added to relatively strong forces at the individual level, can change the resultant selection vector sufficiently to cause evolution towards an alternative adaptive peak than that which might be achieved by individual selection acting alone. The richer dynamics and greater dimensions of a multi-dimensional phase space produces the possibility of a wide variety of interactions among selection forces at different levels. These interactions are surely poorly understood at this time.

But we cannot hope to understand them if we don't even detect them. We fail to do so, it appears, because of biases which are almost perceptual in character.

## 8. HEURISTICS AND THEIR BIASES: SOME AMELIORATIVE REMARKS

The simplifications Wade discusses were almost all (with the possible exception of the fifth) made to improve analytical tractability. Why their biasing effects should not have been noticed is a difficult and probably multifaceted problem. I will here discuss some of the considerations which seem to me to be most salient:

(1) *Inertia*. Some assumptions have a time-honored status, in that they have been made by almost all past models. As a result, unless the model or phenomenon in question itself seems to immediately point to the need for relaxing one of these assumptions, it will be taken for granted, especially since each of these assumptions almost invariably involves advantages of increased analytical tractability. This may relate to the anchoring bias of Tversky and Kahneman (1974), but it seems equally likely that 'anchoring' and 'inertia' are broad phenomenological categories that cover a multitude of sins committed for a variety of reasons. Particularly well entrenched is the assumption of panmixia (equivalent to the mating pool assumption, and *sufficient to guarantee that the mode of inheritance at any higher level of organization is a form of extreme blending*). An assumption that is well entrenched in the theoretical structure is one that is widely used in and at least apparently essential to the derivation of many other results. These dependency relations can lead to a reluctance to give up an assumption even if it is widely known to be problematic, or even if it is generally believed to be false. The assumption of the transitivity of preferences in decision theory, or of the transitivity of fitness in evolutionary theory, both known to be false when choice or fitness is a function of a number of variables, seems to me to be such a case. The assumption of panmixia is almost certainly another. A third, similarly entrenched assumption in quantitative genetics is the assumption that variance in fitness is totally additive. Both this and the assumption of panmixia are absolutely inimical to the existence or significance of higher-level units of selection.

(2) *Perceptual focus*. Given the centrality of reproductive rate in virtually all evolutionary models and in the structure of evolutionary theory, Wade's third simplifying assumption seems almost unintelligible in almost any

circumstances. The only way that I can think of to rationalize it is as follows: model-building activity is performed against a background of presumed mechanisms operating in the interaction of presumed units. If the presumed units are very well entrenched in a given area, there is a strong tendency to describe and to think about even phenomena at other levels of organization in terms of these units. In traditional evolutionary theory, and even at present, the most obvious unit is the individual organism − the unit which our every-day thought and our perceptual apparatus naturally predisposes us to consider. Most of our everyday interactions (as well as those of most other organisms) that call for voluntary action are pairwise interactions with one other organism at a time, so this bias is evolutionarily well founded. Consequently, there is a strong tendency to see and to talk about groups of organisms as *collections of individuals, rather than as unitary entities.* This is true even for colonies of social insects, whose interdependencies extend even to reproductive speciali-zation, making the metaphor of the colony as an organism perhaps more revealing in evolutionary terms than the view of it as a collection of organisms. Hull (1978) has found similar biases in his arguments that species must often be conceived of as individuals in evolutionary contexts. A quick review of Williams (1966) reveals that even in the context of discussions of group selection, groups are usually described as collections of individuals, and it is my impression that this tendency is widespread throughout the literature. But, *in the context of the group selection controversy, description of an assemblage of units as a collection is a theory-laden description, since it suggests that it is an aggregate. The only time it is appropriate to describe it as an aggregate is when the fitnesses of its components are context-independent* (see Section 6) *or are additive. But as we saw above, this is a sufficient condition for their not constituting a higher-level unit of selection!* Perceptually, the focus on individual organisms prevents us from at the same time seeing the groups as individuals. If we do not see the groups as individuals, then we do not see that assuming that each group contributes an equal number of migrants is equivalent to assuming that there is no variance in reproductive rate. The bias, I suspect, is a perceptual one.

Similar remarks apply to seeing panmixia as equivalent to a form of blending inheritance at the group level. Blending inheritance is traditionally viewed as applying only at the individual level. But once the group is seen as an individual, a view that emerges particularly strongly in Wade's work, even in his earliest paper (Wade, 1976, 1977), the analogy is immediate. I saw the analogy when I read an early draft of Wade (1978) at the same time as I was teaching Darwin's blending theory and its criticisms. The phenomenon was reproducible: three months later, Ross Kiester was teaching Darwin and independently pointed out the analogy to Wade. (Seeing the analogy also requires, I suppose, more than a passing acquaintance with blending inheritance.)

The recent focus on the gene (rather than the individual) as the unit of selection has introduced its share of perceptual problems and biases. Williams (1966) and Dawkins (1976) are full of 'perceptual shifts' back and forth from the genetic to the individual level, but both try to maintain the primacy of the former with consequent biases in their description of the complexities of gene interaction.

Wade (1979) has shown a marvelous case of perceptual bias in the foundational work of Hamilton on kin selection theory. In this theory, selection is seen as maximizing an individual's 'inclusive fitness', in which contributions to the fitness of relatives who share genes are included, weighted by their degree of relationship. Hamilton is reported to have described this as "the gene's eye view of evolution," and in any case the description fits. Hamilton's theory involves the assignment of selection coefficients to individual genes, and makes no reference to the genotypes in which they occur, or to which genotypes mate with which to produce offspring. Wade (1979) advances one- and two-locus models in which this minimum amount of detail is added, and in which he is able to prove an analogue of Fisher's fundamental theorem of natural selection, according to which the rate of evolutionary change is proportional to the *between family variance in fitness*. This result shows that *in a more realistic model of kin selection, the unit of selection is not the individual gene, but the mating pair, Wade's 'atomic family'*. This result could not have been derived or even seen in Hamilton's model, because the simplifications of the environment of the gene were such that Hamilton had no structure in his model on which to hang the upper-level phenomena. Mating pairs do not occur in his model, and there is no way of putting them into it. This is a paradigmatic example of Bias 5 in Section 5 above, but also shows the perceptual or quasi-perceptual character of the way in which this bias operates.

(3) *Perceptual reinforcement*. A factor discussed above is the reinforcement of different biases or heuristics. When a force must be greater than a given magnitude to have an effect, several biases tending to underestimate this force may have a cumulative effect that none of them could have alone. But biases can be mutually supportive in another way also suggested there. One bias may act in such a way as to hide the fact that another bias *is* a bias, and conversely. Detailed documentation of these cooperative effects of biasing assumptions in population genetics remains largely a task for future analysis,

but a promising one. The assumptions of panmixia and additivity of fitness variance are plausible candidates for inspection, both because of their ubiquity and also because each carries with it a relatively easily intuited picture of the unstructured environment of the unit of analysis. The assumption that fitnesses are constant in space and in time, and independent of the density or of the relative frequency of the relevant units (usually assumed to be genes) also demand closer inspection, as each is a source of potential systematic bias against recognizing higher level units. Near decomposability assumptions are important tools which are easily misused, here as elsewhere. Some further biases which may be relevant in this context are discussed in Wimsatt (1980), but most of the work is yet undone.

How, aside from analyzing specific cases of bias, and identifying biases more generally can something be done to correct for their effects? There are at least two plausible candidate procedures:

(1) The first is what Levins (1966) has called the search for robust theorems. To counteract the biases of any given model, he suggests building families of alternative models of a given phenomenon which differ in their simplifying assumptions. The models will vary in their consequences and predictions, reflecting the variety of their assumptions, but there may be consequences which are true across all of the models. These Levins calls 'robust theorems', results which are independent of the details of any particular model. Thus, says Levins, "Our truth is the intersection of independent lies" (1966, p. 126; see also his 1968, pp. 6–8, for further elaborations of this approach).

This method has two disadvantages, neither fatal, but both of which are worthy of note. The first is that we are often in situations in which we do not have even one model, much less two or more, to compare their results. Population genetics and ecology has a richness of mathematical models, unlike some other disciplines, but even here, many areas do not have models with sufficient overlap to be able to compare the results. A further contributing problem arises when the assumptions made in deriving the models are not explicit, since it is then not possible to tell how much one can trust the robustness of a result, basically for the second reason.

A second caution is necessary because it is not always possible to tell when the models are *in fact* independent. If a number of models each make a possibly well-disguised but in any case unnoticed assumption, any theorems in common may simply be a relatively direct consequence of the shared assumption or assumptions. In this case, the results will generally be assumed to be quite robust — illegitimately as it turns out. Indeed, just this seems to have happened in the group selection controversy. Until Wade's review, where the assumptions the various models had in common and their consequences were made explicit, the one result which seemed to be robust was that group selection could be efficacious only rarely and under very special circumstances, and probably the majority of evolutionary biologists still believe this to be the case.

Another approach, generally consistent with robustness analysis (see Levins, 1968, pp. 6–8) and suggested by the heuristic use of means-end analysis and identifications in inter-level reductive explanation (Wimsatt 1976a, pp. 231–237, 1976b, Section 8) could be called 'multi-level reductive analysis'. Even if perceptual focussing may leave one blind to the biases of the nine heuristics of Section 5 as applied at any one level, these same heuristics will have biases leading to *different* simplifications if they are applied to the same system at a different level of organization, *simply because the system-environment boundary has changed*. Simplified models of group selection may thus suggest particular structural features of the environment of the individual which should be included in models of individual selection, just as, *e.g.*, an analysis of the structure of the genotype and/or mating system of the individual may suggest important internal constraints on models at the group level. Simultaneous multi-level modelling may thus eliminate the biases of proceeding at only one level, but a final caution is required: this is likely to work only to the extent that the phenomena and entities of a given level are taken seriously in their own right. Seeing them merely as an extension of another level, be it lower or higher, will merely preserve the perceptual focus of that other level, and most biases will go undetected. This is merely another expression of a view I have argued for before (Wimsatt 1976a, 1976b, 1978). Now for pragmatic as well as for theoretical reasons, reduction in science is better seen as the attempt to understand the explanatory relations between different levels of phenomena, each of which is taken seriously in its own right, than as an unending search of firm foundations at deeper and deeper levels in which, as Roger Sperry so aptly put it (quoted in Wimsatt, 1976a), " ... eventually everything is explained in terms of essentially nothing."

*Department of Philosophy,*
*Committee on Evolutionary Biology, and*
*Committee on Conceptual Foundations of Science*
*The University of Chicago*

## ACKNOWLEDGMENT

This work owes a great deal more to many people than is reflected in the footnotes: To Richard Lewontin for his leading me past the most elemental understandings (and sometimes misunderstandings) I brought to my study of population genetics as a post-doctoral fellow in his lab, and for his early advocacy of the importance of gene interactions and higher level units of selection in evolution, which, as with many other things he has taught me, had a formative effect on my present views. To Richard Levins I owe a similar debt, for his tutelage in mathematical ecology and the analysis of complex systems, for his friendly tolerance of my early naive and principled reductionism, while he gradually taught me elements of a richer vision. To George Williams, whose book was an education to me and to many others, formulating for the first time a clear reductionistic vision and many arguments supporting that view, which I continue to respect, even though I now believe it to be mistaken. More recently, I have learned a great deal from discussions with Ross Kiester and Mike Wade about model building and experimental design, as well as a lot of fascinating and useful biology, and Wade in particular has done a lot to shape my views on group selection. I have profited also in particular from discussions with David Hull, but also with Bob McCauley, Bill Bechtel, Bob Richardson, Elliot Sober, and Mary Jane West-Eberhardt, the participants of a Midwest Faculty Seminar on Evolution at Chicago, and seminar or conference participants at the Leonard Conference at Reno, Denison University, Ohio State, the University of Wisconsin at Madison, and the University of Colorado at Boulder, where versions of this work were presented in 1978–79.

I would like to thank Columbia University Press for permission to reprint Figure 23 and 24 from Lewontin (1974), pp. 274 and 280, and to adapt Table 56 from p. 283; and the *Quarterly Review of Biology* for permission to reprint Figure 1, p. 109, and to quote from p. 103 of Wade (1978). Finally, all of this would have occurred substantially later or perhaps not at all without the support of the National Science Foundation (Grant SOC78–07310 research).

## NOTES

[1] Actually the term 'heuristic' was used earlier and perhaps introduced by G. Polya in his book *How to Solve It* in 1945. The idea of a heuristic procedure has however developed substantially further, and has become one of the central theoretical concepts of artificial intelligence. See, *e.g.*, Nilsson (1971) and Winston (1977) for some more recent discussion of heuristic programming.

[2] On the most obvious reading, in which genotypes are the subsystems containing genes, which interact to produce fitness, which is a property of genotypes which affects the multiplication ratios of the genes they contain, selection and mating would be treated as an intersystemic interaction, with segregation and recombination as intrasystemic interactions. Thus the analogy is not quite exact. In this case, with mating assumed to be at random and the genotypic frequencies in linkage equilibrium, the 'long range' behavior of the system involves the interaction of an intrasystemic force (segregation) and an intersystemic one (selection). This particular decomposition into subsystems (at variance with that required for easy analysis of near decomposability) is necessitated

by the particular structure of Mendelian genetics, which, through mating and differential reproduction, inextricably combines inter- and intra-organismic forces.

Other ways of breaking up the system (*e.g.*, into loci as subsystems, which might be suggested by Williams's remarks) produce similar problems: then recombination (a strong force if there is substantial linkage disequilibrium) is intersystemic, rather than intrasystemic, as it 'should' be. Nonetheless, the partitioning of forces into strong and weak, characteristic of near decomposability analysis is found here also, so there remains an important (and probably the most important) ground of analogy.

[3] The biologist's use of tautology here is looser than the philosopher's, and means roughly a relation which has no empirical content because of the way in which it is used. As such it is related to the vernacular use of tautology as in 'covert tautology' rather than to the logician's sense. I will here use the term as the biologist does.

[4] These judgments are themselves an important source of error, associated with heuristics for cutting the world up into entities, using the robustness or overdetermination of boundaries. These heuristics and examples of their application and misapplication are discussed at length in Wimsatt (1980) where they are particularly relevant in understanding the nature and origin of functional localization fallacies.

[5] These make no mention of the geometric rate of natural increase of organisms and the consequent inevitability of competition for resources (Malthus's observation). But this was a subsidiary argument employed by Darwin to establish the second principle — that different types of organisms had different fitnesses. Darwin needed this *a priori* argument because he had no direct observations of the occurrence of natural selection in nature.

[6] Lewontin applies these principles on a genetic micro-evolutionary scale, and points out that for a population in equilibrium of gene frequencies, however temporary, conditions 2, 3, or both are not met (1970, p. 1). And obviously, if there is only a single allele at a given locus in a population (violating condition 1), no change in gene frequency (or micro-evolution) is possible at that locus.

[7] I have analyzed these conditions and their ramifications in much greater detail in a book manuscript now in process and tentatively to be called *Reductionism, Sociobiology, and the Units of Selection*. Further excellent discussions of related issues can be found in Hull (1978), Sober (1979), and, less directly, Cassidy (1978).

[8] Mike Wade felt that this did not emphasize sufficiently strongly that whether an interaction was additive or epistatic is a function of the relation of the system to the environments in which it is studied. He feels that many studies which purport to show that variance in fitness is additive rather than epistatic suffer from looking at a restricted environment (usually in the laboratory) or range of environments, and that investigation of the system in a wide range of environments would show that many or most of the supposedly additive interactions are in fact epistatic.

It is worth pointing out that the term 'epistasis' is traditionally reserved for interactions between genes *within a given genotype*. But the discussion here naturally suggests an extension to interactions between higher level complexes of genes. Thus, when one speaks, as Wade does, of a *group* phenotype, it becomes natural to describe nonadditive interactions between individuals in the group as epistatic.

[9] The notion of additive variance has an implication that speaking of context-independent variance does not. Speaking of additive variance implies the context of a larger unit in which more than one of the smaller units which contribute to fitness will co-occur, so

that their contributions will add. Thus genes which show additive variance will occur in genotypes. In the case of two allels at one locus, this condition is met if the fitness of the heterozygote *Aa* is exactly halfway between the fitness of the two homozygotes, *AA* and *aa*. Talking about groups as units of selection may not imply a larger conspecific group whose fitness they contribute to, and in this case (and other similar cases), it is preferable to talk about *context-independent fitnesses* of groups rather than of *additive contributions to fitness* of groups.

[10] Within the 'traditional' models, the record was even worse. Five out of seven of the models made all five of the assumptions. The structure of the intrademic models, a newer development widely heralded as improving the case for group selection, required dropping one of the assumptions.

[11] Thus, population size is a case of a functional or supervenient property, in the sense widely discussed in philosophy of psychology and recently in biology, by Rosenberg (1978), in that the same value of population size can be realized by a variety of underlying states or mechanisms involving lower-level variables.

[12] Blending inheritance does not require fusion of the hereditary particles. The blending refers to the character traits, not to their underlying determining factors. Unconstrained random mixing of a large number of factors having an additive effect will produce the main effect of blending, a loss of genotypic variance, until Hardy-Weinberg equilibrium is achieved. For this reason a process importantly analogous to blending in the individual case can occur even with Mendelian genetics and can be significant in the loss of genotypic variance for additive multilocus traits.

[13] These are explored in substantial further detail in the manuscript mentioned in Note 7.

[14] This conclusion must be moderated when the effect of finite population size and sampling error are taken into account. As Wade (1978, p. 110) points out, sampling error increases variance (so $\alpha$ would in effect be greater than 1) but at a cost of lowered group heritability. Some simple models taking account of sampling error have been constructed and will be discussed in our forthcoming paper.

[15] See Note 14.

[16] The effects are ʌeversed for survival times of the order of one generation or less and small reproductive rates, but survival times and reproductive rates and both inclined to be appreciably larger under a wide variety of circumstances.

## BIBLIOGRAPHY

Ando, A., F. M. Fisher, and H. A. Simon: 1963, *Essays on the Structure of Social Science Models*, MIT Press, Cambridge.

Bantz, D. A.: 1976, *Does Physics Explain Chemistry?*, unpublished Ph.D. Dissertation, Committee on Conceptual Foundations of Science, Univ. of Chicago.

Bogaard, P. A.: 1979, 'A Reductionist's Dilemma: the Case of Quantum Chemistry', in I. Hacking and P. D. Asquith (eds.), *PSA 1978*, Vol. 2, Philosophy of Science Association, East Lansing, Michigan.

Boyd, R.: 1972, 'Determinism, Laws and Predictability in Principle', *Philosophy of Science* 39, 431–450.

Campbell, D. T.: 1974, 'Evolutionary Epistemology', in P. A. Schilpp (ed.), *The Philosophy of Karl Popper*, Open Court, La Salle, Illinois, Vol. 1, pp. 413–463.

Campbell, D. T.: 1977, 'Descriptive Epistemology: Psychological, Sociological, and Evolutionary', The William James Lectures, Harvard, Spring, 1977, unpublished ms.

Cassidy, John: 1978, 'Philosophical Aspects of the Group Selection Controversy', Philosophy of Science 45; 575–594.

Causey, R. L.: 1972a, 'Uniform Microreductions', Syntheses 25, 176–218.

Causey, R. L.: 1972b, 'Attribute-identities in Microreductions', Journal of Philosophy 69, 407–422.

Causey, R. L.: 1977, Unity of Science, Reidel, Dordrecht, Holland.

Cohen, R. S., C. A. Hooker, A. C. Michalos, and J. van Evra (eds.): 1976, PSA 1974 (Boston Studies in the Philosophy of Science, Vol. 32) Reidel, Dordrecht, Holland.

Darden, L., and N. Maull: 1977, 'Interfield Theories', Philosophy of Science 44, 43–64.

Dawkins, R.: 1976, The Selfish Gene, Oxford Univ. Press, New York.

Dawikins, R.: 1978, 'Replicator Selection and the Extended Phenotype', Zeitschrift für Tierpsychologie 47, 61–76.

Dresden, M.: 1974, 'Reflections on Fundamentality and Complexity', in C. Enz and J. Mehra (eds.), Physical Reality and Mathematical Description, Reidel, Dordrecht, Holland, pp. 133–166.

Fisher, R. A.: 1930, The Genetical Theory of Natural Selection, Oxford Univ. Press, London.

Glassman, R. B.: 1978, 'The Logic of the Lesion Experiment and Its Role in the Neural Sciences', in S. Finger (ed.), Recovery from Brain Damage: Research and Theory, Plenum Press, New York, pp. 3–31.

Globus, G., G. Maxwell, and I. Savodnik (eds.): 1976, Brain and Consciousness: Scientific and Philosophic Strategies, Plenum Press, New York.

Gregory, R. L.: 1967, 'Models and the Localization of Function in the Central Nervous System', reprinted in C. R. Evans and A. D. J. Robertson (eds.), Key Papers: Cybernetics, Butterworths, London, pp. 91–102.

Hull, D. L.: 1973, 'Reduction in Genetics – Biology or Philosophy?', Philosophy of Science 39, 491–498.

Hull, D. L.: 1974, Philosophy of Biological Science, Prentice-Hall, Englewood Cliffs, New Jersey.

Hull, D. L.: 1976, 'Informal Aspects of Theory Reduction', in Cohen et al. (1976), pp. 633–652.

Hull, D. L.: 1978a, 'A Matter of Individuality', Philosophy of Science 45, 335–360.

Hull, D. L.: 1978b, 'The Units of Evolution: A Metaphysical Essay', presented at the University of Aarhus symposium 'Evolution: Its Philosophical and Methodological Aspects', in August, 1978. A later version of this paper will be given at the 1980 meetings of the Philosophy of Science Association, and published in the proceedings.

Jenkin, Fleeming: 1867, 'Review of Darwin's Origin of Species', The North British Review, reprinted in David L. Hull (ed.), Darwin and His Critics, Harvard Univ. Press, Cambridge, 1973.

Levins, R.: 1966, 'The Strategy of Model Building in Population Biology', American Scientist 54, 421–431.

Levins, R.: 1968, Evolution in Changing Environments, Princeton Univ. Press, Princeton.

Levins, R.: 1970, 'Extinction', in Some Mathematical Questions in Biology: lectures on mathematics in the life sciences, American Mathematical Society 2, 75–108.

Lewontin, R. C.: 1970, 'The Units of Selection', *Annual Review of Ecology and Systematics* 1, 1–18.

Lewontin, R. C.: 1974, *The Genetic Basis of Evolutionary Change*, Columbia Univ. Press, New York.

Lewontin, R. C., and M. J. D. White: 1960, 'Interaction between inversion polymorphisms of two chromosome pairs in the grasshopper, *Moraba scurra*', *Evolution* 14, 116–129.

Lorenz, K. Z.: 1965, *Evolution and the Modification of Behavior*, Univ. of Chicago Press, Chicago.

Maull, N. L.: 1977, 'Unifying Science Without Reduction', *Studies in History and Philosophy of Science* 8, 143–162.

Maull-Roth, N.: 1974, *Progress in Modern Biology: An Alternative to Reduction*, unpublished Ph.D. Dissertation, Committee on Conceptual Foundations of Science, Univ. of Chicago.

Maynard Smith, John: 1975, *The Theory of Evolution*, 3rd ed., Penguin, London.

Maynard Smith, John: 1978, *The Evolution of Sex*, Cambridge Univ. Press, London.

McCauley, D. E., and M. J. Wade: 1979, 'Group selection: the genetic and demographic basis for the phenotypic differentiation of small populations of *Tribolium castaneum*', *Evolution* (submitted).

McCauley, Robert N.: 1979, *Explanation, Cross-Scientific Study, and the Philosophy of Mind: An Examination of the Methodological Foundations of Transformational Generative Grammar*, Ph.D. Dissertation, Department of Philosophy, Univ. of Chicago.

Moss, R. Y.: 1979, 'Designing Reliability into Electronic Components', Hewlett-Packard, Palo Alto, California.

Nagel, E.: 1961, *The Structure of Science*, Harcourt, Brace, New York.

Newell, A., and H. A. Simon: 1972, *Human Problem Solving*, Prentice-Hall, Englewood Cliffs, New Jersey.

Newell, A., J. C. Shaw, and H. A. Simon: 1957, 'Empirical Explorations with the Logic Theory Machine: A Case Study in Heuristics', reprinted in E. A. Feigenbaum and J. Feldman (eds.), *Computers and Thought*, McGraw-Hill, New York, 1963.

Newell, A., and H. A. Simon: 1961, 'GPS, a Program that Simulates Human Thought', reprinted in E. A. Feigenbaum and J. Feldman (eds.), *Computers and Thought*, McGraw-Hill, New York, 1963.

Nickles, T.: 1973, 'Two Concepts of Intertheoretic Reduction', *Journal of Philosophy* 70, 181–201.

Nickles, T.: 1976, 'Theory Generalization, Problem Reduction, and the Unity of Science', in Cohen *et al.* (1976), pp. 33–74.

Nilsson, N. J.: 1971, *Problem-Solving Methods in Artificial Intelligence*, McGraw-Hill, New York.

Rosenberg, A.: 1978, 'The Supervenience of Biological Concepts', *Philosophy of Science* 45, 368–386.

Roughgarden, Jonathan: 1979, *Theory of Population Genetics and Evolutionary Ecology: An Introduction*, Macmillan, New York.

Ruse, M.: 1973, *The Philosophy of Biology*, Hutchinson University Library, London.

Ruse, M.: 1976, 'Reduction in Genetics', in Cohen *et al.* (1976), pp. 633–651.

Salmon, W. C.: 1971, *Statistical Explanation and Statistical Relevance*, Univ. of Pittsburgh Press, Pittsburgh.

Schaffner, K. F.: 1967, 'Approaches to Reduction', *Philosophy of Science* 34, 137–147.
Schaffner, K. F.: 1969, 'The Watson-Crick Model and Reductionism', *British Journal for the Philosophy of Science* 20, 325–348.
Schaffner, K. F.: 1974, 'The Peripherality of Reductionism in the Development of Molecular Biology', *Journal for the History of Biology* 7, 111–139.
Schaffner, K. F.: 1976, 'Reduction in Biology: Prospects and Problems', in Cohen *et al.* (1976), pp. 613–632.
Schaffner, K. F.: 1978, *The Logic of Discovery and Justification in the Biomedical Sciences*, unpublished manuscript, Department of History and Philosophy of Science, Univ. of Pittsburgh.
Schaffner, K. F.: 1979, 'Theory Structure in the Biomedical Sciences', *Journal of Medicine and Philosophy* 3, 331–371.
Shimony, A.: 1970, 'Statistical Inference', in R. G. Colodny (ed.), *The Nature and Function of Scientific Theories*, Univ. of Pittsburgh Press, Pittsburgh, pp. 79–172.
Simon, H. A.: 1957, 'A Behavioral Model of Rational Choice', in H. A. Simon, *Models of Man*, Wiley, New York, pp. 241–256. See also introduction to Part IV, pp. 196–206.
Simon, H. A.: 1966a, 'Thinking by Computers', in R. G. Colodny (ed.), *Mind and Cosmos*, Univ. of Pittsburgh Press, Pittsburgh, pp. 3–21.
Simon, H. A.: 1966b, 'Scientific Discovery and the Psychology of Problem Solving', in R. G. Colodny (ed.), *Mind and Cosmos*, Univ. of Pittsburgh Press, Pittsburgh, pp. 22–40.
Simon, H. A.: 1968, *The Sciences of the Artificial*, MIT Press, Cambridge.
Simon, H. A.: 1973, 'The Structure of Ill-Structured Problems', *Artificial Intelligence* 4, 181–201.
Sklar, L.: 1973, 'Statistical Explanation and Ergodic Theory', *Philosophy of Science* 40, 194–212.
Sklar, L.: 1976, 'Thermodynamics, Statistical Mechanics, and the Complexity of Reductions', in Cohen *et al.* (1976), pp. 15–32.
Sober, E.: 1979, 'Significant Units and the Group Selection Controversy', unpublished ms., Department of Philosophy, University of Wisconsin at Madison. A later version of this paper will be given at the 1980 meetings of the Philosophy of Science Association and published with the proceedings.
Tversky, A., and D. Kahneman: 1974, 'Decision under Uncertainty: Heuristics and Biases', *Science* 185, 1124–1131.
Valenstein, E.: 1973, *Brain Control*, Wiley, New York.
Wade, Michael J.: 1976, 'Group Selection among laboratory populations of *Tribolium*', *Proceedings of the National Academy of Sciences, U.S.A.* 73, 4604–4607.
Wade, Michael J.: 1977, 'An Experimental Study of Group Selection', *Evolution* 31, 134–153.
Wade, Michael J.: 1978, 'A Critical Review of the Models of Group Selection', *Quarterly Review of Biology* 53, 101–114.
Wade, Michael J.: 1979, 'The Evolution of Social Interactions by Family Selection', *American Naturalist* 113, 399–417.
Wade, Michael J., and D. E. McCauley: 1979, 'Group Selection: The phenotypic and genotypic differentiation of small populations', *Evolution* (submitted).
Wade, Michael J., and W. C. Wimsatt: Forthcoming, 'The Blending Theory of Group Inheritance'.

Wallace, B.: 1968, *Topics in Population Genetics*, Norton, New York.

Williams, G. C.: 1966, *Adaptation and Natural Selection*, Princeton Univ. Press, Princeton.

Wilson, E. O.: 1975, *Sociobiology: The New Synthesis*, Harvard Univ. Press, Cambridge.

Wimsatt, W. C.: 1974, 'Complexity and Organization', in K. F. Schaffner and R. S. Cohen (eds.), *PSA 1972*, (Boston Studies in the Philosophy of Science, Vol. 20), Reidel, Dordrecht, Holland, pp. 67–86.

Wimsatt, W. C.: 1976a, 'Reductionism, Levels of Organization, and the Mind-Body Problem', in Globus *et al.* (1976), pp. 199–267.

Wimsatt, W. C.: 1976b, 'Reductive Explanation: A Functional Account', in Cohen *et al.* (1976), pp. 671–710.

Wimsatt, W. C.: 1978, 'Reduction and Reductionism', in H. Kyburg, Jr., and P. D. Asquith (eds.), *Current Problems in Philosophy of Science*, Philosophy of Science Association, East Lansing, Michigan.

Wimsatt, W. C.: 1980, 'Randomness and Perceived-Randomness in Evolutionary Biology', *Synthese* 43, 287–329.

Wimsatt, W. C.: 1980, 'Robustness and Functional Localization: Heuristics for Determining the Boundaries of Systems and their Biases', in M. Brewer and B. Collins (eds.), *Knowing and Validating in the Social Sciences: A Tribute to Donald T. Campbell*, Jossey-Bass, San Francisco.

Winston, P. H.: 1977, *Artificial Intelligence*, Addison-Wesley, Reading, Massachusetts.

Wright, Sewall: 1968, *Evolution and the Genetics of Populations*, Vol. 1, Univ. of Chicago Press, Chicago.

EDWARD MacKINNON

# THE DISCOVERY OF A NEW QUANTUM THEORY

The debate, some fifteen years ago, on the logic of scientific discovery, did not lead to any definitive or generally accepted solution. Yet it did sharpen some of the issues involved. Both the logical positivists and their erstwhile opponent, Karl Popper, insisted that scientific discovery was outside the scope of logic.[1] N. R. Hanson's vociferous opposition seems, in retrospect, more a rhetoric about the need for a logic of discovery than a developed logic.[2] Though the process of scientific discovery, especially when rationally reconstructed, inevitably evidences some structural relation between the originating question and the eventual answer, the relation definitely does not seem to be logical in the sense of being governed by formal rules of general import.

As Koestler (1964) manifested in a somewhat anecdotal fashion, and Blackwell (1966) argued in a more systematic way, the crucial issue is not the logic but the *rationality* of scientific discovery. If human rationality develops dialectically and is manifested especially through problem solving, then scientific discovery must be considered among the more striking examples of rationality. Such a conclusion, however, still leaves us on the level of bland clichés. The pertinent question is: What specific form does rationality take in scientific discovery?

Such a question is, I believe, best answered by a detailed reconstruction of particular scientific discoveries. Here, theoretical discoveries, which depend upon a sustained process of reasoning, are more revealing, though often less spectacular, than experimental discoveries. These may depend more on luck or on the exploitation of technological breakthroughs. Elsewhere, in a series of technical articles, I have attempted a detailed reconstruction of the paths that led Louis de Broglie, Werner Heisenberg, and Erwin Schrödinger[3] to the development of a new quantum theory in the mid-nineteen twenties. Here I wish to give a simple descriptive summary of the process involved in each case, paying particular attention to Heisenberg and Schrödinger; for their published papers concealed, more than revealed, the process of discovery. Then I will add a few brief reflections on some common features these discoveries exhibit.

*T. Nickles (ed.), Scientific Discovery: Case Studies,* 261–272.
*Copyright © 1980 by D. Reidel Publishing Company.*

## 1. HEISENBERG: MATRIX MECHANICS WITHOUT MATRICES

Before Heisenberg wrote his paper initiating quantum mechanics, he had written some sixteen technical papers, chiefly concerned with giving an atomic account of distinctive features of atomic lines through calculations based on the Bohr-Sommerfeld model of the atom. Heisenberg's initial breakthrough here was an ingenious adaption of Sommerfeld's core model, a specialized version of the general Bohr model. The core comprised the nucleus and all but the outermost electron, which was thought to be in an encompassing orbit around this core. The basic chemical and spectroscopic properties of the atom were attributed to this outermost, or optical, electron. What Heinsenberg did was to postulate a novel type of interaction between the optical electron and the core. Though this violated the established rules for the assignement of quantum numbers by introducing half-integral quantum numbers, it was helpful in explaining such puzzling phenomena as the anomalous Zeeman effect.

In Copenhagen, where he became a fellow in Niels Bohr's Institute, Heisenberg worked on some new problems concerning the interaction of radiation and atoms. Here the core model did not work, particularly in the problem of dispersion. To everyone's surprise, a virtual oscillator model, based on Lorentz's turn of the century work, fared much better. This came at a time, 1924, when the old Bohr theory was obviously breaking down. No one was more acutely aware of this than Bohr himself. He accepted — and in fact formally introduced — the virtual oscillator model. Yet he did not think of this as a model proper to individual atoms. It was, rather, a means of extending the viable features of the classical treatment of radiation through the guidance of the Bohr Correspondence Principle.

Apart from Bohr, the man who then had the strongest influence on Heisenberg was Wolfgang Pauli, Heisenberg's closest correspondent and severest critic. Pauli outdid Bohr's scepticism concerning the realistic significance of models. He demonstrated that the core model Heisenberg had used was inconsistent with the established results of the special theory of relativity. This strengthened his insistence on abandoning any reliance on models and model-dependent reasoning. Heisenberg wrote Bohr that "Pauli grumbles about everything, especially about my atomic physics."

Early in 1925 Heisenberg wrote yet another paper on the anomalous Zeeman effect, reinterpreting his earlier work in the light of Pauli's criticism and Pauli's new rules for assigning quantum numbers. At the end of the paper, he indicated the path that he thought should be followed. The virtual oscillator model, which did not rely on a descriptive account of atomic orbits, was something of a halfway house between the older descriptive models and the new insistence, coming in different ways from Bohr and Pauli, that it was impossible to give any true space-time description of the motion of electrons within the atom. Heisenberg, accordingly, suggested using the virtual oscillator model to calculate the energy levels proper to the optical electron and then the modifications due to its interaction with all the other electrons.

The calculations involved proved too difficult, even for the simplest case of the hydrogen atom. So Heisenberg took a problem which was physically unrealistic but mathematically tractable, the anharmonic oscillator. Though his mathematics will not be treated here, I will indicate one crucial aspect of how he used the virtual oscillator model. He could treat transitions directly in terms of a jump from a state $n$ to, for example, a state $n - 2$. He could also treat this as a virtual transition from state $n$ to state $n - 1$ followed by another virtual transition from state $n - 1$ to state $n - 2$. A comparison of the two expansions involved led to formulas like,

$$b_2\ (n, n-2) = (1/a^2)a_1\ (n, n-1)a_1\ (n-1, n-2). \tag{1}$$

Because of the interpretation given this formula through the virtual oscillator model, it was necessary to keep the terms on the right-hand side in the order given. The electron cannot make a transition from the state $n - 1$ to the state $n - 2$ until after it has made a transition from state $n$ to state $n - 1$. Or, mathematically,

$$a_1\ (n, n-1)a_1\ (n-1, n-2) \neq a_1\ (n-1, n-2)a_1\ (n, n-1). \tag{2}$$

Max Born, who received a preprint of Heisenberg's paper, recognized equation (2) as a formula found in matrix multiplication. Heisenberg had effectively discovered matrix mechanics without knowing what a matrix is.

With this new method Heisenberg worked out a very detailed solution for the energy levels proper to the anharmonic oscillator. He was also able to solve the same problem in a way that did not depend on his new method. This involved solving the *simple* harmonic oscillator problem by standard methods, and then treating the anharmonic term as a correction to this basic solution. Earlier Heisenberg had assisted Max Born in working out the perturbation methods necessary for such calculations. When both methods gave exactly the same results, Heisenberg was convinced that his new method worked.

He was convinced, but could he convince others? The procedure he followed is manifested through his correspondence.[4] He thought that if he could find an explanation for his new method acceptable to Pauli, whom

Heisenberg called 'the master of criticism', then he could convince the physics community at large. Within the next month he worked out an acceptable justification, one that supressed any overt dependence on the virtual oscillator model. In its place was a combination of Pauli's doctrine, that theories should be geared to observable phenomena, and a hypothetical-deductive approach, relying on formulas developed either by extending and transforming classical formulations or by simply postulating formulas derived from the virtual osciallator model. This Pauli accepted. It is rather ironic that Heisenberg, whose fundamental breakthrough was due to a skilful use of models applied to virtual processes, or processes which are unobservable as a matter of principle, became the symbol for the success of a scientific method which rejects model-dependent reasoning and relies exclusively on observable quantities.

## 2. DE BROGLIE'S DISCOVERY OF MATTER WAVES

Before the discovery of the Compton effect in 1923, few physicists accepted Einstein's hypothesis of light-quanta, later called photons, and the dual wave-particle properties it attributed to light.[5] A notable exception was the young French physicist, Louis de Broglie. He not only accepted light-atoms, but even developed a doctrine of multi-atomic light molecules as a basis for justifying Planck's radiation formula. On his thirty-first birthday, in 1923, he conceived the idea of extending this wave-particle duality to electrons, an idea that was given a complex and developed justification in the elaborate theory developed in his doctoral thesis.

This theory, for all its suggestiveness, was soon superseded by Schrödinger's wave mechanics. A critical examination shows that de Broglie's argument rests on a misidentification of two fundamental technical relations. As a result, the only really viable aspect of de Broglie's work proved to be the concept of extending wave-particle duality to electrons, and the formula $\lambda = h/p$, relating wave-length to momentum. This celebrated formula appeared only one time in de Broglie's thesis, and then only as a nonrelativistic approximation valid for standing waves in a gas. For this reason, we will leave de Broglie and pass on to the more fruitful discovery of wave mechanics due to Schrödinger.

## 3. SCHRÖDINGER'S ROAD TO WAVE MECHANICS

The text of Schrödinger's papers, developing wave mechanics, has, until recently, supplied virtually the only guide to this historical development. Here, as in Heisenberg's work some six months earlier, the finished text was written to convince contemporaries rather than to enlighten historians and philosophers interested in the process of discovery. Fortunately, recent scholarship has clarified this significant breakthrough.[6]

Erwin Schrödinger, fourteen years older than Heisenberg, had made significant, though not outstanding, contributions to various branches of physics through some forty three technical papers when, in the mid-nineteen twenties, he turned his attention to the problem of ideal gases. The work he did here, which led to his wave mechanics, is best understood as his contribution to a dialogue involving Max Planck and Albert Einstein.

The classical theory of ideal gases, based on the statistical methods developed by J. C. Maxwell and Ludwig Boltzmann, could not accomodate the phenomenon of degeneracy. This is a kind of thermodynamic collapse which the Nernst heat theorem, the third law of thermodynamics, predicted must occur at very low temperatures. Max Planck, the conservative physicist who made the radical discovery of quantum theory, found a way to accomodate degeneracy by what was essentially a mathematical trick. Einstein, as might be expected, made a more daring and significant breakthrough. A young Indian physicist, Satyandra Nath Bose, developed a new derivation of the Planck radiation formula and sent it to Einstein. His key idea was that, instead of counting photons or radiant energy distribution, one should count occupation cells in the abstract phase space that represents the system. Einstein, recognizing the potential this new approach had, translated Bose's article into German, had it published, and then wrote the first of two papers adapting Bose's way of counting phase-space cells into a new method of treating ideal gases.

After finishing the first ideal gas paper Einstein received and read a copy of de Broglie's thesis, which Pierre Langevin had sent him. He realized that de Broglie's idea of matter waves could be adapted to the parallel Einstein himself was using between the statistical treatment of radiation and ideal gases. The wave properties Einstein then attributed to molecules led to a peculiar coupling which is particularly significant for a degenerate gas.

During this period Schrödinger wrote three papers on ideal gases and carried on an extensive correspondence with Planck and Einstein on the issues involved. In the Fall semester of 1925 he also, at the suggestion of Pieter Debye, conducted two seminars in Zurich on de Broglie's thesis. Here we will consider only his third gas paper, 'On Einstein's Theory of Gases'. In that paper he accepted the parallel between radiation and ideal gases that Einstein had used, but rejected the Bose-Einstein method of counting cells in phase space. On this point Schrödinger followed Planck in treating the gas as a whole. Instead of counting molecules, he based his statistics on a count of energy states proper to the gas as a whole. On this basis he was able to reproduce the results Einstein had achieved through his new statistics.

The parallel between Einstein's and Schrödinger's treatment brings out the conceptual option then available. It was not possible to retain both the traditional way of doing statistics and also the traditional idea of molecules as isolated, independent units and then explain the gas degeneracy proper to Bose-Einstein statistics. Einstein changed the statistics. Schrödinger retained the Boltzmann method of doing statistics, but modified the idea of a molecule. The parallel between radiation and ideal gases which Schrödinger used, together with his method of treating a gas almost as an organic whole, suggested that a molecule be treated as a wave crest in a system of waves. This suggestion, a modification of de Broglie's idea, presented difficulties which Schrödinger explicitly cited in the conclusion of this ideal gas paper. A wave-packet disperses; a molecule retains particulate properties. Schrödinger expressed the hope that a more sophisticated mathematical treatment of wave-packets might overcome this difficulty.

In the discussion following Schrödinger's second seminar on de Broglie's thesis, Debye suggested that the proper way to treat wave motion is to look for a variational principle. Schrödinger followed this suggestion and, like de Broglie, first sought a relativistic wave equation. When this did not seem to work he sought a nonrelativistic approximation. This, now known as the Schrödinger wave equation, led to correct results for the energy states and frequencies of the hydrogen atom, and a host of other problems, now familiar from textbook summaries.

It is interesting to compare Schrödinger's way of developing wave mechanics with the way Heisenberg had developed quantum mechanics some six months earlier. Heisenberg deliberately supressed his dependence on the virtual oscillator model. Before the Copenhagen Interpretation was hammered out, Heisenberg was relying on the doctrine that a physical theory is essentially a mathematical formalism correlated with observable data. Schrödinger initially followed a similar method of presentation. The model of electrons as waves, which suggested the Schrödinger equation, is not given any foundational role in Schrödinger's first two quantization papers. The emphasis there is on a mathematical formalism and its observable consequences. After Schrödinger proved the mathematical equivalence of his wave mechanics and Heisenberg's matrix mechanics, he attached considerably more importance to the physical interpretation which differentiated his system from

Heisenberg's prior system. This led him in his third, and especially in his fourth, quantization paper, to develop a rather new physical interpretation of his $\psi$-function and to accord this interpretation a foundational role. Since Schrödinger was never able either to develop his new interpretation in a consistent and adequate way or to prove its superiority to the competing probabilistic interpretation of the $\psi$-function developed by Max Born, his efforts proved to be counterproductive. After Heisenberg's indeterminacy paper, undercutting Schrödinger's stress on intuitiveness, Schrödinger withdrew from active participation in the developments he had done so much to establish.

## 4. PATTERNS OF DISCOVERY

There are, to be sure, various approaches to the problem of explaining scientific discovery. Though I think it helpful, indeed indispensible, to approach the problem through theories of scientific discovery, that approach has, in the past, too often manifested a peculiar limitation. If one begins with a definitive list of the forms rationality must take in science, such as deductive, inductive, and probabilistic reasoning, then many significant examples of scientific discovery are found not to fit into this procrustean bed without extreme cutting and stretching of joints. The convenient way out is to dump discovery into an all purpose grab bag labelled 'Irrational', 'Superrational'. or 'Serendipity Strikes Again', and then seek an account in terms of unconscious inference, subliminal bisociation of consciously blocked matrices, or through some superrational intuition possessed only by certifiable geniuses.

The cases we considered manifested human rationality in a highly developed and strongly socialized form. Each of the three men considered had a fairly clear idea of the same underlying problem, the inadequacy of the Bohr-Sommerfeld theory of the atom. Each also had a fairly clear idea of what would constitute a solution, a new theory in which established experimental results could be derived and new results predicted through the application of consistent rules, without the *ad hoc* hypotheses, patchwork corrections, and the rather arbitrary mixture of classical and quantum laws that characterized the final stage of the Bohr theory. Finally, the men involved, particularly Heisenberg and Schrödinger, developed their positions through a sustained critical dialogue with such men as Albert Einstein, Niels Bohr, Wolfgang Pauli, Max Planck, Pieter Debye, Hermann Weyl, Pierre Langevin, and H. A. Lorentz. No Platonic dialogue ever contained such talented interlocuters.

Yet none of the theoretical discoveries considered can, in my opinion, be explained in terms of any established pattern of deductive, inductive, or probabilistic reasoning. Under the circumstances, the best approach would seem to be a quasi-phenomenological one. One begins with a detailed description of the complex process each of these men followed. This is essentially a matter of painstaking historical reconstruction. After that one should work from the descriptive account towards underlying structures. Such a search is, to be sure, shaped by the hypotheses one has concerning possible forms of explanation. It may be shaped, but it is not predetermined. Though hypotheses may propose, history must dispose.

I hope to attempt such an analysis at a later date. Here I will simply suggest, without even a sketchy justification, some of the structural elements that I think such an analysis should include. For the historical examples considered, and for others like them, there are two basically different reasoning structures involved, which we will call the mathematical formalism and the physical framework. Because of its technicality and abstruseness, the mathematical formalism looks like the more formidable component. It is, in fact, the more manageable. Learning how to set up and solve the Schrödinger wave equation, or the equivalent matrix equation, is essentially a question of mastering a technical skill, a skill acquired, for the most part, by studying textbooks and doing assigned problems. The reasoning is deductive, and the rules governing it are farily well understood.

The physical framework is basically the language we use to describe the world of which we are a part plus some specialized extensions proper to different disciplines. It is difficult to isolate, because it is impossible to get away from it. One point needs stressing. This distinction is quite different from the earlier distinction, made by the Logical Positivists, between an observational language and a theoretical language. My purpose is not to impose an epistemology, but to suggest a framework that is at least descriptively adequate to the developments considered.

The interrelation between the physical framework and the mathematical formalism is most clearly seen in the case of measurement. This may be generally described as a homomorphism of a physical structure, or of a structure embedded in a a language used to describe some physical system, into a mathematical structure.[7] Thus a domain of elements with an ordering relation that is asymmetric, transitive, and connected may be mapped into the ordinal numbers. For a mapping into the cardinal numbers, one needs a unit, a method of concatenation, and a scale.

What is basic in measurement need not be conceptually basic. In quantum mechanics, for example, 'state of a system' is a foundational concept. This is given through the maximal number of observables characterizing a system that admit of simultaneous specification. Since states obey a superposition principle, one uses as the appropriate mathematical system a Hilbert space, whose vectors also obey a superposition principle.

This homomorphism constitutes the minimal correspondence between the physical framework and a mathematical formalism. To go beyond this minimal connection it is helpful to distinguish three stages in the development of a scientific theory: a formative state; a stage of deductive unification; and a stage of axiomatic reformulation. Since the first is our present concern, I will only indicate the features of the other two that distinguish them from the formative stage. For the developments we have been considering, and for many similar ones, the formative stage is characterized by two distinctive features. The first is an ontological luxuriance. When a standard way of doing physics has reached a dead end, researchers are willing to postulate, on a trial basis, new entities, hidden processes, various types of models and analogies, to try anything that offers any hope of a breakthrough. The second, closely related feature, is a rather loose fit between the physical framework and a mathematical formulation. A model, such as the virtual oscillator model, wave-particle duality, or a gas as a quasi-organic whole, suggests formulas which are developed and tested, though their relation to established mathematical formulations is not at all clear.

After one formulation proves successful, there is a tendency to prune the luxuriant ontology, minimize any dependence on physicalistic reasoning, and concentrate on a deductive unification. Thus, instead of reasoning in terms of various models proper to particular problems, as was the fashion in 1923–1925, one could simply set up and solve a Schrödinger equation. An axiomatic reformulation differs from this deductive unification in that it is the physics of the philosophers rather than the physics of the functioning sceintists. In functioning physics a theory is a conceptual tool used to understand and control some domain of physical reality. In the type of rational reconstruction philosophers are wont to indulge in, a theory is detached from its normal interpretative background and becomes an object of study in its own right. These stages will not be treated here except to bring out one significant contrast. Any discussion of normative forms, of justification, and related themes, generally occurs in discussions of deductively unified theories, their axiomatic reformulations, or the ideals thought to be proper to such formulations. When we discuss the formative stage, we are not attempting to impose any norms or seek any justifications. The goal is simply to find a generalized descriptive framework adequate to handle the developments we considered and others like them.

In the developments we discussed, various types of models were used as stepping stones towards mathematical formulations which proved successful. To make sense out of these, and even more out of the subsequent debates, it is necessary to distinguish two different types of models. *Iconic*, or visualizable models, are the type of models in which something relatively familiar, or relatively well understood, serves as a model for something relatively unfamiliar, usually because of a presumed structural resemblance. *Conceptual* models are more difficult to isolate because they are embedded in the language we speak. Here, I am using ideas developed in different ways by Kant and by Wilfrid Sellars, basically the idea that concepts embody laws and so supply a basis for material rules of influence. Because something is an $X$ it has property $Y$ or relation $Z$ when these are entailed by the concept of being an $X$. An attempt to justify this position involves problems familiar, at least since Hume. Material rules of inference rest on presuppositions, which are themselves in need of justification. Here, however, we are attempting to describe, not to justify, to treat issues internal to the development considered, not external questions.[8]

Within this general framework, we may consider the discoveries discussed earlier by using the distinction Kordig (1978) introduced between initial thinking, plausibility, and acceptability. Plausibility and acceptability ultimately require fitting a new discovey into a coherent overall account. Initial thinking, of the type we have been considering, generally involves introducing something inconsistent with the established theories. The contrast between Heisenberg and Pauli brings this out most clearly. Both were trying to develop a mathematical formulation of electronic motion which would supply a unified basis for deducing allowed transitions, their frequencies, intensities, and polarizations. Pauli was attempting to achieve this through a critical search for justifiable general laws. Heisenberg succeeded, where Pauli did not, by using different models as conceptual tools applied to manageable cases. Both de Broglie and Schrödinger were like Heisenberg in developing successful mathematical formulations on the basis of models which, in spite of valiant efforts, could not be retained as realistic images of the structures and processes in the atom.

The iconic models proved to be mere temporary props. The conceptual models had a more complex career. In the ensuing discussions in Copenhagen, Bohr rejected both Schrödinger's insistence that his wave model must be accepted as describing something real within the atom and Heisenberg's insistence that. one simply discard all models and simply think of a theory as a mathematical formulation coupled to observable data. Though the problems and difficulties this entailed are quite different from those proper to the originating discoveries, both show the same basic need. The framework required for a minimally adequate account of the different components and complex relationships that go into the discovery, development, and interpretation of new theories, is quite different from anything supplied by logical reconstructions. The development of such a minimally adequate framework remains an outstanding task for philosophers of science.

*Philosophy Department*
*California State University, Hayward*

## NOTES

[1] The most succinct statement of this position has been given by Karl Popper (1959, p. 31): "The initial stage [in a scientific development], the act of conceiving or inventing a theory seems to me neither to call for logical analysis nor to be susceptible of it."
[2] Hanson developed this position in various places such as (1958a, 1958b, 1961) and in some of his Yale seminars.
[3] See my (1976), (1977), and (forthcoming a) for de Broglie, Heisenberg, and Schrödinger, respectively. An overall interpretative history of these developments will be presented in my *Scientific Explanation and Atomic Physics, Vol. I: Historical Development* (forthcoming b).
[4] This scientific correspondence is available on microfilm. A detailed list of such material may be found in Kuhn *et al.* (1968).
[5] See Stuewer (1975).
[6] The account given here has drawn heavily on Wessels (1975) and (forthcoming) and on Hanle (1977a), (1977b), and (forthcoming). I have also benefited from corresponding with both authors.
[7] The treatment of measurement I accept as basic is that given in Krantz *et al.* (1971).
[8] I have attempted to clarify the significance of this distinction in my (1975).

## BIBLIOGRAPHY

Blackwell, Richard: 1966, *Discovery in the Physical Sciences*, Univ. of Notre Dame Press, Notre Dame, Indiana.
Hanle, P. A.: 1977a, 'The Coming of Age of Erwin Schrödinger: His Quantum Statistics of Ideal Gases', *Archive for History of Exact Sciences* 17, 165–192.
Hanle, P. A.: 1977b, 'Erwin Schrödinger's Reaction to Louis de Broglie's Thesis on the Quantum Theory', *Isis* 68, 606–609.
Hanle, P. A.: forthcoming, 'The Schrödinger-Einstein Correspondence and the Sources of Wave Mechanics', *American Journal of Physics*.
Hanson, N. R.: 1958a, 'The Logic of Discovery', *Journal of Philosophy* 60, 1073–1089.
Hanson, N. R.: 1958b, *Patterns of Discovery*, Cambridge Univ. Press, Cambridge.

Hanson, N. R.: 1961, 'Is There a Logic of Scientific Discovery?', in H. Feigl and G. Maxwell (eds.), *Current Issues in the Philosophy of Science*, Holt, Rinehart and Winston, New York, pp. 20–35.

Koestler, Arthur: 1964, *The Act of Creation*, Macmillan, New York.

Kordig, Carl R.: 1978, 'Discovery and Justification', *Philosophy of Science* 45, 110–117.

Krantz, D. H., *et al.*: 1971, *Foundations of Measurement*, Vol. I: *Additive and Polynomial Representations*, Academic Press, New York.

Kuhn, T. S., *et al.* (eds.): 1968, *Sources for the History of Quantum Physics: An Inventory and Report, Memoirs of the American Philosophical Society*, Philadelphia.

MacKinnon, Edward: 1975, 'Discussion: A Reinterpretation of Harré's Copernican Revolution', *Philosophy of Science* 42, 67–79.

MacKinnon, Edward: 1976, 'De Broglie's Thesis: A Critical Retrospective', *American Journal of Physics* 44, 1047–1055.

MacKinnon, Edward: 1977, 'Heisenberg, Models, and the Rise of Matrix Mechanics', *Historical Studies in the Physical Sciences* 8, 137–188.

MacKinnon, Edward: forthcoming *a*, 'The Rise and Fall of the Schrödinger Interpretation', in P. Suppes (ed.), *Foundations of Quantum Mechanics: The 1976 Stanford Seminar*.

MacKinnon, Edward: forthcoming *b*, *Scientific Explanation and Atomic Physics*, Vol. I: *Historical Development*.

Popper, Karl R.: 1959, *The Logic of Scientific Discovery*, Science Editions, New York.

Stuewer, R. H.: 1975, *The Compton Effect: Turning Point in Physics*, Science History Publications, New York.

Wessels, Linda: 1975, *Schrödinger's Interpretation of Wave Mechanics*, unpublished Ph.D. dissertation, Indiana University, available from University Microfilms, Ann Arbor, Michigan.

Wessels, Linda: 1977, 'Schrödinger's Route to Wave Mechanics', *Historical Studies in the Physical Sciences* 10, 311–340.

WILLIAM T. SCOTT

# THE PERSONAL CHARACTER OF THE DISCOVERY OF MECHANISMS IN CLOUD PHYSICS

The field of research on clouds and rain is a fruitful and hitherto untapped source of illustrations for the growth of knowledge in science. I wish particularly to use examples from this field in a discussion of Michael Polanyi's (1958) insistence that personal elements are necessary rather than ancillary to any account of knowing and reasoning, and of Rom Harré's (1970) assertion that the primary aim of science is the elucidation of mechanisms and structures rather than the generation of deductive systems of propositions. I do not mean to imply that Polanyi is the only writer concerned with personal elements, but I do mean that he has given especial attention to the commitments whereby scientists proceed with confidence in the face of the unspecifiability of many of their operations of observing and theorizing. I also do not suggest that Harré alone takes up the mechanism-and-structure point of view, or that I endorse all that he says about it; but I find his view a congenial approach to the complex taxonomy of explanations and consequently the wide variety of kinds of discovery.

Michael Polanyi's contribution to epistemology is a claim that the process of getting and holding knowledge involves personal elements in a fundamental way, that is, not merely as sources of limitation and distortion but as essential to observation and theory construction. I hope to make clear by examples the sense meant here of the term 'personal', but I should say at the outset that I refer to an individual's concern for truth and rationality and not to subjective biases or efforts at self-aggrandizement.[1] The personal elements that are fundamental to knowledge involve in the first place the integration of clues and cues into a perceptual or conceptual Gestalt, such as a visible cloud or a microscopic cloud droplet. It is obvious in ordinary practical experience that the particular subsidiary elements on which we rely for perceiving a whole are not generally specifiable, but Polanyi goes further in claiming that subsidiaries of which we are aware in their character as pointing to a whole and as participating in its unity will lose these characteristics if examined directly. The knowing we have of unspecifiable subsidiaries is tacit knowing. Furthermore, the rapid and unconscious nature of the integration process puts it also quite beyond specifiability − just consider how fast you can recognize a friend whom you catch sight of in a crowd of people, or how quickly you can detect

273

T. Nickles (ed.), Scientific Discovery: Case Studies, 273−289.

the cogency of a novel argument. Polanyi identifies this rapid tacit integration
of clues into a whole as the proper meaning of the term 'intuition', and
because it leads to conclusions on which we rely, even though not by a rule-
controlled process, he also calls it 'tacit inference'. The unspecifiable nature
of perception requires us to account for it in terms of the person doing the
integration, as over against an account of the integration itself.

Perception of things and relations is also personal in the sense that it involves an agent. The key word describing this agency is attention. We attend
focally *to* a whole while attending *from* subsidiaries to that whole. The subsidiaries to a perception are those elements on which we rely for our focal
attention.

Because attention comes from interest and interest can shift, the process
of integrating clues into wholes is clearly transitive. If we shift attention from
a whole to larger or more complex wholes, focally-seen objects become
subsidiaries, with consequently altered appearances, to the perception of other
objects. For example, we may perceive a single cloud by utilizing clues of
color, shape, background of sky, outline, and many items of previous acquaintance stored in memory. We may then shift to observing a field of clouds and
the accompanying sunshine and wind to see whether it is a good day for
sailing. The clouds and the clues to them change their character in this higher-
level perception.

Polanyi's claim that all specifiable knowledge rests on an unspecifiable,
tacit base amounts to a denial of any hope for establishing the validity of an
observation or of the attaining of an insight in terms of logical relations
between specifiable particulars and consequent wholes. Such a denial leads to
the necessity of a non-specifiable accreditation of our tacit personal powers
of perception and integration. Accrediting one's own or another's tacit intellectual powers means the acceptance of these skills as valid in just the same
way that we accredit our abilities to walk or swim or read or write, and this
accrediting, as an act of trust or faith, is fundamentally personal. As a theoretical cloud physicist, I am aware that my meteorologist friends are more
skilled than I in perceiving and classifying clouds and weather patterns, and I
am more skilled than they in the mathematics of cloud formation.

Dudley Shapere's account (1979) of scientific beliefs gives accreditation to
scientists' ability to believe, and Marx Wartofsky (this volume) accredits our
ability to make judgments. However, accrediting one's powers of judgment
inevitably involves acceptance of the possibility of error. We accept for ourselves and others the results of observation and reasoning that however skillful
are not strictly proven and so are always open to possible doubt. Polanyi

handles the central problem of doubt in the philosophy of knowledge by asserting that all knowledge is held by personal commitment, rather than by impersonal and detached indubitable proof. That is, we accept and take into our mental existence results that we *do* not doubt even if logically we *could*.[2] In my research on cloud droplet growth, I commit myself to vast amounts of physics and physical chemistry which underlie my subject and which I believe to have been reasonably but not absolutely established. I also commit myself to each aspect of droplet growth of which I become convinced either from my own work or from published results of others.

An important type of commitment is to an argument's being logically correct. It requires a skilled tacit integration of clues to be assured that the terms and symbols have been correctly used, that each step is logically correct, and that the steps follow each other without hiatus. For a complex argument, there is always the possibility of overlooked alternatives, hidden assumptions, and errors in reasoning, so the commitment of a skilled scientist to logical deductions is made in the awareness that they *could* be mistaken along with the conviction that they *are* not. The acceptance of the correctness of a logical proof is personal, but that which is proved is seen, personally of course, to be the most objective, universal and impersonal of entities. The structure of such commitments shows clearly the significance of Polanyi's claim that the essentially personal transcends the subject-object dichotomy. Wartofsky's account of judgment (this volume) does essentially the same thing.

It is a fact of all intellectual life that any given scholar contributes only a small amount to knowledge in her or his field and thus must rely on — and commit himself or herself — to a vast body of the results of others, as Shapere has so well described. In the more elementary subjects like physics and chemistry, the others are almost always in the same subject, but for meteorology and even more for the life sciences, many of the others are in related, supporting fields where the reliance is harder to justify. Accrediting the skills of these other researchers and trusting in the processes of refereeing, editing, scholarly publishing, and textbook writing are essential if we are to trust the background within which any individual's work can find significance. In short, participation in a community of knowers, along with its disciplinary matrices, controversies and confusions, is another essentially personal aspect of the knowing process.

To the personal coefficients of tacit knowing, self-accreditation, commitment, and trust in community, we must add the personal longing for and contact with reality. Polanyi's grounding of the mind in the body puts the person in direct contact with her or his world, in contrast to epistemological views in which the inferring mind is remote from the *Ding an sich* and must use elaborate methods of inference to get information about the world. Polanyi's view is that our bodies — better, our mind-body organisms — are those systems which we know almost exclusively in the subsidiary mode, as we rely on our sensory organs and motor abilities for attending to what we perceive and do. Retinal images are among the clues we use for perception[3], and the sensation of a pen in hand typifies the clues on which we rely for the use of a tool for writing or whatever. We extend ourselves into our tools and even our language and all our conceptions, dwelling within them as parts of our bodies and meeting the world thus embodied and being in intimate contact with it.

A person from earliest childhood reaches out with eyes and hands and all the body to the world of real things and happenings and persons. As we grow, our passion deepens into one for rationality within the world of the real. The intellectual passions of caring for evidence and logic, for intersubjective testability, and for successful prediction are developed expressions of this primary concern for what is real. Is a cloud really made of hundredth-millimeter size droplets? Are these really held up by rising air passing up through the cloud, forming ever-new droplets as the air reaches the region of 100% humidity? Or are these assertions merely good explanations of observations without bearing on reality? The reality questions are the ones we ask in our science and answer for each other and for our students. As Shapere, Wartofsky and others at this conference have said, we have discovered, recognized, and had revealed to us, many aspects of nature.

An epistemology of personal knowledge thus requires an ontology. In a world observed by transitive tacit integration, the restriction of the term 'reality' to mean something hidden behind phenomena, some essence or collection of elementary particles or systems of universal fields, will not do. It is far more pertinent, and much simpler, to say with Polanyi that we take as real any entity which we expect to meet again in unexpected ways. A cloud once recognized by the pilot of a research airplane may be detected again in terms of unexpected dimensions and properties recorded on instruments as the plane flies through, and it may be watched on the ground through the last wisps before it evaporates after a half-hour's life. The good sailing conditions may be experienced by those in boats in terms of breezes whose direction and strength they could not have predicted.

To recognize a case of rationality in nature is to commit oneself to an aspect of reality. The relation of the height to which a cumulus rises and the temperature variation with altitude is law-like and approximately explainable in rational, theoretical terms. So also is there a rational but not strict account of the roughly half-hour life of a small cumulus from first condensation to the final evaporating wisps. These accounts tell us a bit of what Nature is like.

It is the character of a real thing or a rational relationship which makes it possible for us to meet it or experience it over and over in new ways that convinces us of its reality. The mere repetition of a given experiment is not nearly so convincing — as well as being not publishable — as the carrying out of a different experiment to test the same relationship or existential claim. Our ability to anticipate unexpected meetings is both a general characteristic of all our conceptualizations and an indication of the personal powers by which innovation and discovery are made. For instance, each new cloud droplet measuring device reinforces but also modifies our previous understanding of the distribution in sizes. The terms and structures of our language have the well-known 'open texture' by which our terms are regularly given new instantiations. What we say is nearly always, except for play-acting and recitation, something new we have never said before. Furthermore, we accredit ourselves confidently in our ability to say what we mean today but had no idea of yesterday. Without such confidence our speech would be totally blocked.

Is a new instantiation of a term logically implied by our previous understanding of it? If this were so, we could, in principle at least, retrace our steps from the new use to the initial givens, as we can for ordinary deductions. But the integration of tacitly-known clues into a whole, whether familiar or unfamiliar, is an irreversible process. Once we have perceived a Gestalt, the clues to it have the character of pointers to that particular entity; beforehand some of them may have been seen tentatively as pointing to something else, or as not pointing at all. Only in rare cases can we reconstruct the state of mind we had before having seen that $x$ is present or $y$ is so. As we accept a new instance or meaning or perception and commit ourself to it, our mind changes, grows a wee bit, and we are not the same as we were before. When a pilot sees a thunderstorm ahead, his attention and flight plan are inescapably changed. Irreversibility is another of the essentially personal characteristics of the knowing process, both in the sense of a logical gap between the familiar and the new and in the sense of the resulting change in the person.

The recognition of a hitherto unknown entity or relationship is a discovery, a term which I take to cover all degrees of innovation rather than just applying to those cases of considerable surprise and substantial novelty at the tail end of the spectrum of advances in knowledge. One might think that the personal elements I have sketched out represent the psychological rather than the logical aspects of discovery; alternatively, perhaps they belong to the so-called context of discovery as contrasted with the context of justification.

However, an experience that has the character of meeting something real or true involves both the psychological elements of perception, motivation, acceptance and so on, and propositional content — those cottony clouds *are* collections of tiny droplets and there *is* a quantitative relation between the height of cloud base and the relative humidity. Furthermore, the clues by which a new entity is first seen are among those reflected on when a second look or further thought is given to the question of whether one really did see *x* or not. Justification and discovery utilize the same standards of the recognition of rationality in nature by which we become convinced; or to say it otherwise, discovery requires justificaton for the discoverer before he or she is fully committed to having made a discovery.[4]

However, there remains an important sense for the distinction between the two contexts. What is found is valid independently of the person finding it. Not infrequently the same discovery is made by several individuals without knowledge of each other. Different clues and different integrations must have been carried on by the different investigators. One cannot conclude however from the variability in personal participation among discoverers that discovery somehow must be describable in impersonal terms, for the fact remains that *any* exercise of judgment as to the truth of a proposition or the existence of an entity is a responsible and personal act. But there are many aspects of these personal acts, including chance matters of individual history and temperament, that remain in the context of an actual discovery (by one person) but do not belong to its justification (within a community).

Polanyi distinguishes between the term 'personal' and the term 'subjective', the latter having to do, for instance, with aspects of the individual that interfere with the effort to find truth. It is obvious that his description of the personal in knowing is an ideal one. We do not always act in such ideal ways. Sometimes scientists make discoveries with little awareness, as if sleepwalking, to borrow from the title of Arthur Koestler's book (1959) on Copernicus, Kepler and Galileo. Scientists sometimes consciously and more often unconsciously steal ideas and results from others, fight over prestige and priority, and do a number of other nasty things. However, when we take 'discovery' to mean an advance in knowledge that convinces others and becomes accepted as part of established truth, we intend thereby to overlook or discount these subjective failures.

A scientific discovery is such only within an accepted paradigm of explanation and body of scientific beliefs. In cases in which a discovery shifts a paradigm, it will only be within the new paradigm that it is recognized as a discovery. Some account of scientific explanation is therefore needed to fill out a Polanyian view of discovery. I take it that Frederick Suppe's account (1977) of the demise of the 'Received View of Science' — the two-languages-and-logic view — is accurate. My alternative is to combine personal knowledge with Rom Harré's view that the fundamental aim of science is to develop statement-picture complexes describing the mechanisms of nature, using the broadest meaning of the term mechanism. According to Harré's view, the formulation of law-like generalizations and deductive systems is an assistance to the basic aim, and often a powerful one, but not the aim itself.

The mechanism of cumulus cloud formation, whereby a parcel of air is warmed by heated ground, rises in Archimedean style through surrounding cooler air, cools as it rises by adiabatic expansion, achieves 100% humidity at a certain level, and develops droplets above this level as condensation occurs on appropriate sizes of dust particles, is an explanation that makes sense of what skilled observers report about clouds. The elements of this explanatory conception have all been observed in sufficient detail to make convincing both their separate existences and the validity of the statement-picture complex of their joint operation in the formation of cloud droplets. The mechanism as a whole has been established. On the other hand, while certain quantitative law-like generalizations have been found approximately valid, these generalizations have little of the precision which has come to be expected of atomic and nuclear physics. For instance, the quantitative thermodynamics of the cooling of rising air is well established and corroborated, but so far the only attempts mathematically to model the motion of rising parcels of air involve radical idealizations of the shape and uniformity of these parcels. They are idealized into cylinders or other figures of revolution, or represented by variables at a small number of grid points in space. Cloud physicists generally accept the conception that actual parcels are more complex and varied than the models and beyond the power of any mathematics to represent precisely.

How do we make discoveries about the mechanisms and laws of nature? How do we make nature speak persuasively to our colleagues who share our scientific beliefs? Most scientific discoveries are the outcomes of deliberately chosen problems. Occasionally a discovery is a surprise, a bit of serendipity, but even in these cases, the chance finding sets up a problem, perhaps just the problem of the verification that something indeed has been found, but more often the problem of conditions, properties, and structures in that which was observed.

Hence we can ask, what is a problem? When Socrates announced his problem of determining in what virtue consists, Meno asked him the basic question of human agency in discovery: how could he possibly search? If he did not know the answer he would have no criterion for recognizing it, yet if he knew the answer he had no problem. Polanyi's (1966) reply to this dilemma is to describe a problem as a tacit integration of clues that point only vaguely to the as-yet-unseen coherence. In fact, finding a good problem in science is itself a discovery. An example from our laboratory is Jim Telford's problem of the relation between the history of a cumulus cloud and characteristics of the process whereby dry air is mixed into it. As we proceed to experiment and calculate concerning such a possible coherence-yet-to-be-found, the clues change, more clues are added, and the coherence either sharpens into a recognition of rationality in nature or perhaps disappears if the problem turns out to be empty. (One of the important skills of a scientist is the ability to know when to stop an investigation that has proved a non-problem, or an insoluble one.) All the personal elements of tacit knowing, accreditation of one's own and another's tacit powers, commitment, participation in a community of trust, passion for reality and rationality, recognition of unexpected consequences of one's conceptions, and irreversibility in the acquisition of knowledge, are needed for a proper answer to Meno. They jointly make possible an account of discovery in spite of the unspecifiability of the tacit. Strategies and maxims can be described, but no algorithms can be specified.[5]

Discovery consists in finding, coming upon, becoming convinced of, some aspect of the mechanisms of nature responsible for some given set of phenomena. The existence of enormous varieties of mechanisms in nature, taken with the view that understandings of mechanisms are what we are about, leads to the expectation of finding a wide variety of types of discovery. The inner workings of a mechanism may be found from observational clues, or deduced from known theories, or found by tracking down discrepancies, or arrived at by brilliant speculation from a combination of new observations, established theories, and general scientific background. The discovery itself may be of a structural element in a process, or a quantitative law for a measurable aspect of a process, or a new mathematical method for making a difficult computation. In addition to differing varieties of structures and processes, the demands of a given subject matter, variants in the coherence or lack thereof in scientific background, chance occurrences, and more specifically the personality of the discoverer and her or his relations to colleagues, all serve to generate a broad spectrum of discovery types. Thus a taxonomy of types is more likely to be insightful than attempts at widely-applicable logical generalizations. Without even attempting a systematic taxonomy in my present research field, I should like to illustrate the variety and the personal coefficients by reference to half-a-dozen cloud physics examples, interspersed with comments on reductionism, on frameworks of commitment, and on false starts in problem solving.

The quantitative theory of condensation growth is an example of *discovery by construction from known theories belonging to lower levels of complexity, following an integrative appreciation of the mechanism.* The chemistry of the solubility of cloud nuclei and the resulting effects on the vapor pressure of small droplets, along with the ordinary properties of liquid-to-vapor equilibrium and the effects of surface curvature, lead to a prediction of the rate of growth of a droplet when its immediate surroundings are slightly more humid than equilibrium conditions. The theories of vapor diffusion and heat conduction allow a computation of the relation between the conditions at the drop, the conditions far away in the average or so-called ambient behavior of the region between droplets, and the rate of inflow of vapor toward a growing drop and the corresponding rate of outflow of the released latent heat. H. Köhler (1921) saw that one could establish a simple formula for the radius of a droplet which was just in equilibrium for a given humidity, when the added evaporation due to surface tension just balanced the restriction due to dissolved salts. Howell (1949) made the first extensive calculation of the growth of a group of droplets based on the deviations from equilibrium and competition for vapor, and Squires (1952) worked out a precise quantitative formulation involving skilled judgments as to which of many side effects are to be included and which to be neglected. The results are believed to be a reasonable approximation not because of any reliable experimental tests but on the grounds of coherence with known physics and chemistry and the logic of deduction when contingent disturbances can be reasonably argued to be negligible.

Laboratory experiments on the growth of particles from the dust phase about a hundred-thousandth of a millimeter in diameter to water droplets a hundredth of a millimeter in size are extremely difficult, and the effects of droplet fall-out in laboratory chambers without benefit of the kilometer-sized updraft that nature provides underneath clouds have just about prevented any direct corroboration. However, the possibility of making such tests has now arisen, in the proposal for the orbiting NASA Space Shuttle to carry aloft a gravity-free cloud chamber and associated apparatus. The Atmospheric Sciences Center research group at Nevada's Desert Research Institute has a major share of the responsibility for the design of these experiments. I am fortunate to have a small part in the improvements of the theory to a point where its expected precision corresponds to the 2 to 3% accuracy hoped for in the new experiments. The opportunity has thus arisen for a highly systematized testing of a theory already believed to be substantially correct, with the possibility of some disconfirmation which might lead to a genuine discovery (*sometimes* falsificationism is an appropriate doctrine).

The theory of the condensation process may seem to be reducible in the sense of its being a logical conjunction of its various underlying laws of physics and chemistry. However, this is not the case. The logic of the independence of upper levels from lower in a hierarchy of explanation has been given by Polanyi in his concept of boundary control.[6] The logic of this principle seems so clear to me that I wonder why it had not been noticed before — it is quite a different type of argument from the Polanyian personal knowledge thesis. In the case of droplets, the argument is that a droplet may fail to grow or may even evaporate without any of these underlying laws failing to be instantiated. The laws of physics and chemistry are all capable of being written in the subjunctive conditional mode, if so-and-so then such-and-such: if the gradient of water vapor were to have a certain magnitude outward from the drop, vapor would flow in at a certain rate. It is the conditions represented by the 'ifs' that matter here. Failure to have a droplet grow on a cloud nucleus means that the air is not sufficiently saturated. It is obvious that boundary conditions cannot be made parts of the laws of chemistry and physics, because the latter would then lose their generality as subjunctive conditionals. It is not so obvious that boundary conditions in complex cases may become sufficient conditions for the existence of a structure or mechanism, conditions which may readily fail of instantiation. In such cases, the boundary conditions may be said to control the existence or functioning of a level of organization that in conventional metaphor is called 'higher' with respect to those physical processes that are thus controlled.

Another way to put it is that the laws of physics and chemistry cannot distinguish between a growing droplet and a dry or drying-out nucleus, so that they cannot possibly lead to a deduction concerning the existence of these structures. The underlying laws are of course *necessary* for an explanation, but they are clearly not *sufficient*. The sufficiency is provided by the organizing principles of the mechanisms in nature which Harré says are what we are looking for anyway.

An especially good example of the hierarchical principle of boundary control is the mechanism of cloud formation that was introduced earlier. This is a case of the *discovery of an organizing principle by the independent development of the necessary underlying laws and their almost obvious synthesis.* The common and most reasonable scientific view of cumulus clouds in the seventeenth century was that they float aloft because they are made of hollow, lighter-than-air bubbles. It was the change of scientific beliefs in the nineteenth century in respect to six aspects of physics and chemistry that made the present mechanism self-evident. Stokes brought forward the law of the terminal velocity of small falling spheres, the thermodynamics of heat conduction and convection was made clear by Clausius and others, Poisson showed that rising air cools adiabatically, the distinction between air and water vapor and the concept of the molecular mixing of air and water molecules were made evident by Dalton's chemistry, the conception of relative humidity was derived by Lavoisier and Dalton, and the necessity of dust nuclei on which water drops could form above 100% relative humidity was shown experimentally by Coulier and Aitken.[7] I do not know who first consciously pulled all these elements together into a discovery of how clouds are formed, although there are records of many who took part in applying one or another of these insights to clouds. Perhaps we could say in Shaperean or Kuhnian style that the paradigm of what constitutes explanation and what type of research is appropriate had shifted with the development of a new set of scientific beliefs, of commitments of wide generality as to what nature is really like. However, each of these developments was relatively 'normal' in Kuhn's terms. The paradigm shift for the constitution of clouds and for the appropriate kinds of research programs developed as a consequence of the joint meaning of various discoveries, few of which were themselves motivated by dissatisfaction with an old paradigm. Nevertheless, the modern view is decidedly novel in the sense of not being derivable or even intuitable from the old interpretive framework.

The realist point of view I am following implies that a discovery is felt to be one by the discoverer, and accepted by those capable of critical appreciation of it, only if it is considered to have bearing on reality. Coherence with what is known — with commitments already held as to what is real — is one of the basic features of a conviction that something new has been discovered. But nearly all of what any scientist believes about the field of his expertise he has learned from others, others whom he trusts to have responsibly reported their findings, or to have accurately transmitted research results in textbooks, review articles, university lectures and reports at meetings. The frame of reference of any discovery is not merely a paradigm of accepted methods, but a system of beliefs about what is the case, exactly as Shapere has said. Meteorologists do believe in the molecular theory of gases and a great deal more of physics and chemistry. Many of these beliefs have a tentativeness about them, as being the best available results but subject to possible modifications, but nevertheless are accepted and operated with as relatively certain.

All of a scientist's commitments are accepted with a framework of more general commitment. This has been one of Polanyi's principal claims and has led some philosophers to attack him as a relativist. Polanyi appears to say that since every communally accepted set of commitments has the same structure, every one is equally valid. However, say his critics, there are surely rational and objective grounds for accepting some sets of commitments, or at least the general scientific set, over others, such as witchcraft, occultism, or primitive nature worship. I find as a scientist that there are indeed rational and objective grounds for accepting the commitments I and my colleagues make. In fact, the personal aspects of knowing on which I insist are those by which our search for rationality and objectivity may be carried on. However, rationality, objectivity, and the openness to new truth, values held personally and passionately by scientific researchers and students, are themselves part of the frame of reference. These values, of course, are also held in nonscientific parts of our culture. There is no intellectual ground outside science and therefore outside rationality on which I could objectively compare scientific beliefs with any others.

The scientific community as a whole consists of many overlapping and interlocking communities of men and women doing research in specialized areas, such as the cloud physics field in which I now make my home and the particle physics area to which I have long belonged. The overlapping judgments which scientists can make of neighboring fields lead to a single interconnected fabric of science as a whole. No so-called revolution in science can break, or break away from, this fabric, so that even if some basic abstractions, such as those on absolute time or perfect determinism, are changed, shared commitments are maintained about a host of real entities at all levels of the hierarchy of complexity — molecules, diffusion processes, droplets, updrafts, clouds, wind fields, polar fronts, global circulation patterns, and on and on.

To get back to examples, the accepted method of calculating droplet growth by condensation, as mentioned above, is a case in which the theory is believed because of its derivation from accepted physics and chemistry along with its own organizing principle, without there being as yet any reliable experimental verification. An opposite situation is that of the terminal velocity of fall of small droplets, *a case in which many experimental results are combined into a believable formula without an accompanying theoretical validation.* Stokes's famous calculation was based on the neglect of inertial forces as compared to viscous forces and only applies to the smallest droplets, less than say 1/100 millimeter or 10 microns in diameter. For larger droplets, the inertial forces become important and the necessary mathematics is intractable. The only theoretical result available is the established scaling relation according to which the ratio of the actual drag force to the Stokes value is the same for all cases of the same Reynolds number, namely the product of speed, liquid density, and diameter, divided by the coefficient of viscosity. Experiments with falling spheres in vertical wind tunnels have provided the now-accepted results for the relation between drag force and Reynolds number without there being any substantial theoretical vertification.[8]

Condensation growth theory and the terminal velocity empirical relation are developments of relatively little surprise, not often dignified by being entitled 'discoveries'. A clearly innovative example is the first discovery of a mechanism of rain formation from cloud droplets, *a case in which discovery combined a chance observation with astute speculation grounded in previous skill and training*. In 1922, the Norwegian meteorologist Tor Bergeron was aware of the problem of rain formation: each rain drop has roughly the volume of a million cloud droplets, and cloud droplets seem to be so uniform in size and so little subject to mutual collision that it was a puzzle as to how one drop in a million could ever collect the rest. He also knew the result, pointed out by Wegener in (1911), that the vapor pressure of supercooled water drops is higher than that of ice crystals with a maximum difference at about $-15°C$. Furthermore, Bergeron had spent many hours over many years in observations of clouds that rained and clouds that did not. His chance observation came as he took several morning walks in a winter fog at a hotel in the mountains above Oslo. He noted that when the temperature was below freezing, but only then, the pathways in the woods were free of fog and at the same time the tree branches were coated with ice. With the repressed memory of Wegener's result as undoubtedly one of his clues, he recognized right away ("il me vint alors de l'ésprit") that the fog droplets had evaporated and their vapor had been taken up by the ice on the trees. The reason the fog outside the wooded area was still there was that there were no ice crystals in or near it to start collecting water vapor; he knew that the individual fog droplets would not freeze spontaneously until the temperature dropped to $-30°$ or $-40°C$. Bergeron put this observation in the back of his mind during his work with the Norwegian Weather Service in the development of frontal analysis and gradually came to see clearly that if a cloud which was high enough to be at below-freezing temperature contained a few particles on which ice could form — only one in a million in proportion would be needed — the resulting ice crystals would grow and each take up the water from many droplets until the crystal was big enough to fall through the cloud and collect still more water. As the crystal fell into the lower, warmer regions, it would melt and become a raindrop. By 1928 Bergeron was convinced he had made a genuine discovery and so published a paragraph suggesting this precipitation process in his doctoral dissertation, Bergeron (1928).

It was not until 1933 that Bergeron presented a full paper on his conception of the releasing mechanism for precipitation at an international conference in Lisbon.[10] As he wrote in 1972, his enthusiasm for his discovery was great enough both to lead him to work out its conceptions in detail and also to become committed to the idea there could be no other process, so that his claims were exaggerated. Nevertheless, as he pointed out, to get a new idea accepted it must be believed in and fought for.

It has now turned out from both field observations and theoretical considerations that there are other processes — some warm clouds contain a small fraction of extra large droplets which fall and collect others until they are of raindrop size, and in some clouds which are mostly ice, larger ice particles collide with others into graupel particles which melt as they fall. However, after some important quantitative work by Findeisen (1938), the Bergeron-Findeisen process remains as probably the most prevalent of the rain-producing mechanisms. Bergeron's mechanism was accepted because of his arguments from terrestrial observations, as well as on the basis of general, large-scale cloud observations, before microphysical observations in clouds aloft could directly verify it.

*A case of a microphysical process observed in the attack on a carefully chosen problem but not yet clarified theoretically* is that of the Hallett-Mossop ice multiplication process (1974). Several observations had been made in which prospective ice nuclei were taken from beneath or beside a cloud in a large bag of air and tested in the laboratory for their ice-crystal-producing ability. The density of these particles was thus compared with the crystals actually in the upper part of the cloud (with due regard for time lags). In a number of cases, there were 1000 to 10,000 times as many crystals high in the cloud as could be accounted for by the contents of the air that moved up into the base of the cloud. The problem was then to account for the discrepancy. Careful examination was made of the instruments from which the discrepancy was reported, but in spite of the generally low precision of such observations, no convincing case could be made against the difference being real. A study was made of the known facts of crystal growth and of possible ways in which crystals could splinter into smaller ones and multiply themselves by chain reaction; however, no convincing mechanism could be found which fitted the facts.

In 1975, John Hallett from the University of Nevada Desert Research Institute and Stanley Mossop of the Commonwealth Scientific and Industrial Research Organization in Australia set up apparatus in a large tent within which they could form clouds of droplets of controllable sizes in an atmosphere of controllable temperature and humidity. Within this atmosphere they rotated a small rod on which ice could grow, at controllable speeds of motion relative to the droplets so as to resemble a falling ice crystal. They found a combination of variables, at a temperature of only -5°C or so (much warmer than had been suspected) when they could see and count large numbers of tiny ice crystals, scintillating in their tent-enclosed cloud. The ice on the rod grew long needle-like spikes at this temperature, and presumably some mechanism of collision with both small and large droplets, along with a Bergeron-like process with accreted droplets, led to crystallites breaking off. The existence of the mechanism was clearly shown and found to fit the observations and account for the discrepancy. However, the mechanism itself remains to be elucidated.

The work of Hallett and Mossop illustrates a case in which discovery was deliberately sought, the innovative feature being an ingenious experiment which allowed a range of variables and a method of observation that had been previously unavailable. It is a good example for examining the relation of discovery and justification. Justification here means a carefully argued case that indeed there is such a multiplication process operating under such-and-such conditions. But until Hallett and Mossop had repeated the experiment enough and had checked all aspects of it − i.e. had justified their results − they were not convinced that they had made a discovery.

There are surely many cases in which the initial intuition of a possible discovery proves false. Much of my research has involved the search for ways to solve equations representing physical processes or relationships. For instance, I recently found a hundred-times-more-rapid-than-usual method for computing the growth of a group of droplets, using mathematics expressive of the principal physical effects (Scott, 1978). This was *a case of an intuitive scanning of plausible starting points, followed by a chain of reasoning along with numerical experimentation.* In this type of work, some of my more intuitive efforts are carried on at night in bed during periods of wakefulness. There have been several occasions in which the imagining, intuiting, and reasoning I carry on, focussed on general principles rather than on details for which writing is needed, leads me to a method I believe will work. In the morning, I write formulas representing my hoped-for discovery and carry on mathematical deduction to find the consequences. Sometimes I am thus led down a path of reasoning to an interesting and unsuspected conclusion. More often than not, I find a logical flaw, unnoticed in my nocturnal state, that makes the method untenable. Then what happens is that the tacit integration of clues which pointed to a solution quite disappears, being replaced by a sense of a collection of clues that points nowhere. I can scarcely remember the bright nocturnal idea. The intuition was not a discovery.

How often it must be for all kinds of theoretical and scientific work that a temporary vision of discovery disappears when something to contradict it is discovered. Once this happens, that pulling together of subsidiaries that is governed by our obligation to the truth collapses, and the subsidiaries are no longer the same. The discoveries we wish to study in the philosophy of science are those that in fact are justified, at least by the finders.

Actually, a discoverer has at least three stages of personal justification, one when she or he is convinced of having found something, another when it is described to one's local colleagues, and a third when it is written up for publication. The occurrences of acceptance by editors, by referees, and by the scientific community constitute external stages. However, it is important to keep in mind that the only true outside justifiers are those in the same or closely related fields whose skills and training make them capable of valid falsification and therefore of valid positive criticism. But all the justifications are commitments to the discovery of a reality that will be met again in un-expected ways.

Let me conclude with some of Polanyi's own words on discovery:

... The true sense of objectivity in science ... [is] the discovery of rationality in nature. ... The kind of order which the discoverer claims to see in nature goes far beyond his understanding, so that his triumph lies precisely in his foreknowledge of a host of yet hidden implications which his discovery will reveal in later days to other eyes. (1958, p. 64)

*Department of Physics*
*University of Nevada, Reno*

## NOTES

[1] It is in this sense of 'personal' that we must construe Burian's (1980) account of the human-centered aim of philosophy.
[2] Shapere (1980) gives a good account of the allowance for special but not general doubt.
[3] See James J. Gibson (1968).
[4] See Polanyi (1958, p. 13–14).
[5] *Cf.* Marx Wartofsky's contribution to this volume.
[6] See Marjorie Grene (1969).
[7] W. E. Knowles Middleton (1965) gives extensive references for both the seventeenth century ideas and the nineteenth century changes.
[8] See Beard and Pruppacher (1969).

[9] Professor Bergeron described this discovery to me in Uppsala in 1970. See his (1972).
[10] See Bergeron (1928).

## BIBLIOGRAPHY

Beard, K. V., and H. R. Pruppacher: 1969, 'A Determination of the Terminal Velocity and Drag of Small Water Drops by Means of a Wind Tunnel', *Journal of the Atmospheric Sciences* 26, 1066–1072.

Bergeron, T.: 1928, 'Ueber die dreidimensionale verknüpfende Wetteranalyse', *Geofysiske Publikasjoner* 9, 29–31.

Bergeron, T.: 1935, 'On the Physics of Cloud and Precipitation', *Proc. 5th Assembly, Sessions of the Meteorology Association, International Union of Geodesy and Geophysics*, Vol. 2, p. 156. Lisbon.

Bergeron, T.: 1972, 'L'Origine de la Théorie des Noyaux de Glace Comme Déclencheurs de Precipitation, un Cinquantaire', *Journal de Recherches Atmospheriques* 6, 49.

Burian, R.: 1980, 'Why Philosophers Should Not Despair of Explaining Discovery', in T. Nickles (ed.), *Scientific Discovery, Logic, and Rationality*, D. Reidel, Dordrecht.

Findeisen, W.: 'Die kolloidmeteorologischen Vorgänge bei der Niederschlagsbildung', *Meteorologische Zeitschrift* 55, 121.

Gibson, James J.: 1968, *The Senses Considered as Perceptual Systems*, George Allen & Unwin, London.

Grene, Marjorie (ed.): 1969, *Knowing and Being, Essays by Michael Polanyi*, Routledge & Kegan Paul, London.

Hallett, J. and S. C. Mossop: 1974, 'Production of secondary ice particles during the riming process', *Nature* 249, 26.

Harré, Rom: 1970, *Principles of Scientific Thinking*, Univ. of Chicago Press, Chicago.

Howell, W. E.: 1949, 'The Growth of Cloud Drops in Uniformly Cooled Air', *Journal of Meteorology* 6, 134.

Koestler, Arthur: 1959, *The Sleepwalkers*, Macmillan, London.

Kohler, H.: 1921, 'Zur Kondensation des Wasserdampfes in der Atmosphäre', *Geofysiske Publikasjoner* 2, No. 1, p. 3 and No. 3, p. 6.

Middleton, W. E. Knowles: 1965, *A History of the Theories of Rain*, Oldbourne, London.

Polanyi, Michael: 1958, *Personal Knowledge: Towards a Post-Critical Philosophy*, Univ. of Chicago Press, Chicago.

Polanyi, Michael: 1966, *The Tacit Dimension*, Doubleday, Garden City, N.J.

Scott, W. T.: 1978, 'Fast Method for Computing Near-Equilibrium Droplet Growth Rates', *Bulletin of the American Physical Society* 23, p. 20, paper A12.

Shapere, Dudley: 1980, 'Rational Change of Standards in the Development of Science', in T. Nickles (ed.), *Scientific Discovery, Logic, and Rationality*, D. Reidel, Dordrecht, Holland.

Squires, P.: 1952, 'The Growth of Cloud Drops by Condensation', *Australian Journal of Scientific Research, A*, Vol. 5, pp. 59 and 473.

Suppe, Frederick (ed.): 1977, *The Structure of Scientific Theories*, 2nd ed., Univ. of Illinois Press, Urbana.

Wegener, A.: 1911, *Thermodynamik der Atmosphäre*, J. A. Barth, Leipzig.

Wartofsky, Marx: 1980, 'Scientific Judgment: Creativity and Discovery in Scientific Thought', this volume, p. 1.

DAVID A. BANTZ

# THE STRUCTURE OF DISCOVERY: EVOLUTION OF
# STRUCTURAL ACCOUNTS OF CHEMICAL BONDING

## 1. INTRODUCTION

In their important 1927 paper, Werner Heitler and Franz London provided the first qualitative explanation of the chemical bond on the basis of quantum mechanics. Their application of Schrödinger's equation to the simplest molecule appeared to elucidate the physical basis of the covalent bond, and thereby initiated 'quantum chemistry'. Although the numerical results (for bond length and bonding energy) were only roughly in agreement with the experimentally determined values, the theory was regarded as a great success and, according to Slater (1975, p. 93), "produced a sensation among physicists." The recognition given to the work of Heitler and London by both chemists and physicists, the extent to which subsequent work in the quantum structure of molecules (particularly valence bond theory) is based upon the methods introduced (or at least made use of) in the Heitler-London theory, and the continued use even of the terminology Heitler and London employed in their interpretation of approximate solutions for the bound state of the electrons and the binding energy ('overlap integral', 'exchange energy', for example) all point to the significance of the Heitler-London theory of the hydrogen bond as a theoretical discovery. In this paper I am going to take this theory, and the path to it, as an example of a discovery of a theoretical description or account which at the same time bridges two fields of science (or points toward a 'unification' of those fields or 'incorporates one field into' another) and is the basis of a great deal of further scientific work on similar problems (the source of a 'research program' or minor 'scientific paradigm').

291

*T. Nickles (ed.), Scientific Discovery: Case Studies*, 291–329.
*Copyright © 1980 by D. Reidel Publishing Company.*

A study of the development of ideas on the physical explanation of chemical bonding may thus be relevant to the unity of science or reduction (at least the questions of how such unifications or reductions are arrived at, or what kinds of scientific reasoning support reduction claims); it may also provide a useful illustration of the manner in which a solution to a problem (even, as in this case, an approximate and partial solution) serves as a model or at least a foil for subsequent work on the same and similar problems. It will become clear, I hope, that these two problem areas in the philosophy of science are intertwined with the elucidation of the process of *discovery*, and indeed that more or less 'orthodox' views in these areas have helped obscure the understanding of that process. A false image of scientific theories, and of the relation of theories to evidence, can create the illusion that theories are, and even *must* be, all of a piece, and consequently that discovery or invention in the theoretical realm can have no structure. Rather than argue in general terms, I will illustrate this fact presently with respect to the Heitler-London theory; to provide at least a preliminary analysis of the stages of reasoning in the discovery of a theoretical account of a subject matter will require examination of questions not addressed (or even raised) by orthodox accounts of scientific theories. This is of course one reason the orthodox accounts have made an analysis of discovery seem a will-o'-the-wisp. These questions will in turn require an examination of various models of molecular structure proposed by both chemists and physicists and the relations between them, and the use to which approximate and even recognizably false descriptions can be and are made use of in constructing more adequate accounts (the way in which, in Popper's phrase, scientists 'learn from their mistakes', but learn more than merely *that* they are mistakes, as Popper's account of testing suggests). Within the space of this paper I will only be able to sketch a rather coarse grained picture of the structure of the discovery process, directed toward depiction of the progression of major ideas and directions of research rather than the processes leading to the contribution of particular scientists; but even this coarse grained picture will illustrate some general features of discovery and progress within a limited field of science, features which are inadequately and misleadingly thought of under the rubrics of hypothesis testing or applications of theory. While many of the features to be indicated are apparently generalizable to discovery in other fields, some may characterize differences between the methodology of chemistry (or of the branch of chemistry concerned with the explanation of the basic phenomena of bonding and molecular structure) and that of other fields.

## APPENDIX: THE HEITLER-LONDON THEORY

The outlines of the Heitler-London theory of the homopolar (covalent) bond in the hydrogen molecule, as developed in their 1927 paper, can be indicated as follows: (1) The hydrogen molecule is a stable system of two hydrogen atoms, each atom in turn composed of a single proton and (lighter) electron (see Figure 1).

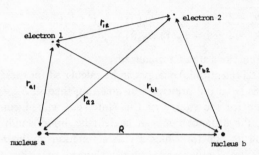

Fig. 1. Schematic representation of $H_2$; notational convention of Heitler and London's treatment.

(2) These particles interact electromagnetically, other interactions being neglected in this treatment. (This means the potential energy, $V$, is

$$e^2 \left[ \frac{1}{R} + \frac{1}{r_{12}} - \frac{1}{r_{a1}} - \frac{1}{r_{a2}} - \frac{1}{r_{b1}} - \frac{1}{r_{b2}} \right]$$

where $e$ = electron charge.

(3) The framework for a quantum mechanical description of a system of interacting particles is a series of state functions ($\psi_n$) which are functions of the coordinates of all the particles, the Hamiltonian operator ($H_{op} = \frac{h^2}{2m} \nabla^2 + V$) which can be written from the classical expression of the energy of the system, and the time-independent Schrödinger equation relating these functions and the energy ($E_n$) of the state, $H_{op} \psi_n = E_n \psi_n$.

(4) Writing the Schrödinger equation for the hydrogen molecule (making use of (1) and (2) above) is a straightforward exercise. Its solution is not, and Heitler and London had to devise simplifications and approximations to show that the lowest energy state, $\psi_o$, with energy $E_0$ describes a bound state.

(5) At large separations of the nuclei we have in effect two hydrogen atoms, and the state representing the two electrons is just the product of the

two *atomic* wave functions, so that in this limit the molecular wave function can be written down: $\psi = \psi_a(1)\,\psi_b(2)$. In the molecule, electron 1 might just as well be on nucleus $b$ as nucleus $a$, so that an equally good state would be one with the coordinates of the two electrons interchanged: $\psi = \psi_b(1)\,\psi_a(2)$. More generally, the molecular state can be written as some combination of these two states. Heitler and London used arguments similar to those Heisenberg had used in the case of helium to show that the only allowed combinations were

$$\Psi_+ = \psi + \bar{\psi} = \psi_a(1)\,\psi_b(2) + \psi_b(1)\,\psi_a(2) \text{ and}$$
$$\Psi_- = \psi - \bar{\psi} = \psi_a(1)\,\psi_b(2) - \psi_b(1)\,\psi_a(2).$$

(These wave functions are unnormalized.)

(6) The total (electronic) energy can be calculated for these (approximate) wave functions from the principles of quantum theory: $E_\pm = \langle \Psi_\pm | H_{op} | \Psi_\pm \rangle$. The expression for the energy of this state, $E_\pm$, will of course involve the parameter $R$, the distance between nuclei (the 'bond length'), and a bound state then will be one for which $E_\pm$ as a function of this separation has a definite minimum. Heitler and London showed that the state $\Psi_+$ had a minimum at about the correct (empirically determined) separation and that $\Psi_-$ led to a repulsive interaction. They thus interpreted these two states as representing bond formation and elastic collision, respectively (see Figure 2).

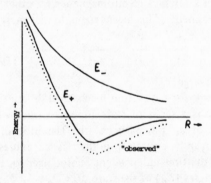

Fig. 2. Calculated energy of $\Psi_+$ and $\Psi_-$ as a function of internuclear distance, $R$. (after Heitler and London, 1927, p. 462).

(7) The expression for the energy was interpreted as follows: since the molecular wave function is formed from the hydrogen *atomic* wave functions, the energy can be broken into components from different combinations of

these functions. The points emphasized by Heitler and London can best be illustrated if we use the functions $\psi$ and $\bar{\psi}$ defined above and write $H_{op}$ as the sum of the operators which appear in the hydrogen *atom* problem, $H_a = \hbar^2/2m\nabla^2 a - e^2/r_{a1}$ and $H_b$, plus additional terms arising from the additional interaction of the two atoms, $V_i = e^2 [1/R + 1/r_{12} - 1/r_{b1} - 1/r_{a2}]$. The total energy of the bound state is easily shown to be the sum of terms which appear in the hydrogen problem, that is $E_a$ and $E_b$, the calculated energies of the hydrogen atom in the (electronic) state $\psi_a(1)$, plus further terms arising from the atomic interaction, $V_i$. The energy in the bound state, $E_+$, is given by

$$E_+ = \langle \Psi_+ | H_{op} | \Psi_+ \rangle = E_a + E_b + \frac{\langle \psi | V_i | \psi \rangle}{1+S} + \frac{\langle \bar{\psi} | V_i | \bar{\psi} \rangle}{1+S}$$

where $S = \langle \psi | \bar{\psi} \rangle$, the *overlap integral*. The product $\langle \psi | V_i | \psi \rangle$ is referred to as the *coulomb integral* and $\langle \bar{\psi} | V_i | \bar{\psi} \rangle$ the *exchange integral*. The coulomb integral is smaller than the exchange integral. Both quantities *lower* the overall energy of the bound state, but Heitler and London emphasized the importance of the exchange integral as providing the stability of the bound state.

(8) Finally, Heitler and London linked their discussion to the Pauli principle by noting that the state $\Psi_+$ is *symmetric* in the coordinates of the electrons while $\Psi_-$, which represents a repulsion (elastic collision) of the two atoms, is antisymmetric. Symmetric (antisymmetric) wave functions describe electrons with antiparallel (parallel) spins. Hence, the covalent bond is formed between atoms with antiparallel electron spins; if the spins are parallel, no bond is formed.

## 2. FRAMEWORKS FOR DISCOVERY

The Heitler-London theory of the hydrogen bond is evidently a significant scientific discovery, but what kind of framework can be used to understand that significance — understand it, that is, as more than merely 'something new'? Familiar accounts of the nature of scientific theories as a set of postulates or fundamental assumptions plus correspondence rules suggest two possibilities. These two suggestions — (A) *application* of a fundamental theory and (B) an empirical *hypothesis* connecting the concepts of physics and chemistry — will be shown inadequate; the questions they fail to answer (or even ask) suggest a more adequate framework: (C) the solution of chemical problems in terms of the physical structure of atoms and molecules.

(A) The interest in the Heitler-London paper shown by physicists (as opposed to chemists), and its appearance during the period of rapid development and generalization of the quantum theory suggest an interpretation of this theory as an application of the newly formulated general principles of quantum theory to a specific subject area. The demonstration that even an approximate solution to the Schrödinger equation represents a bound state (a state in which the component hydrogen nuclei are separated by a finite distance, $R$, approximately the same as the experimentally determined bond length) indicates the *applicability* of the general theory to a 'molecular environment'. (Both classical physics and the old quantum theory had failed to account for the stability of molecules, although, as we shall see, there had been some ingenious attempts.) Thus within this interpretive framework, the significance of Heitler-London is (1) the nature of the special assumptions which allowed them to construct a solution to the Schrödinger equation in this particular case (the writing of the general equation itself being, as I noted, trivial) and which might be expected to be generally useful given their success in this case, and (2) the confirmation or corroboration of the principles of quantum theory which were the basis of their calculations. On this interpretation the work of Heitler and London can be seen as a rational extension of Heisenberg's solution of the other problem involving two electrons: the helium atom; and in fact the starting point of Heitler and London's solution is to use the same combinations of atomic wave functions from which Heisenberg constructed wave functions for the helium electrons.

This framework is obviously consistent with the orthodox view of scientific theories as general calculi to be supplemented by correspondence rules or interpretive postulates for application to a particular domain; it is also consistent with widely accepted doctrines of the unity of science and the fundamentality of physics, according to which the *most* general laws and properties are those dealt with by physical theory, and which are restricted or specialized in particular contexts by the assumption of additional 'constraints' or 'boundary conditions'. This framework brings some structure into an account of the discovery of Heitler and London, as the rationale in the previous paragraph indicated, but it also leaves much about this discovery (indeed of any discovery of the kind I am dealing with) mysterious: Why, for example, should *chemists* manifest a professional interest in a theory in which the quantitative results have errors of the same magnitude as the quantities themselves, far greater than experimental error in the techniques of physical chemistry for the same quantities? Also, it is *not* the case that the existence of bonding was deduced as a consequence of quantum theory plus 'boundary conditions'. Instead, the results of *chemical* studies showing the importance of valence electrons in bonding, together with the known configuration of nuclei, suggested several simplifications and approximations to Heitler and London. In what ways does this fact affect our understanding of what was discovered by Heitler and London and its relation to other fields of knowledge? Also, since the formulation of the Schrödinger equation, the solution of which yields the state or wave function, and the energy of that state, is trivial, is not the contribution of Heitler and London directed toward a mere *computational* problem: the problem, that is, of finding an approximate solution (based on empiricial knowledge of the system described) of an otherwise intractable mathematical equation?

(B) If the orthodox view of the structure of scientific theories appears to reduce the significance of the Heitler-London theory to a computational trick of interest to quantum physicists wishing to extend the range of applicability of quantum theory, perhaps the real significance of the discovery is as an empirical hypothesis accounting for ('explaining', 'reducing') the formation of the chemical bond. This view, the framework of hypothetico-deductive reasoning, would account for the chemists' interest in the Heitler-London theory despite its poor quantitative predictions: for if arguments can be given that the discovery reveals the mechanism or cause of chemical bonding (at least in this paradigm case) in terms of the electronic structure of the molecule and the basic equations of quantum mechanics, this is the solution to a *chemical* problem. Within this interpretative framework, then, we are led to ask for the specific hypotheses relating the chemists' characterization of bonding with aspects of the quantum-mechanical description. Heitler and London pointed to the importance of the *exchange integral* (defined in the Appendix to Section 1) in their expression for the energy of the bound state, and for some time it was common for physical chemists to refer to 'exchange forces' as a nonclassical (*i.e.*, purely quantum-mechanical) cause of bonding. The significance of the Heitler-London theory (or indeed of any of the physical explanations of bonding) would then lie in (1) the identification of (or reduction of) the chemical bond to specific physical effects, as embodied in one or another hypothesis relating the chemical characterization with physical descriptions of molecular structure, and (2) the explanation or justification of specific *chemical* theories or descriptions of bonding (*e.g.*, Lewis's theory of valence and electron pairing).

But the apparent explanation of the significance of a theory such as the one under discussion on the basis of hypothetico-deductive reasoning will not stand scrutiny. If we judge the Heitler-London theory (or even later, 'better' theories of bonding) solely on the accuracy and relevance of its deductive consequences, it is difficult to see why the theory should be regarded as a significant discovery. The theory only purports to give any sort of explanation in the case of the very simplest molecule; the grounds, if any, for supposing the explanation for more complex molecules is the same, or even 'similar', are not consequences of the theory itself; nor indeed, are the improvements or corrections obviously required to bring an approximate theory into agreement with experimental data. For what reasons, one might even ask, is the chemist *qua chemist* interested in a structural description of molecules from the point of view of fundamental physics? What specific, scientific reasons, that is, rather than merely vague and hopeful attitudes of 'unity of nature'? This question becomes more pressing when it is recognized that many of the chemists' conceptions of the interaction of valence electrons and the structure of molecules (conceptions developed in light of chemical evidence) are not obviously translatable (or even consistent with) the physicists' language of superpositions, probability density and the like.

I have sketched the frameworks for scientific discovery that emerge from orthodox interpretations of scientific theories as axiom systems and of scientific reasoning as hypothetico-deductive. But essentially the *same* alternatives are suggested by what may be called the orthodox heterodoxy of Kuhnian paradigms. For within this scheme we may regard Heitler and London (A) as working within the newly established paradigm of quantum theory, and their contribution as a piece of normal science (*i.e.*, showing how the general features of the paradigm may be applied to a case considered problematic within an older paradigm); alternatively, (B) the Heitler-London theory might itself be seen as the origin of a (minor) paradigm within chemistry. In case (A) we again have a view of the discovery as merely 'working out the calculational difficulties': fundamentally uninteresting and of minor consequence (particularly to the concerns of chemists). If we attempt to regard the Heitler-London theory as the discovery of something new and interesting (case B), we are faced with the explicit denial (by Kuhn and also writers who, paradoxically, profess an interest in discovery, such as Hanson) of any structure to the discovery process. (Both Kuhn and Hanson claim such discovery is analogous to the gestalt switch that occurs when a new pattern is recognized; and both writers emphasize that this occurs 'at once', that is, that no analysis into stages of development or reasoning is possible.)

I have argued that the frameworks of applying a fundamental theory (A) and of a new causal hypothesis (B) are inadequate or at least incomplete because each leaves important questions unanswered. Even more serious, however, are questions left unasked: that is, questions which find no place in the frameworks above, but which arise naturally if we put aside for the moment philosophical accounts of scientific theories and scientific progress. How and why, for example, does the perceived need for an explanation of bonding arise? In particular, why was this *not* a pressing question to most chemists of the 19th century, or of previous centuries? Quantum mechanical theories of the chemical bond are important partly because they join physics and chemistry, and numerous questions arise from this connection. Why should the explanation of bonding be in terms of the physical structure of atoms and molecules? Is this in some way suggested or presupposed by the formulation of the problem of chemical bonding? Under what circumstances, and to what extent, will evidence and theories from the field of physics force revision or abandonment of the chemists' conception of molecular structure? Conversely, is the role of those chemical conceptions in the development of theories of bonding *merely* heuristic or suggestive, or, in light of the success of those conceptions within chemistry, is there a more central — even necessary — role of the chemists' picture? (Palmer, 1965, p. 125, attributes to Heisenberg the view that prior knowledge of the chemical concept of valency is *required* for the quantum mechanical account of bonding.) Heitler and London obviously made use of previously articulated models and theories of both physics and chemistry despite their inadequacy to account for the problem of bonding; and these models in turn were based on earlier, less adequate theories. How is this plain fact to be accommodated? How can scientists learn to *make use of* the inadequacies in a theory, rather than simply *rejecting* the theory as false? Is there any rationale or structure to the sequence of (inadequate) theories or models? Beside these questions concerning the utility of inadequate or 'false' theories we have to pose analogous questions concerning the deficiencies of successful or 'partially correct' theories. The Heitler-London theory is (like most theories) approximate, tentative, based on some unrealistic assumptions, and applicable only to the simplest case of interest. In what sense can this be considered good enough to rank as an important scientific discovery? Clearly, part of the merit of the theory is that scientists were able to improve upon the results obtained, and extend the range to which the conceptions of the theory apply. How is this done? (Even approximate solutions of the quantum mechanical problem of describing a molecule can be derived for only a few of the simplest cases, yet quantum chemists frequently claim that quantum theory explains quite generally the behavior and properties of molecules, and in particular the formation of bonds. What kind of reasoning supports such an extrapolation?)

(C) The questions posed in the last paragraph revolve around (1) the formulation of a problem and (2) the way in which scientists produced a *series* of more or less adequate solutions of that problem (in this case the solutions are descriptions of molecular structure in which formation of the chemical bond can be seen as a consequence of that structure). Thus a more illuminating framework for understanding the structure of a theoretical discovery may be in terms of the formulation, attempted solution, partial solution, and reformulation of scientific problems. Herbert Simon (1977 and elsewhere) has argued for the application of such a framework to the process of discovery generally, but has developed his ideas with respect to the detailed processes of solving a more or less well defined 'puzzle' by an individual. Simon has emphasized for example the significant role of 'heuristics' induced from repeated attempted solution of problems sharing important features; but it is not obvious that heuristics in Simon's sense can play a large role in the historical development of theories of bonding. Nevertheless, the analysis of the discovery process I am about to offer for this case bears strong analogy to Simon's suggestion. I will begin by indicating some of the ways in which the problem before us differs from the paradigm cases of puzzle solving discussed by Simon (*e.g.*, designing a chess program, or finding a pattern in data presented according to a standard format). This is the subject of the next two sections, in which specific aspects of the 'complexity' of the problem of understanding chemical combination are discussed; it will be seen that the appropriate *description* of a subject area, and the *formulation* of problems within it are themselves part of the discovery process: scientific descriptions and problem formulations are themselves products of scientific reasoning.

### 3. COMPLEXITY AND THE LANGUAGE OF PROBLEM FORMULATION

The explanation of chemical bonding is far more complex than the cases usually taken as paradigmatic of problem solving or, indeed, of scientific explanation, and several aspects of that complexity confront us at the outset. These concern (1) the language in which the problem is stated, (2) the 'connections' of physics to the object of study, and (3) the fact that we deal here with three levels of organization: fundamental physics, atomic and molecular structure, and chemical characterization.

(1) First, the language of chemical bonding is itself the result of extensive scientific work, and of course incorporates important results of that work. The development of this descriptive vocabulary, and the way in which it suggests or even demands explanations — in other words, the way in which

the domain of chemical bonding has evolved — is an important aspect of the relation between chemistry and physics. The manner in which chemists use this descriptive vocabulary to describe chemical reactions, the properties of molecules, and so on, presents the physicist who would explain chemistry with a set of concepts and problems; that is, it is the chemist who has determined what statements require explanation, and what concepts have to be justified, or shown dispensable (or whatever it is the physicist does when he explains chemical bonding). Thus what the physicist has to explain is itself the result of a chain of scientific reasoning, and an understanding of the sense in which chemistry is explained by physics requires us to examine this reasoning by which the domain is specified.

There exists a refined vocabulary or conceptual scheme for describing cases and kinds of bonding developed over a long period of experiment and observation. This is (roughly) the language of atoms, valence electrons, covalent bonds, ionic bonds, binding energy, bond lengths, bond angles, double bonds, and so on. It is this vocabulary, or rather its referents, that is an object of study of quantum mechanics. As an object of study, the objects described within this scheme (or, for the sake of brevity in some cases, the corresponding vocabulary) may be called the 'phenomenological level.' However, we must realize that the notion of a phenomenological level is a relative one: for the practicing chemist (especially in the first quarter of the twentieth century), the phenomenological level *and* the chemical bond theory became an object of theoretical study, and thus together form what I shall refer to as the phenomenological level for quantum chemistry. When I wish to refer to the *chemists'* atomic theory of bonding (*i.e.*, to their model of atomic structure and explanation of behavior at the phenomenological level), I will speak of the 'chemists' picture' of bonding. The orthodox empiricist interpretation of theories would identify what I have called the *phenomenological* vocabulary with an 'observation vocabulary'. Now, in the case of (textbook presentations of) the kinetic theory of gases this view seems *prima facie* to make sense: terms like 'pressure' and 'volume' seem at least reasonably close to being 'theory-free', 'uninterpreted' ideas based on direct experience. This can, of course, be pressed and questioned; but my point is simply that the claim has an initial plausibility — much more, as we shall see, than in the case of the phenomenological level of chemical bonding.

For the terms 'atom' and 'bond' as they are used in chemistry cannot plausibly be claimed to be free of theoretical presupposition or to be 'observational', at least on the accounts of observation usually given by philosophers. These terms are, of course, the results of a long process of development of theoretical ideas on chemical combination. They presuppose (rather, incorporate) an answer, for instance, to the scholastic question of the perdurance of elements in a compound, by speaking of the 'elements' (in this context, the constituent atoms or radicals) as retaining their identity but linked by bonds to one another in specific physical relations. When combined with the symbolic notation for compounds developed in the nineteenth century as an elaboration of the notation of Dalton and Berzelius – the familiar alphabetic designation of elements joined by lines representing bonds – the terminology of bonds incorporates the law of multiple ratios and the concept of valency, and suggests that compounds have a definite geometrical structure, with directional bonds. The fact that 'atom' and the uses of that term are likewise the result of extensive theoretical reasoning, and are incompatible with some other possible interpretations of particular experimental work is well known and requires no further justification here. The extent to which the framework of atoms and bonds transcends naive observational evidence on the nature of compounds is perhaps indicated by a partial catalogue of the terms no longer employed in that context: 'substantial forms', 'remission of the qualities' of elements, 'affinity', and 'type' have all been supplanted (at least in their original senses) by the terminology of 'atoms' and 'bonds', and descriptions employing this terminology are incompatible with descriptions in the older terms.

Chemists can describe chemical combination in terms of atoms and bonds and make measurements of various properties of bonds while remaining agnostic on *any* physical explanation *of* chemical bonds. In deliberately introducing the descriptive term 'bond' in the mid-nineteenth century, Frankland recognized both the theoretical import of the term and the absence of any definite account of the physical basis of bonding, or of the stability or specificity of bonds:

I intend merely to give a more concrete expression to what has received various names from different chemists, such as atomicity, an atomic power, and an equivalence . . . . It is scarcely necessary to remark that by this term I do not intend to convey the idea of any material connection between the elements of a compound [this would in any case merely replace the problem of chemical combination of elements with an equivalent one of the combination of elements with bonds – D.A.B.], the bonds actually holding the atoms of a chemical compound being, as regards their nature, much more like those which connect the members of our solar system. (Quoted in Russell, 1971, p. 90)

The analogy to gravitational attraction was appealing and recurs in the literature on chemical bonding, but even this very broad physical characterization could be dispensed with in speaking of bonds and determining some of their properties, and Frankland did so four years later, modifying the concluding sentence of the passage above:

... the bonds which actually hold the constituents together being, as regards their nature, entirely unknown. (*Ibid*.)

It would be as incorrect to label the vocabulary of atoms and bonds as 'theoretical' as it would be to regard it as 'observational'. Terms like 'atom' and 'bond' and their cognates, and the conventions regarding their usage, provide a very general framework within which theories of, say, the structure of benzene, the action of an enzyme, or the physical basis of stable bonding may be formulated. The justification for retaining the label 'phenomenological' rests on the practice of chemists, especially before the development of quantum mechanics, of (i) employing such terms consistently, without any reference to a theory of bonding; (ii) determining, in particular cases, that there are bonds formed between particular (kinds of) atoms, and classifying bonds into groups based on experimental studies of those compounds (*e.g.*, into ionic, covalent, single, and double bonds); (iii) making quantitative comparisons of the strength and stability of bonds, using techniques independent of any particular theory of bonding (for example, determining relative bond strengths by the amount of amount of heat produced in an exothermic reaction).

It's not merely that the theory/observation distinction is blurred or that the phenomenological vocabulary is partly theoretical or 'theory laden'. Rather, in the sense of presupposing a wide range of facts which evidently might have been otherwise, and describing phenomena in a way incompatible with a wide variety of conceivable theories (and in fact many past theories) of the nature of compounds, *the phenomenological vocabulary is unequivocally theoretical*. But this does *not* entail (what is suggested by the continuum view) that such descriptions 'bias' the stated results of research in a way that makes comparison or testing of alternative theories of bonding one-sided, or that the reported results are in some way 'created' by or made intelligible by the theories in question. On the contrary, chemists were able to refer to specific types of bonds, measure and describe them in some detail without presupposing or suggesting one or another theory of bonding, yet in such terms as allowed those results to be used in the comparison and testing of alternative theories of the nature of those bonds; and in this sense the phenomenological vocabulary within which this is done is *unequivocally observational*.

(2) What is the relation between the concepts in a problem solution or explanatory hypothesis and the elements of the problem description or data?

In Simon's models of problem solving, the concepts are specified by the formulation of the problem. In scientific discoveries, new concepts or new relations between concepts can arise. In the case (to draw on a favorite example again) of early kinetic theories of gases, the connections between macroscopic quantities in the phenomenological vocabulary and characteristics of the components arose almost 'automatically'. Gas pressure, for instance, is force exerted on a container per unit area; once the supposition that a gas is a collection of atoms (or molecules) in motion is granted, the background theory of Newtonian mechanics makes it quite natural to identify this (macroscopic) force with change in momentum of the gas atoms upon collision with the container walls. The identification is 'natural' here because the fundamental dynamical theory employed for those atoms, Newtonian physics, identifies force quite generally with change in momentum. (If we write Newton's second law in differential form it reads $F = dp/dt$, that is, net force is the total time derivative of momentum.) While the question of just how such identification should be construed has attracted some attention from philosophers (i.e., whether such identifications are genuine identities, and if so, whether 'synthetic', 'necessary', or otherwise), there seems to be little point in this case to detailed discussion of how or why one macroscopic quantity is identified as (with?) one or another characteristic of the physical components. On the other hand, as we shall see below, in the case of quantum mechanical explanations of chemical bonding, such identifications are problematic and an object of some theoretical interest and effort on the part of chemists. This is one reason a hypothetico-deductive account appears a plausible reconstruction of the reasoning leading to kinetic theory but is radically incomplete for the discovery of theories of bonding. In the absence of well established identifications of bond types as quantum mechanical (QM) properties, should we conclude that in fact QM does not provide an explanation of chemical bonding? To do so would be counter to both the statements of physical chemists and the entrenched view that physics explains chemistry. To see in what sense QM does explain chemical bonding in spite of this requires an examination of interactions between and development of chemists' and physicists' theories of atomic and molecular structure.

(3) There are different directions from which the problems of chemical bonding have been attacked, arising from fields of chemistry and physics. On the one hand, chemists have devised models or theoretical conceptions of atomic and molecular structure with the primary aim of systematising chemical behavior, characterized in the phenomenological vocabulary. Carried on largely independently of rigorous quantum mechanical treatment, these

inventions relate the properties of bonds to the structure of atoms, particularly the outer electron 'shells'. This evolving conception of structure I call the 'chemists' picture'. On the other hand there are attempts to describe atomic and molecular structure on the basis of fundamental physics, only indirectly providing an account of *chemical* behavior or of bonding. When directed toward determining physical and chemical properties of atoms and molecules, I will refer to *this* evolving conception as the 'physicists' picture'. Of course, quantum mechanical descriptions of atomic structure, which have been carried out in some detail for at least some cases, have in turn greatly influenced the heuristic atomic models used by chemists.

Thus we have three aspects of the physical explanation of bonding to examine: (1) the account of bonding based on successively more sophisticated models of atomic structure and the behavior of electrons; (2) the modifications of those accounts dictated by quantum mechanical descriptions of atomic structure and the behavior of electrons; (3) attempts to account for bonding directly in terms of characteristics of solutions or approximate solutions of Schrödinger's equation, or of general features of any such solution.

We can begin our examination by indicating, though only briefly, the reasons adduced for representing a compound as a specific geometrical arrangement of atoms linked by (directional) bonds, and the domain of phenomena this representation was supposed to elucidate.

## 4. THE NEED FOR A THEORY OF BONDING

In a 1908 review of the 'Development of General and Physical Chemistry During the Last Forty Years', Walther Nernst suggested a threefold classification of chemical properties, which he called *molar*, *additive*, and *constitutive*. Constitutive properties are those which depend on the precise composition and structure of the molecule. Additive properties are, as expected, those which can be understood as a combination of the properties of the constituents (paradigmatically, but certainly not necessarily, the numerical measure of that property equalling the sum of the measures of the same property of the constituents). Molar properties are characterized as those the measurement of which "render(s) possible the immediate deduction of the values for the molecular weights." This last characterization is surprising, since it seems that molecular weight is paradigmatically additive. Nernst's examples of such a property are vapor density and osmotic pressure, both then-common methods of molecular weight determination. Essentially, the first of these methods attempts to determine the density of a gas of the compound in question in the region where the behavior is very nearly that of an ideal gas (very low pressures), or to extrapolate to zero pressure by empirically determining the constants in van der Waals's equation of state. Then, according to Avogadro's hypothesis, the density of the gas is directly proportional to the molecular weight, because equal volumes of gas contain equal numbers of molecules. As a means of determining relative molecular weights the method is, as Nernst remarks, very direct. The use of osmotic pressure to determine molecular weights is similar, but applicable to dilute solutions. (Osmotic pressure is the pressure of the dissolved substance on a barrier permeable to the solvent but not the dissolved substance; the barrier separates the solution from pure solvent.) This method depends on an extension of Avogadro's hypothesis by van't Hoff: the osmotic pressure of a substance in dilute solution depends on the concentration of that substance in molecules per unit volume (in this respect, a substance in dilute solution is like an ideal gas).

With the example in mind, we can see that in both cases the simplicity of the deduction of (relative) molecular weights depends on the starkness of the *connection* between the macroscopic measurement and atomic properties — in both cases, details of the constitution of the molecule are irrelevant. All gases, regardless of constitution, contain the same number of molecules in a given volume (for a single temperature and pressure); osmotic pressure depends on the number of molecules in a given volume of solution only, regardless of constitution. *The significance of what Nernst calls molar properties is that they in no way depend on the constitution of the molecule, but depend only on the number of molecules present.* The connection with molecular weight determinations is that these properties facilitate the determination of (relative) average weights by allowing one, in essence, to count molecules. The existence of such properties is an argument for the existence of molecules. With this understanding (or perhaps distortion, for Nernst's meaning is unclear) of molar properties, we can reformulate the characterization of the other two classes of properties in a parallel manner:

($p_1$)     *Molar* properties are those which do not depend on the constitution of the molecule, but only on the existence of a number of molecules. [Examples which fit this (revised) criterion of molar properties are the pressure and volume of a gas under conditions where intermolecular forces are negligible, and the 'Brownian motion' of small particles.]

($p_2$)     *Additive* properties are those which depend on the nature of the components (*i.e.*, atoms, ions, radicals or whatever) but not on the detailed arrangement or structure of those components; such properties (it is claimed) can be seen as the 'addition' of properties of the individual components. [Examples of these are molecular weight, refractive index (which depends on the number of electrons), and rate of diffusion (which depends on the square root of the molecular weight).]

($p_3$)     *Constitutive* properties are those which depend on *both* the nature of their components and the detailed arrangement or structure of those components; such properties cannot (it is claimed) be seen as the sum of the properties of the individual components. [Examples of these are optical activity, which depends on an asymmetric (*i.e.*, left-handed or right-handed) structure, chemical reactivity with other compounds, and characteristic absorption bands − *i.e.*, color.]

Note that, since molar and constitutive properties are specified in terms of the existence of molecules, and determinate structures of those molecules, the instrumentalist must deny that there are any properties irreducibly molar or constitutive. Conversely, the instrumentalist position suggests that any property which *appears* to be either molar or constitutive ought to be reducible to an additive property; such a reduction would − or so it seems on the instrumentalist view − provide a more fundamental explanation of the property in question. This implicit program contrasts strongly with the suggestions of Nernst on the ultimate nature of the properties of compounds, which are compatible with the implicit program of realism:

In the last analysis all these [molar and additive] properties are probably constitutive also, and their interpretation as either purely molar or purely additive is only a more or less close approximation. (p. 248)

Subsequent results would seem to bear out this suggestion. Whether or not this can be construed as an argument in favor of a realistic interpretation of science generally, it does show, contrary to what is sometimes claimed, that realistic and instrumentalist interpretations of science can lead to sharply contrasting research programs.

*If there are genuinely constitutive properties, then a theory of the spatial arrangement of atoms in the molecule*, i.e., *a theory of bonding, is required.* Now Nernst himself gave an example of how one might transform a constitutive property into an additive one by, as we should say, reifying the (hypothetical) bonds between atoms:

we may attribute to the double bond a determinate amount of refraction [of light] and take into account with a high degree of approximation, the influence of its constitution, by a return to the additive method. (p. 247)

It is, as I have indicated, in keeping with an instrumentalist interpretation of atomic theory to search for such reductions of constitutive properties, and Ostwald (to take a well known example) apparently felt that this could be done successfully by paying special attention to the different amounts of energy contained in a composition. There is, however, one class of phenomena that apparently cannot be understood without reference to constitutive properties: that associated with the contrasting properties of isomers — compounds with identical composition (*i.e.*, same elements in the same ratio) but different chemical or physical properties. Ostwald expressed the view that even differences between isomers could be understood in terms of the different energy contents of the isomers. It is true, of course, that in many cases there is a difference in, for example, the heat of formation of the two forms of a polymorphic compound, but Ostwald was unable to parlay this into any real theory of the different properties of pairs of such compounds. Conspicuous by its absence in Ostwald's discussion (*e.g.*, in 1909, Chapter 10) of isomerism is the case of optical activity, first studied by Biot about 1815, but better known from the extensive work of Pasteur (Berry, 1954, p. 94; Freund, 1904, pp. 571–576).

Optically active substances rotate the plane of polarization of incident light. Pasteur was able to show that ammonium sodium racemate which is optically *in*active consists of equal portions of two isomers, each optically active, but which rotate the plane of polarization in opposite senses (that is, to the right in one case, to the left in the other, viewed from the direction of the incident beam of light). Here then was a case of the same compound (*i.e.*, by other tests chemically uniform) occurring in two forms distinguished only by a small cluster of physical properties (crystals of the two forms are distinguishable on the basis of the asymmetric arrangement of the faces, again left and right 'handed' — this was the basis of Pasteur's most famous method of separating the two forms):

All that can be done with one [isomer of the] acid can be repeated with the other under the same conditions, and in each case we get identical, but not superposable products; products which resemble each other like the right and left hands. The same [crystalline] forms, the same faces, the same angles, hemihedry in both cases. The sole dissimilarity is in the inclination ... of the facets, and in the sense of the [optical] rotary power. (Quoted in Freund, 1904, p. 576)

Pasteur reasoned that the macroscopic asymmetry reflected an asymmetry in the geometrical structure of the molecule:

Are the atoms of the right acid arranged along the spiral of a right handed screw, or placed at the corners of an irregular tetrahedron, or disposed according to some particular asymmetric grouping or other? We cannot answer these questions. But it cannot be the subject of doubt that there exists an arrangement of the atoms in an asymmetric order, having a non-superposable ['handed'] image. (*Ibid*.)

Others doubted it nevertheless; but Wislecenus, le Bel and van't Hoff, in separate work, developed the specific conception of tetrahedrally directed bonds of a central carbon atom. If four different atoms (or groups) are attached at the corners of a tetrahedron, they can be arranged in two geometrical arrangements, one the mirror image of the other. If any two (or more) of the bonded atoms are identical, no asymmetric structures are possible, and if this is indeed the basis of optical activity, one expects no optical activity in such compounds. This is in fact what is observed: compounds with the general formula $CH_2R'R''$ do not display isomerism, while there are in general two (optically active) isomers when the carbon atom has four different atoms bound to it (Russell, 1971, pp. 159–162).

The conception of tetrahedrally directed bonds of the carbon atom thus provided an early version of a specific *geometrical* model of molecular structure, including 'bonds' with a specific direction in space, fixing bound atoms in relative position. Van't Hoff actually made small tetrahedral models to represent such structures; furthermore, he supposed that double bonds could be represented by two tetrahedra with a common edge, and triple bonds by two tetrahedra with a common face.

The tetrahedral model of carbon bonding reinforced the conclusions that were to be drawn from the work of a number of chemists on the composition of organic substances, especially concerning substitutions of one atom or group for another in a larger organic compound. Among those conclusions were that (i) atoms, in particular the carbon atom, have a fixed combining power or *valence* (an atom of carbon "always combines with four atoms of a monatomic, or two atoms of a diatomic element; that generally, the sum of the chemical units of the elements which are bound to one atom of carbon is equal to four" – Kekulé; quoted in Russell, 1971, p. 58); (ii) substitutions of one atom for another in organic compounds can result in a little-changed compound; particularly striking were the results of Dumas and Laurent which showed that in many cases the substitution of chlorine for hydrogen results in a derivative quite similar to the original compound (see below); and (iii) that partly as a consequence of (i) and (ii), as we shall see, the 'dualistic' view of chemical combination developed by Davy and Berzelius, according to which bonding was an electrostatic attraction of (charged) atoms, is untenable.

While thus 'overthrowing' the dualistic view of chemical combination, these results at the same time made the need for *some* theory of the process of chemical combination all the more apparent; for the theory of Davy and Berzelius offered what the developing ideas of stereochemistry demanded but failed to give: an account of the nature of bonding (or 'affinity'). Under the influence of his experiments with electrolysis, which implicated electrical forces in the interaction of atoms to form compounds, Davy had proposed, and Berzelius had elaborated, an account of chemical combination (anachronistically, of bonding) based on the attraction of positively and negatively charged atoms. For diatomic molecules the theory is straightforward; for complex molecules it had to be assumed that larger groups also possessed a characteristic charge, so they could combine to form still larger groups, thus accounting for these molecules as hierarchial structures, with each level of the hierarchy constituting two combining groups oppositely charged (or becoming charged upon encountering each other). A molecule like zinc sulfate, then, which we should write as $ZnSO_4$, was thought of as consisting of two groups $(ZnO)^+$ and $(SO_3)^-$, and in turn, $ZnO$ as a combination, $Zn^+$ and $O^-$.

Attractive though it was, this theory could hardly be defended in the light of the near equivalence of compounds with the strongly negative chlorine substituted for the positive hydrogen. How, on the dualistic theory, could such a substitution even occur? An equally fundamental problem for the theory was to account for the formation of diatomic molecules such as $O_2$, $N_2$, or $H_2$: since the two atoms are identical it seems pretty implausible that one is charged positively, the other negatively.

At the beginning of the twentieth century, then, the need was recognized for (i) the description of the geometrical arrangement of atoms within compounds (as well as chemical composition) for detailed explanations of the properties and behavior of molecular compounds; (ii) an account of how such arrangements are stabilized by the formation of (directional) chemical bonds; and (iii) a theory of the nature of the chemical bond to replace the untenable dualistic theory. Perhaps surprisingly, there seems to have been no real alternative proposed before Lewis's theory of shared electron pairs (conceived as early as 1902, but not published until 1916). In part this may have been due to the limited success of a descendant of the dualistic theory in a restricted area of application: electrolytes in dilute solutions. Faraday was one of those strongly influenced by Davy's electrochemical theory, and his experiments leading to the laws of electrolysis can be seen as extending and underscoring the importance of electrical forces in chemical combination. Certainly Helmholtz saw it this way:

I think the facts leave no doubt that the very mightiest among the chemical forces are of electric origin. The atoms cling to their charges, and opposite electric charges cling to each other . . . " (Quoted in Russell, 1971, p. 266)

Arrhenius extended the range of phenomena for which the dissociation of electrolytes into ions must be assumed, thus showing that ionization was intimately connected to chemical forces. Essentially, molar properties of electrolytes (which, it will be recalled, depend only on the number of particles) in dilute solution indicate more molecular species present than the supposition of undissociated molecules would require. This fact, combined with the results of electrolysis, suggest something very like the dualistic theory of bonding of the two ions.

Other than Arrhenius's version of the electrochemical theory for electrolytes, the only attempts to specify the nature of the chemical bond in this period were rather vague and hesitant comparisons of the 'chemical force' with the force of gravity, and corresponding analogies between the solar system and the constitution of a molecule. We have already seen Frankland's tentative suggestion of the analogy and his later deletion of the remark in favor of protestation of ignorance. Nernst, in the review cited earlier, retains the Newtonian framework of forces between the atoms of a compound, but, recognizing the differences between the *universal* force of gravitation and the species-specific attractions apparently required to account for the articulated structure of molecules, concludes that there must be uniquely chemical forces, of which, however, nothing is known:

we are obliged to admit that during the period under consideration [c. 1865–1905] there has been no answer to the question (of the nature of chemical bonding) which really tells us anything more than we can see with our own eyes. It seems reasonably certain that we should admit the existence, not only of electrical and therefore polar forces, but of nonpolar natural forces somewhat of the nature of Newtonian gravity. (1909, p. 251)

The distinction made here between polar and nonpolar (later, covalent) bonding was to remain a fundamental dichotomy until the unification made possible by quantum theory.

The accounts of bonding to this point have been nearly independent of the structure of the atoms bonded. Not surprisingly, more fruitful theories of bonding have arisen only in conjunction with theories of the electronic structure of atoms.

## 5. CHEMICAL BONDING AS CONSEQUENCE OF ATOMIC STRUCTURE: LEWIS'S THEORY OF COVALENT BONDING

G. N. Lewis was the central figure in the development, roughly from 1902 to 1923, of the shared electron pair of the covalent (*i.e.*, nonionic) bond, which forms the basis of modern electronic theories of bonding. It was also the starting point of the more elaborate and physically detailed theories of bonding in terms of quantum mechanics developed by London, Pauling and others. Lewis self-consciously developed his theory of bonding in the light of developing ideas on the structure of atoms.

The popular, and popularized, 'plum-pudding' model of the atom created by J. J. Thomson forms a conceptual bridge to the theory of Lewis. Thomson's model assumed a uniform sphere of positively charged matter, within which moved much smaller electrons. Thomson calculated that *stable* structures could be constructed if the electrons were grouped in spherical *shells* concentric with the positively charged sphere. Thomson himself placed this model in the service of the 'dualistic' or electron transfer theories of bonding: chemical bonding occurred, he supposed, by the transfer of electrons from the outermost 'shell' of one atom to that of another. The supposition was backed up with calculations (based on classical electromagnetic theory and his model) which showed that atoms with between fifty-nine and sixty-seven electrons would have an outer shell holding up to eight electrons, an atom with fifty-nine electrons having an empty outer shell, and one with sixty-seven having its outer shell of eight completely filled. Thus Thomson's model could provide a rough accounting for the valences across a row of the periodic table, interpreted according to the electron-transfer view of bonding. Thomson did consider the possibility of two atomic spheres sharing an electron, but argued that such an arrangement would be metastable: even a small disturbance of equilibrium condition would result in a transfer of the electron to one of the two atoms.

Lewis was able to combine two developing strains of thought relevant to the structure of molecules: (1) physical theories of the electronic structure of atoms, including inferences from the chemical periodicities summarized in the periodic tables as well as physical theories *per se*, such as those of Thomson, Rutherford, and Bohr; (2) the increasing recognition of the inadequacy of polar or electron transfer theories of bonding for organic molecules.

Thomson's 'plum-pudding' model unwittingly provided the first step toward models of nonpolar bonding, suggesting the possibility of atoms 'sharing' electrons.

Lewis's earliest advocacy of nonpolar bonding, however, was not presented in terms of a structural theory. Instead he pointed out the failure of polar bonding theory to account for a wide range of phenomena which it might reasonably be supposed to explain. This failure had become more striking, not less, with the revival and elaboration of polar bonding theory by followers of Thomson; for although one could, on the supposition of polar bonds, account for the properties of a large class of inorganic compounds, for many other compounds, particularly organic compounds, nonionizability, low dielectric constants, and lack of reactivity argued against the presence of polar bonds between charged atoms. In addition, Lewis argued that the simple attractive electromagnetic forces could not explain the stability of complex geometrical arrangements in nonpolar molecules:

The nonpolar molecule, subjected to changing conditions, maintains essentially constant arrangement of the atoms; but in the polar molecule the atoms must be regarded as moving freely from one position to another . . . like the bits of glass in a kaleidoscope. (Lewis, 1913, p. 1449; quoted in Kohler, 1971, p. 357)

This is perhaps unfair to the sophistication of, for example, Thomson's calculations on stable orbits of electrons, but it does emphasize the basic incompatibility of the clearly directional and rigid character of bonds in organic molecules and symmetrical electromagnetic forces between ions invoked by polar theory. Lewis's objections to polar or 'dualistic' theory of bonding (at least if extrapolated to all chemical bonds, as was customary) indicated *that* something more was required to explain bonding in nonpolar molecules, but he was unable to give an account of *how* such a nonpolar bond might be formed, or *what* such a bond could be like, other than, of course, being directed and 'nonpolar'. It was identified, in other words, merely as the cause of — that responsible for — chemical combination in compounds where the polar theory was clearly inadequate. At this initial stage, then, it was suggested that there were two entirely distinct types of chemical bond, polar and 'nonpolar', but of the latter little could be said positively.

Here we have an instance of significant reformulation of a problem based on a partial solution of that problem. The dualistic theory had provided a tentative answer to the problem of explaining the formation of molecules from atoms. As Lewis showed, however, when this solution is pushed to account for such features as the directionality of bonds, or the rigid structures of organic compounds revealed by stereochemistry, it clearly fails. Nevertheless, in certain cases the dualistic theory is supported by evidence of the atomic interactions; the features of bonding in general (directional bonds, *etc.*) which seem incompatible with this partial solution then become the focus of attention, and in this case the basis of a fundamental classification of bonding types (*i.e.*, into cases which are fairly well described by this partial solution, and those where incompatible features are prominent − in other words into polar and nonpolar bonds). This reformulation even suggests the direction of research into a more adequate theory of nonpolar bonding: the dualistic theory (based as it is on a picture of atoms as simple charged particles) is incompatible with the increasingly detailed picture of molecular structure *because* charged particles do not exert specific, directional forces, while that is what is required between atoms. Atoms, then, for the purpose of explaining bond formation *cannot* be described merely as centers of some net force; that is, the account of atoms must be more detailed.

If the properties of substances could not be explained by the mere assumption of charged atoms, might they not be explicable if we should no longer regard the atom as a unit, but rather if we might ascertain where the charge or charges resided within the atom itself? (Lewis, 1923, p. 74)

Lewis already *had* a model of electronic structure in atoms, based largely on the periods of eight in the 'periodic law' and the implication of electrons in bonding. This was the model of electrons stationed at the vertices of a cube. Between 1902 and 1916, when he published the theory of the shared pair bond, there were other attempts to describe the nonpolar bond in terms of atomic structure, and these (apparently) suggested the way in which Lewis could resurrect the cube model, which he tells us, he had "ever since [1902] regarded as representing essentially the arrangement of electrons in the atom" (Lewis, 1923, p. 29). The cube model (see Figure 3) accounted for a period of eight elements, the characteristic valences of those elements (in ionic bonding), and the inertness of the noble gases (which had a complete octet),

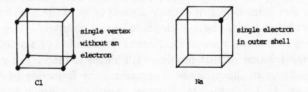

Fig. 3. Lewis's "cubic atoms" of Na, Cl.

but there was no obvious way to represent a nonpolar bond within the scheme. Other contemporary theories of nonpolar bonding proposed a mechanism of bonding in which two kernels (*i.e.*, the nucleus and inner electrons) both interacted strongly with an electron or a pair of electrons (Kohler, 1971). Thomson's model proposed electrons tied by 'tubes of force' to a positively charged nucleus; if there were fewer than eight electrons in the outermost shell, the electrons could interact with a second nucleus. If a transfer of electrons (supposed to occur in *polar* bonding) is energetically favored because of some intrinsic stability of the *completed* octets (and this was part of the original scheme, which was so well adapted to *polar* bonding), then the concept of kernels sharing electrons suggests that a nonpolar bond can be represented, in a manner reminiscent of the tetrahedral carbon model, by cubes sharing an edge. It is intrinsic to this representation that an electron *pair* is always shared. Double bonds were represented as cubes sharing an entire face (and hence sharing two electron pairs). The importance of *pairs* led Lewis to abandon the interpretation of the cube as representing the spatial distribution of electrons within an atom in favor of a tetrahedral arrangement, with a pair of electrons at each vertex.

Having justified, in terms of a specific atomic model as well as by the general arguments adduced earlier, the existence of a distinctly nonpolar bond, Lewis was able to effect a reunification of the two types of bond. (That is, covalent and polar bonds were merely the extremes of degrees of sharing the electrons; equal sharing constituted a covalent bond, complete transfer a polar bond.) Electron pairs might be, on Lewis's developed theory,

between two atomic centers in such a position that there is no electric polarization, or it may be shifted toward one or the other atom in order to give that atom a negative, and consequently to the other atom a positive charge. But we can no longer speak of any atom as having an integral number of units of charge, except in the case where one atom takes exclusive possession of the bonding pair, and forms an ion (1923, p. 83).

Lewis, in his book, *Valence*, and Langmuir, in a series of papers published after Lewis's initial statement of the theory in 1916, elaborated the theory, giving at least qualitative explanations of a great many chemical phenomena. These explanations were based on Lewis's schematic model and the various 'rules' for the formation of stable structures, most importantly, the original 'rule of eight', which attributes saturation and particular stability to complete octets, and the 'rule of two' which lays the basis of chemical bonding on the creation of stable pairs of electrons, which are shared by the two bonded atoms. (Langmuir, by the way, introduced the terms 'covalent' and 'electrovalent' to characterize the two extreme forms of the bond.)

## 6. CHEMISTS' ATOMS AND PHYSICISTS' ATOMS

The theory of Lewis and Langmuir presents a lucid and suggestive 'picture' of the formation of chemical bonds in terms of a theory of atomic structure, and the rules of formation of stable combinations of electrons. Appealing and fruitful though this picture was, there were reasons for supposing that this picture, and the fundamental details of the atomic picture it relied on, were incorrect. Lewis's original cube model had supposed electrons were merely stationed at the vertices of a cube, and even the modified tetrahedral model was pictured with the electron pairs neatly sandwiched between the two kernels. Such a fixed arrangement undoubtedly was originally thought necessary to account for the chemical stability and geometrical rigidity of the molecule. But such a picture of the atom was in conflict with the growing weight of physical evidence, and physical theory, culminating in the Bohr model of the atom, which pictured the atom as the arena of rapidly moving electrons in discrete orbits. More sophisticated quantum theories made the 'chemists' atom' seem only more removed from physically reasonable atomic structure.

Despite its utility in providing an explanation of chemical combination, the chemists' model of atomic structure provided no clues to the interpretation of atomic spectra, and it was largely this area that seemed to demand explanation by a less stable or rigid atomic structure. "Physicists who studied spectra wanted very active extranuclear electrons; chemists were inclined to let them loaf," but the failure to provide any hint of the explanation of spectra forced the abandonment, as a physical interpretation, of such static structures: "the loafing electron was permanently enlivened into the circulating" (Gregory, 1931, pp. 204–205).

Lewis (1923) recognized the differences between the conceptions of atomic structure based alternatively on the periodic laws and inferences from chemical composition and on physical models to account for spectra. He devoted an entire chapter to the 'reconciliation' of the two views, in which the loafing electrons of the initial chemical model were replaced by electronic orbits and average electronic positions. His hope (and it was really not much more than that – the theory is entirely qualitative) was that the chemical models could be retained with slight modifications by interpreting the positions of electrons and electron pairs in those models as their average position (or the position of the orbit) in Bohr's model. Using a conception of electrons reminiscent of Parson's magnetons, Lewis constructed a general scheme of electronic structure and from it derived the electronic structures of elements in each of the rows of the periodic table. Our interest lies not in the details of his derivation of these structures but in the general framework used to present those structures.

This scheme is important in the development of physical explanations of chemical phenomena, for it presented in a clear, intuitive and suggestive way the basis of the regularities in the periodic table in the electronic structure of elements. This scheme is essentially that now used in nearly every textbook of chemistry. Even Lewis's notation for representing the build-up of elements of higher atomic number from kernels isoelectronic with those of elements with lower atomic numbers has been retained. (Thus, in the third row of the periodic table, *Na*, the first element, has its electronic structure represented by '*2-8-1*' indicating a filled innermost shell of 2 electrons, a filled second shell (the first octet), and a single valence electron. *Mg*, second in the row, has the same kernel, with one additional valence electron in the unfilled outmost shell; '*2-8-2*'.) It is the basis of even avowedly quantum mechanical treatments, such as the classic text of Pauling, despite the fact that such a shell structure (in this simple form) is known to be a vast oversimplification of the 'correct' structure in terms of the principles of quantum mechanics applied to many-electron systems.

Regarding electrons within a 'shell' as chemically equivalent can be justified only to the extent that the first, or 'principal' quantum number alone is sufficient to characterize the electron. However, even such an identification of the principal quantum number with Lewis's shells works only for the first three rows of the periodic table. The fourth row, for example, the first with eighteen elements, contains elements with 'valence electrons' in *3d, 4s* and *4p* orbitals. Of course, modifications to the Lewis scheme are possible which take into account additional quantum numbers (discussed further below), but even these, of course, are based (as will be argued) as much on chemical 'sense' as physical theory. This approximative, nonrigorous picture, while still employed by chemists (and for good reasons) is, not surprisingly, an object of derision among some physicists: " ... G. N. Lewis with his *octet theory* which is still popular with some chemists" (Finkelnburg, 1964, p. 396).

Of course the Bohr model was not the last modification in physical thinking on atomic structure, and each of the subsequent developments in physical models again necessitated revisions of the schematic presentation of atomic structure used by chemists. This is especially true of the heavier elements, for which little direct information on structure was available, the focus of physical theories of atomic structure being, understandably, placed on the lighter and presumably simpler atoms. Thus Sommerfeld's extension of the Bohr model to include relativistic effects and elliptical orbits required two quantum numbers to specify a state. This was the first step in breaking down the simple shell model into finer distinctions, leading ultimately to a unique set of numbers for each electron in an atom, expressed by Pauli as an 'exclusion principle': there cannot exist in an atom more than one electron with the same values for each of its (four) quantum numbers. This effectively eliminated the simplistic picture of concentric shells of 2, 8, 8, 18, and so on equivalent electrons, in favor of a more complex picture of shells, subshells, and distinctions within subshells. The order of the different states in atoms heavier than hydrogen (in terms of energy, and hence of 'filling' in moving up the atomic numbers) was determined by what Bohr dubbed the '*Aufbau*' principle: one assumed that the electronic states of heavier atoms were essentially those of the hydrogen atom (that is, starting with the ground state of hydrogen, following with the first excited state (*2s*), and so on). That this method works at all seems barely short of a miracle, for although one can give a rough justification for the use of such a method by assuming that the inner electrons, which might be supposed to drastically affect the possible states of additional electrons, are symmetrically distributed and hence 'screen' the increased nuclear charge (so that an additional electron behaves as though it 'saw' a hydrogen nucleus, and occupies the first unoccupied hydrogen-like state), this assumption is in sharp conflict with the 'shape' of the orbitals calculated. It is not surprising, then, that there are cases where the actual electronic structure is clearly in violation of the *Aufbau* principle. An example is the ions of the 'transition metals':

.... It seems puzzling that for all these ions the *4s* rather than *3d* electrons are gone, particularly since for the most of the neutral atoms the *4s* electrons are lower in energy, more stable than the *3d*. The experimental facts, however, are unequivocal and indicate that in the ions the *3d* level must drop much earlier than it does in the neutral atoms. It seems quite probable that even in neutral atoms shielding of nuclear charge by core electrons is not a one-to-one process and that in the metal ions complete loss of two shielding electrons is sufficient to allow the *d* electron to see the nuclear charge; *the so-called anomalous behavior of neutral chromium and copper are results of highly complex interelectronic forces and cannot as yet be predicted accurately. We can only (as in much of chemistry) search the experimental facts and try to propose reasons for the observed behavior.* (Companion, 1964, pp. 32–33; emphasis added)

Nevertheless, the overall scheme is so useful in formulating a picture of the structure of the heavier atoms that its use is well entrenched, and those cases where the principle leads to incorrect predictions are regarded as violations or exceptions to the scheme, rather than as falsifications.

The acceptance of Schrödinger's wave equation as a fundamental equation determining allowable quantum states, and the probability density interpretation of the solutions to that equation introduced by Born, forced other important revisions of the 'chemists' atom'. The formation of a bond in Lewis's scheme was associated with the formation of stable electron pairs (the 'rule of two'); although it was not necessary, there was a natural tendency to think of the pair on the model of two magnets, that is, spatially localized 'close' to one another; this was the basis of Lewis's preference of the tetrahedron as a representation of the geometry of an atom's valence shell. In the light of wave mechanics, on the other hand, one had to speak less definitely concerning the location of electrons, and bonding was now thought of in terms of the *overlap of electronic orbitals* which, on Born's interpretation of the wave function, represents the probability of finding each electron in the region of overlap [see point (7) of the Appendix to Section 1]. Now it is unclear, when stated in this manner, why this overlap of orbitals should represent a more stable state and hence bonding, since the notion of stable pairs is presumably moot in this context. (This is argued more fully below.) One can give an answer to the question of why bonds form, however, by invoking the quantum mechanical version of the virial theorem: "Bonds can form only if the potential energy of the electrons and nuclei decreases as the atoms come together," and such decreases are to be expected when the (valence) electrons are near two nuclei at the same time − that is, when they have a high probability of being between the two nuclei (Pimentel and Spratley, 1969, p. 72*ff*). This answer needs to be modified somewhat in the light of the possibility of 'antibonding orbitals', but this simple answer shows how *QM* leads to a justification of certain features of the chemists' picture, though not without some modification of that picture.

The abandonment of the original version of the chemists' picture of the atom of 'loafing' electrons, with fixed positions of valence electrons, and the specific modifications made are evidence of the rational way in which theoretical constructs are elaborated in the light of additional evidence and general 'background' theories (in this case fundamental physical theories of atomic structure and the nature of electrons). Having contrasted the picture of atoms obtained by chemists, based on structural theory and the periodic table of elements, with that inferred from physical evidence (particularly spectra and the Bohr model), Lewis remarked that the two views must, ultimately, be reconciled, for "it is the same atom that is being investigated by chemist and by physicist." Which view (or which aspects of each view) are to be modified, and just how? For that matter, why is it necessary that there be a unified picture of atomic structure for both 'chemical' and 'physical' applications? One can, of course, point to vague hopes for 'unification' or 'simplicity' in science, and such hopes have undoubtedly a large role to play, but precisely because of their pervasiveness and generality they can provide no specific direction for the elaboration of one or both of these pictures of atomic structure. What emerges from the revisions proposed (and of course in many other cases as well) is a kind of priority of the *physical* model. The revised picture of the 'chemical atom' is based to a considerable extent on a desire to preserve the shell structure and the rule of eight, which were so fruitful in explaining chemical phenomena; and the approximations and idealizations of the revised model are incorporated with an eye on the original chemical picture. Despite these uses of the original picture of the chemical atom, the modifications are one sided. The features of the original picture which are incompatible with the physical picture are all but abandoned; to the extent they are retained, the 'unphysical' aspects are reinterpreted in the terms of the physical model — for example, Lewis's rather unconvincing reinterpretation of the location of electrons in the chemical picture as *average* positions or orbiting electrons in planetary models of the atom.

The reinterpretation is unconvincing because the average positions of *all* the electrons on simple planetary models would be at or near the center of the atom, and even if one supposes a more complex picture of the moving electrons would produce *average* positions at chemically 'sensible' positions — for example, on Lewis's theory, at the vertices of tetrahedra — the fact that much of the time the electrons supposed to form a *pair* would in fact be spatially *separated* within the molecule would destroy the classical picture of them as 'linked' or strongly interacting (for example as paired magnetons in the scheme favored by Parson and Lewis). Furthermore, while electrons 'loafing' at the vertices of a tetrahedron or cube held some hope of accounting for molecular shape, it is unclear how such an accounting can be had on the assumption of 'orbiting' electrons with no fixed position. Thus in an important sense, the 'reinterpretation' of the cubic atom by Lewis in fact removed a good many of the advantages of the original model, without supplying any real alternative. In fact, on quantum mechanical principles one can state flatly that electron pairs are *not* the cause of bonding, as the chemists' picture suggests. Lewis's reinterpretation — which attempts to reconcile the chemists' picture of the atom with the physicists' quantum mechanical picture — is faced with the fundamental difficulty of understanding how pairing can result in the formation of a chemical bond. Its resolution requires a serious revision of the chemists' picture: pairing is the result of the Pauli 'exclusion principle', and not the cause of bonding.

Now why should the chemist abandon such a promising model because the results in a different field indicate a different model of atomic structure? Merely attributing the abandonment to the fact that both fields investigate 'the same atom' would be inadequate as an answer; for the fact alone is insufficient to account for the priority given to physics over chemistry in forcing the rejection of central features of this version of the chemists' picture, at the expense of much of its explanatory power. The abandonment cannot be the result solely of the inability of the chemical model to account for spectral lines, or other physical data which formed the empirical underpinning of the physicists' description of atomic structure; after all, the physical models (at the time of Lewis's writing) did not, as presented by Bohr, Sommerfeld, or Pauli, provide any means of understanding chemical bonding, and surely failure in this area should be of more concern to a chemist. Besides, we have learned not to expect a theory at its inception to be able to solve all problems − even important problems − in its domain; theories, we have learned, are not rejected or seriously modified *solely* because they don't explain everything they ought ideally. What other concerns or considerations then are operating in this case? Why is it, to use out-of-favor language, that chemists decided that they had incorrectly identified the cause of chemical combination as the completion of octets in a tetrahedral valence shell? The 'chemists' atom' was plausible on two counts, but rejected on a third:

(i) Those electrons least firmly attached to the atom were certainly implicated in the formation of bonds, as evidenced by the partial success of the ionic theory, so a general account of bonding must certainly give a prominent role to the valence electrons, and perhaps include the ionic theory as one mechanism of bonding or as a special case of a more general mechanism. Furthermore, Abegg's rule, according to which the sum of the highest positive and negative valences of an atom is eight, clearly implicated groups of eight electrons in bonding once bonding is associated with electrons; Lewis's observation of the overwhelming preponderance of an *even* number of valence electrons in stable molecules similarly implicates groups of *two* electrons. All of these 'implicated' features of bonding, together, of course, with the conviction that an adequate explanation will necessarily involve an account of the bond in terms of the (electronic) structure of atoms, lend credence to the chemists' picture because those features are captured in that picture.

(ii) Not only does the chemists' picture of the atom capture general features which have been 'implicated' in bonding, it provides a (postulated) mechanism for the articulation of those features which would produce the chemical bond. Now the mechanism is admittedly *ad hoc*, invoking as it does unexplained stability of electrons in groups of two and eight, and the ability of two atoms, but not more, to 'share' electrons. These hypotheses are *ad hoc* because they (presumably) result from the operation of fundamental physical laws, but no hint is given as to how the mechanism could be explained in terms of physics, and no evidence from other fields indicates the existence of similar mechanisms. Despite the lack of positive evidence, though, even an *ad hoc* mechanism supports rather than undermines the picture, since there was apparently no reason to suppose that such a mechanism was *incompatible* with physical laws or principles, or *couldn't* be ultimately explained in terms of physics. There is of course yet another possibility: the mechanisms of chemical bonding might, as some physical chemists had imagined, be the result of some as yet unknown additional force. On this alternative too, there is no positive reason to reject the picture. Thus the ability to state, even employing *ad hoc* assumptions, a means of production of the chemical bond in terms of the chemical picture is reason to adopt that picture (tentatively) as presenting the cause of chemical bonding.

Two broad considerations, then, support the chemists' picture of the atom and its description of chemical bonding. The fact that the picture is dramatically altered (or, if you will, rejected in favor of another version) in the light of a third consideration suggests that each of the three may be important components of the identification of the cause of a phenomenon.

(iii) The third consideration apparently illustrates the operation of scientific realism in (partially) determining the direction of research programs; the chemists' picture of atomic structure we have been considering was rejected on the basis of its incompatibility with 'direct' physical evidence indicating that the proposed structure was simply incorrect in important ways. The nature of the changes made in the chemists' picture were all in the direction of producing a structure that would satisfy the constraints imposed by the theories and (supported) models of physics. The original chemists' picture of the formation of chemical bonds — the cause of chemical bonding — was incorrect because that picture does not represent the true structure of atoms. This has the ring of the obvious about it, but the rejection of a proposed cause or causal explanation on this basis (when our other two considerations support this identification) is possible only on the supposition that an adequate explanation must be in terms of the 'actual', 'real' or 'true' structure, as determined by the most direct evidence available (in this case atomic physics). This requirement, which demands that chemical models (and presumably those of other fields as well) not deny the complete applicability of physical models (insofar as those are well supported) is far stronger than any desire or goal or presupposition of 'unity' in science.

On what basis then, can we claim the physical picture is the 'correct' one for atomic structure, and in particular for providing the framework within which bonding will have to be explained? First, of course, there is the evidence on which the physical models themselves were constructed, most importantly the Rutherford scattering experiments, which demonstrated the existence of a very small, positively charged nucleus; and spectral emission lines, which formed the basic data for developing and elaborating the Bohr model. It is quite natural to think of the inference to atomic structure from these data as considerably more 'direct' than the chemists' inference from regularities in the periodic table and catalogs of stable compounds. Scattered alpha particles and spectral lines are the direct results of processes within the atom, while the chemists' inferences are based on plausible guesses as to the regularities or features at the atomic level that might produce observed patterns of bonding. The sense of 'direct' here is not easy to explicate, and I will not try to give a detailed treatment of it, but it is relevant to point out that scattered alpha particles (in Rutherford's experiments) and emission lines are detected without physical interactions intervening between the atomic processes which produce them and their detection (by Geiger counter, fluorescent screen, spectroscope, or whatever); on the other hand, the evidence the chemist relies on lacks any such direct causal link to atomic structure or processes. What one would *like* to say is that scattering experiments and the emission spectra are means to 'directly observe' atoms, while chemical compounds and regularities in the periodic table provide 'circumstantial evidence' concerning atoms. This directness leads to (a) more detailed structure, for instance 'splitting' of levels so that electrons in a shell may not be (chemically) equivalent and (b) a contrast with the chemists' picture of atomic structure.

In addition to providing a description of atomic structure within which features of the chemists' picture can be justified, quantum physics corrects that picture in some ways, and removes anomalies by providing qualitative explanations of those anomalies in terms of the physical conception of atomic structure (in some cases this is so even though the theories of structure are unable to *predict* such bonding patterns). These examples typically require consideration of the formation of molecular orbitals — that is, a quantum mechanical treatment of allowable states for electrons in polyatomic environments. There are examples of what would seem, on the Lewis picture, to be stable electronic structures of possible molecules, but the compound is unknown and (presumably) not stable: for example $O_4$ which might be supposed to have the electronic structure:

There are also examples of known molecules without satisfactory electronic arrangements in the Lewis scheme ($I_3^-$, the triiodide ion, for example, seems to require the combination of two structures with already complete octets, $I_2$ and $I^-$ :

For still other molecules, the Lewis scheme does suggest the formation of a stable molecule, but the electronic structure proposed is in conflict with the observed properties. The Lewis structure of $O_2$, for example, ought to be:

but this would not account for the observed paramagnetism of the $O_2$ molecules, which presumably indicates the presence of unpaired electrons. This behavior can be understood on the basis of *ad hoc* arrangements of molecular orbital energy levels, and thus, in some sense explained in the terms of quantum mechanical theory, while the Lewis theory must simply regard the paramagnetism (and the consequent violation of the 'obvious' Lewis structure) as an exception to the rules — that is, no explanation is available within the Lewis scheme.

## 7. CONCLUSION: THE STRUCTURE OF CHEMICAL DISCOVERY

The significant points which emerge from the preceding analysis of discovery within a particular branch of physical chemistry can be stated in a fashion that suggests a structure of discovery. This structure and the attendant reasoning patterns contrast sharply with the framework of theory application and hypothetico-deductive reasoning, which I argued in Section 2 are suggested by orthodox philosophical analyses of science.

(1) The structure of the discovery process is exhibited in the succession of solutions (theoretical explanations) put forth in response to a general problem; but this structure appears only when we recognize the various schemes and atomic models proposed as more than a mere series of attempted solutions to a fixed and clearly stated problem. Rather, there is an interplay and constructive progression between partial solutions, inadequacies, reformulations of the problem, and the results of other fields of science (*i.e.*, physics) which bear on the same objects of investigation. This structure is hidden so long as one relies on a view of the theoretical accounts of a body of facts as a theory (or paradigm) understood holistically, that is, as an elaborated structure upon a stable base of observations, or as an 'interpretation' or 'conceptual scheme' for an entire body of knowledge. Some features of this progression are:

(a)    Succeeding descriptions of the objects of study (atoms and molecules in the case of chemical bonding) are increasingly detailed. More precisely, there is a progression toward *higher resolution descriptions* of the structure of the objects of study.

(b)    Succeeding explanations of bonding attempt to describe a mechanism to account for or 'produce' the effects regarded as problematic or requiring explanation. That is, the succession of problem solutions is a search for causal explanations. (This is amplified in point 2.)

(c)    The problem is subdivided and sometimes reformulated on the basis of partial (if inadequate) solutions, but the structural

features of models or partial problem solutions lead to sharpened
or reformulated problems only after independent evidence
supports the existence of the relevant structures or features of
a model. Thus, for example, the problem of a *directional* bond
can be formulated when geometrical models of molecules are
introduced, but reformulation of the problem of accounting for
chemical combination in terms of supplying a mechanism to
produce a directional bond occurred only after organic chemists
had shown the importance of stereochemical descriptions (and
the geometrical arrangement of atoms was, in turn, taken seriously
by most chemists only after the demonstration of the general
applicability of valency or determinate 'combining power' of
atoms).

(2) What the scientist is in search of in the way of a solution to a problem
is neither the 'general laws' from which laws concerning his subject (*e.g.*,
'laws of chemistry') may be derived, nor the 'boundary conditions' for the
application of general laws. Indeed, it is difficult to state any 'laws of chemistry' concerning chemical combination or bonding for which a derivation or
deduction might be taken to explain the formation of bonds. Theoretical
accounts of bonding based on atomic structure are instead directed toward
providing a general scheme (description of structure and identification of
important causal mechanisms, and features which distinguish cases of relevance
to the initial problem from 'uninteresting' cases) within which the preeminent
facts of chemical combination and in particular the problems high-lighted by
previous theories are 'accounted for' or explained. The sense of explanation
here is the subject of another paper, but enough has been said to indicate:

(a)     The explanation depends upon a structural description of the
        objects of study (atoms and molecules).
(b)     The behavior and interaction of the components identified by the
        structural description are connected in a narrative of plausible
        and (so far as is known) physically possible transformations and
        interconnections. Such a narrative can be called a *pathway*.
(c)     Mere plausibility and the lack of a demonstration of physical
        impossibility leave open innumerable (logically) possible path-
        ways, so that empirical verification is desirable (or perhaps
        required) to substantiate the account (*e.g.*, by confirming the
        existence of some stage of the pathway, perhaps in a particularly
        simple case).

(d)      The explanation *need* not exhibit the facts to be explained as a deductive consequence of general physical laws (or other recognized natural laws).

A scheme of explanation similar to this was suggested for biological explanations by Kauffman (1970); he labelled such explanations 'articulation of parts explanations'.

(3) Every molecule and every chemical bond is of course 'physical' and hence can be described — at least 'in principle' — in terms of fundamental physics. This (presumed) fact — token physicalism — is evident in the roles played by fundamental physics in constraining the pathways in explanations of bonding.

(a)      The 'catalog' of entities and properties and the possible interactions and transformations of these, are those regarded as fundamental by physics or reducible to them.

(b)      Physical evidence of the structure of atoms, and, by implication, molecules (what I called the 'physicists' picture') restricts and sometimes corrects the models of structure used by chemists (the 'chemists' picture') since, as Lewis, noted, each is an attempt to describe 'the same atom'.

Philosophers as diverse as Duhem and Feyerabend have lamented the imposition of restrictions or 'limitations' on the models or hypotheses employed by the scientists. In the case studied here, however, the constraints based on token physicalism play a crucial positive role in the discovery of the explanation of bonding, by focusing the attention of physical chemists. In the absence of those constraints, we can surmise that the physical chemist in search of a solution to the general problem of (covalent) bond formation would flounder in a flood of 'possible' pathways.

(4) The discovery of the physical basis of bonding is based on a more complex and dynamic relation between the fields of physics and chemistry than is usually envisioned (for example in standard accounts of reduction and the unity of science). Repeatedly, the chemists' picture of the atom has been revised or corrected on the basis of the more direct evidence of physics; at the same time however, the chemists' understanding of the molecular properties that atomic structure will have to account for has guided the development of physical reasoning. In my original example of the Heitler-London theory of the hydrogen bond, the approximate solutions which were shown to represent a bound state were formulated partly on the basis of prior 'chemical'

knowledge of the general features of molecular structure of hydrogen. Thomson's atomic model and the (approximate) characterization of the electronic structure of atoms into 'shells' were clearly guided at least partly by the constraint that the physical description must, in the end, conform to what the chemist knows about the combination of atoms into molecules (electron pairing, Abbeg's rule and the like). Differences remain in the conceptions of atoms employed by physicists and chemists (since any such picture is only an approximation to what is ultimately, in a quantum mechanical description, an incredibly complex state function in a many-dimensional space), but this mutual constraining draws those pictures into closer agreement.

*Department of Philosophy and*
  *Program on Science, Technology*
  *and Society*
*University of Illinois*
  *at Urbana-Champaign*

## BIBLIOGRAPHY

Atkins, Peter W.: 1974, *Quanta: A Handbook of Concepts*, Clarendon Press, Oxford.
Berry, A. J.: 1954, *From Classical to Modern Chemistry*, Cambridge Univ. Press, Cambridge; reprint edition, Dover, New York, 1968.
Companion, Audry L.: 1964, *Chemical Bonding*, McGraw-Hill, New York.
Finkelnburg, Wolfgang: 1964, *Structure of Matter*, translated from the 9th/10th edition of *Einfuhrung in die Atomphysik* by W. Finkelnburg and O. Matossi-Reichemeier, Springer, Berlin.
Freund, Ida: 1904, *The Study of Chemical Composition*, Cambridge Univ. Press, Cambridge; reprint edition, Dover, New York, 1968.
Gregory, Joshua C.: 1931, *A Short History of Atomism from Democritus to Bohr*, A & C Black, London.
Heitler, W., and F. London: 1927, 'Wechselwirkung neutraler Atome and homöopolare Bindung nach der Quantenmechanik', *Zeitschift für Physik* 44, 455–472.
Kauffman, Stuart A.: 1970, 'Articulation of Parts Explanations in Biology and the Rational Search for Them', in R. C. Buck and R. S. Cohen (eds.), *Boston Studies in the Philosophy of Science*, Vol. 8, D. Reidel, Dordrecht, pp. 257–272.
Kohler, Robert E.: 1971, 'The Origin of G. N. Lewis's Theory of the Shared Bond', *Historical Studies in the Physical Sciences* 3, 343–376.
Lewis, Gilbert N., 1923, *Valence*, Chemical Catalog Company; reprint edition, Dover, New York, 1966.
Nernst, Walther: 1909, 'Development of General and Physical Chemistry During the Last Forty Years', in *Annual Report of the Board of Regents of the Smithsonian Institution ... 1908*, U.S. Government Printing Office, Washington.

Ostwald, Wilhelm: 1909, *The Fundamental Principles of Chemistry*, translated by Harry Morse, Longmans, Green and Co., New York.

Palmer, W. G.: 1965, *A History of the Concept of Valency to 1930*, Cambridge Univ. Press, Cambridge.

Pimentel, George C., and Richard D. Spratley: 1969, *Chemical Bonding Clarified through Quantum Mechanics*, Holden-Day, San Francisco.

Roby, Keith R.: 1976, 'Mathematical Foundations of a Quantum Theory of Valence Concepts', in W. C. Price, S. S. Chissick, and T. Ravensdale (eds.), *Wave Mechanics: The First Fifty Years*, John Wiley, New York, pp. 38–60.

Ruedenberg, Klaus: 1962, 'The Physical Nature of the Chemical Bond', *Reviews of Modern Physics* 34, 326–376.

Russell, C. A.: 1971, *The History of Valency*, Leicester Univ. Press, Leicester.

Sienko, Michell J., and Robert A. Plane: 1974, *Chemical Principles and Properties*, 2nd ed., McGraw-Hill, New York.

Simon, Herbert A.: 1977, *Models of Discovery*, D. Reidel, Dordrecht.

Slater, John C.: 1975, *Solid State and Molecular Theory: A Scientific Biography*, John Wiley, New York.

RACHEL LAUDAN

# THE METHOD OF MULTIPLE WORKING HYPOTHESES
# AND THE DEVELOPMENT OF PLATE TECTONIC THEORY

One of the more curious aspects of most recent theories of scientific change is that their proponents have left unexplained how scientific change can occur at all. For change implies the existence of alternatives, and neither Kuhn (1962) nor Lakatos (1970) has specified the circumstances under which it is rational to attempt to develop alternatives to a prevailing paradigm or research programme. Indeed, the methodologies of these philosophers provide no rationale for any scientist to explore and develop a new scientific 'maxi-theory'. Feyerabend (1975) has attempted to redress the methodological conservatism inherent in Kuhn and Lakatos, but his own arguments for 'theoretical pluralism' remain unconvincing. Thus the 'logic of discovery' of new maxi-theories remains one of the most neglected parts of contemporary theories of scientific change.

My aim in this paper is to state afresh the case for a form of theoretical pluralism, utilizing the history and philosophy of recent geology as its springboard. I shall begin by discussing a theory of geological methodology first stated by the early twentieth-century geologist T. C. Chamberlin, a methodology that provides a rationale for the simultaneous development of several rival theories. In the body of the paper, I shall show how a methodology of this type was instrumental in generating the plate tectonic revolution in geology in the 1950s and 1960s, particularly at the hands of one of the leaders of that revolution.

It is now widely recognized that between 1960 and 1970 the vast majority of geologists rejected all their previous theories about the origin and structure of the earth's crust in favor of 'plate tectonics' (see Cox, 1975, Hallam, 1973, Marvin, 1973, Takeuchi *et al.*, 1967, and Uyeda, 1978). This theory postulates that the surface features of the crust should be explained by reference to the lateral movement of a small number of thin rigid plates. As might be expected, the discovery of such a sweeping new theory, or 'paradigm' as it is frequently termed by the geologists involved, was not the work of any single individual, but the combination of a number of separate developments. One of the most important of these was the work of the Canadian geophysicist, J. Tuzo Wilson, on the nature of a certain type of plate junction known as a 'transform fault'. For many geologists, the confirmation of Wilson's hypothesis about these

331

*T. Nickles (ed.), Scientific Discovery: Case Studies*, 331–343.
*Copyright © 1980 by D. Reidel Publishing Company.*

faults was one of the major events that clinched their acceptance of plate tectonics. From the point of view of the logic of discovery, Wilson's career is a particularly interesting one to explore, since he only developed the transform fault hypothesis after a long career devoted to pursuing quite different geophysical theories. Here was no brash newcomer to the field able to break with traditional prejudices largely because of his ignorance of them, but a scientist who for most of his life had worked on opposing theories.[1]

Yet although the philosophical literature on scientific change has a great deal to say about why a scientist should continue to work on a given major theory, it has little to contribute on the question of why a scientist might attempt to formulate a new theory. For example, with the publication of his *Structure of Scientific Revolutions*, Kuhn introduced a new classification of scientific theorizing which has implications for any attempt to discuss the discovery of scientific theories. On the one hand, there was normal science. The scientist working within a given paradigm was charged with the task of solving puzzles for that paradigm, and one way in which he was to set about this was the 'discovery' of different specific theories within the framework of the paradigm. On the other hand, except in rare instances, Kuhn suggested that the scientist should not challenge the paradigm itself. He should not attempt to discover alternative maxi-theories or paradigms, for to do so would be to take time and attention away from the activities of normal science, where the real opportunities for scientific progress lay. Of course, at intervals in the development of science, new paradigms were discovered, but since they were incommensurable with older paradigms, the path to this discovery lay in the realms of psychology rather than philosophy.

With Lakatos, the division between the methods of discovery of macro-theories and their specific instances became even more marked. The succession of theories within a research programme was closely constrained by the positive and negative heuristic of the programme. In a sense, the discovery of new theories could be considered almost automatic. However, the initial development of the programme itself was not so·constrained. Although Lakatos (1970, p. 133) argued that "the actual hard core of a programme does not actually emerge fully armed like Athene from the head of Zeus. It develops slowly by a long, preliminary process of trial and error," he never made more detailed suggestions about how this development occurred. Indeed, even in this statement there is considerable ambiguity between developing an existing programme and creating a new one. In either case, Lakatos did not make it clear who was to undertake the development, since the rational way for a scientist to be successful and make progress was to work within a progressive research programme. In doing so, one of his main tasks was to protect that programme from criticism, and hence he was scarcely in a position to develop a rival programme. Thus, although both Kuhn and Lakatos can give some account of the discovery and development of individual theories within programmes or paradigms, the discovery of maxi-theories is a mysterious process which they never really discuss. This is surely an important lacuna in both accounts of scientific change for it leaves unanalyzable some of the most important episodes in the development of science.

In this paper I want to argue that there is more that can be said about the discovery of maxi-theories than either Kuhn or Lakatos allow. In saying this I am not referring to the moment where the scientist first lights on the germ of an idea, but to the process whereby that initial insight is transformed into a program of research that is taken seriously by the scientific community. This is perhaps not discovery in the classical sense, but I believe that it is close to the sense of discovery that many people now adopt. Shapere and his followers, for example, extend the sense of discovery in very much this way (see Suppe, 1977, pp. 687–688). Furthermore, the more complex the theory to be discovered, the less appropriate is the notion of a moment of discovery and the more appropriate the examination of a process.

My suggestion is that one reason Kuhn and Lakatos believe that so little can be said about the discovery of maxi-theories is that they assume that the rational procedure for any scientist is to pick *one* program of research and concentrate on it exclusively. Although this is the way that many scientists have worked, it by no means applies to all scientists, and further, it surely should not. Rather, many scientists have followed the 'method of multiple working hypotheses', and they were right to have done so. In the remainder of this paper, I want to describe the method of multiple working hypotheses as it was developed in the late nineteenth century, relate it to contemporary debates, and then give an example of its use in modern science.

In a series of brief but influential articles published in the last decade of the nineteenth century, that geologists still read and quote, the geologist and educational administrator T. C. Chamberlin argued strenuously for the 'method of multiple working hypotheses'.[2] The name speaks for itself. In the process of research the scientist should compare and contrast a number of hypotheses, rather than pursuing one exclusively. As Chamberlin put it,

the effort is to bring up into view every rational explanation of new phenomena, and to develop every tenable hypothesis respecting their cause and history. (1890, p. 128)

Although at the date he was writing Chamberlin naturally did not distinguish maxi-theories and their specific instances, I believe that it is plausible to claim that the kind of theories he had in mind were closer to the former. For example, in suggesting that the geologist examine a number of hypotheses for the explanation of the origin of the Great Lakes, the examples of possible hypotheses that he gives are excavation by rivers, excavation by ice, and downwarping of the crust. These were the major paradigms available at that time and not specific theories within a paradigm. Had he been comparing those, Chamberlin would have been more likely to mention hypotheses involving (say) the *different paths* of a river, or *different volumes* of water in a river. Thus it seems fair to claim that, translated into modern terms, Chamberlin was urging that an individual scientist be capable of acting, within a short period of time, as if he was pursuing rival paradigms.[3] As we shall see, he also believed that this should be done constantly in science for maximum scientific progress to occur. Thus he drew no distinction between normal science and revolutionary science but believed in a continuous process of comparing maxi-theories.

Chamberlin offered a number of reasons why he believed such a method to be superior to what he took to be the available alternatives, namely the inductive method and the normal form of the hypothetico-deductive method where only one hypothesis was considered. To his mind, one of the most important effects of his method was that the scientist did not become unthinkingly attached to one hypothesis, unconsciously emphasizing its advantages and downplaying its disadvantages. Rather, by adopting his method,

The investigator thus becomes the parent of a family of hypotheses: and by his parental relation to all, he is forbidden to fasten his affections upon any one. (p. 128).

(This is an interesting contrast with the assumption widely held today that only by becoming an ardent advocate of one theory will a scientist pursue it with the enthusiasm necessary for its full development.) But Chamberlin gives other reasons that are more germane to the logic of discovery. If the virtue of any method of hypothesis is that it suggests lines of investigation, then the method of multiple working hypotheses naturally multiplies such virtues. Says Chamberlin,

The value of a working hypothesis lies largely in its suggestiveness of lines of inquiry that might otherwise be overlooked. . . But if all rational hypotheses relating to a subject are worked co-equally, thoroughness is the presumptive result, in the very nature of the case. (p. 128).

This in turn leads to another virtue of his method; not only will empirical investigation be more thorough if all possible hypotheses are pursued, but the hypotheses themselves will be refined and developed more quickly if they are constantly compared with others. As Chamberlin put it,

The reaction of one hypothesis upon another tends to amplify the recognized scope of each, and their mutual conflicts whet the discriminative edge of each. The analytic process, the development and demonstration of criteria, and the sharpness of discrimination, receive powerful impulse from the coordinate working of several hypotheses." (*Ibid*.)

Described in other words, Chamberlin is arguing that only when a number of theories are compared do all their consequences become traced out quickly and their development occur most rapidly. Of course, even Kuhn and Lakatos are more than prepared to grant that there is rivalry between different paradigms or research programmes at certain stages in the history of science, and that this rivalry can force proponents of both theories to make them richer and more sophisticated. But what neither Kuhn nor Lakatos concedes, let alone urges as Chamberlin does, is that the individual scientist *should* work this way. There is yet another advantage of the method of multiple working hypotheses that Chamberlin hints at but does not explore, namely, that elements from one hypothesis can often be amalgamated with elements from another to produce a third, different hypothesis. As I shall show, Wilson frequently followed this procedure.

It is interesting to note that views similar to those of Chamberlin were put forward in England a few years later by F. C. S. Schiller (1921). He stated that "inquiry demands an abundance of hypotheses." He gave a number of arguments for this, including among them "that the 'facts' are always *more or less relative to the hypotheses* which apprehend them, and in terms of which they are described" (*ibid*.), and that therefore a variety of hypotheses would direct attention to a variety of phenomena. He concluded that

science should habitually reckon with a plurality of hypotheses, and eschew the sharp antithesis between the 'right' hypothesis and the many 'wrong' ones. Its concern is really with the relative values of the hypotheses in active service. (*Ibid*.)

Clearly Schiller's method resembles Chamberlin's, although it is not clear whether there was any contact between the two. Both in turn conjure up a similar picture of science to that urged by Feyerabend (1975). However, Chamberlin's (and Schiller's) arguments for the consideration of a plurality of theories are more straightforward and sounder than Feyerabend's. The latter argues that only by exploring alternative hypotheses can scientists reveal some of the anomalies of any given old hypothesis. There seems no good reason why this should be necessarily so. Thus it seems that there is a plausible philosophical case for reconsidering Chamberlin's version of the method of multiple working hypotheses. In the remainder of this paper I want to try to establish an historical case as well.

To this end, I shall examine one case where I believe an individual scientist did follow the method of multiple working hypothesis, namely that of J. Tuzo Wilson. In the brief period between 1959 and 1965, he wrote papers on at least three major geological programs of research, namely the dominant contracting earth hypothesis, the expanding earth hypothesis, and the convecting earth hypothesis, which he helped develop into plate tectonics. He constantly compared and contrasted them, gradually developing testable consequences of each model. The time was one of rapid theory change and development in geology, and by considering a range of theories, Wilson was instrumental in discovering unforeseen consequences of each. It does not seem implausible to claim that the progress of geology would have been slower had Wilson worked solely in one program of research. As it is, he is now recognized as one of the most significant innovators of the period, and one of the handful of scientists who were responsible for the development of plate tectonic theory.

In the mid-1950s, Wilson compared what he then took to be the two major rival theories for explaining the features of the earth's crust, particularly mountain ranges, namely "compression due to thermal contraction and viscous drag of convection currents in the substratum" (1954, p.168). Despite the fact that severe difficulties were known to beset both of these programs of research, and although some forms of both hypotheses were unacceptable, Wilson, like most other geologists, believed that contraction theories looked more promising than convection theories. Even though the latter were favoured by a few authors such as Griggs (1939) and Hess (1948), Wilson (1954) argued that convection current theories failed to account for a number of geological phenomena. For instance, they could not explain the length and continuity of orogenic belts, nor the sharp deflections found in them at intervals; nor were they readily reconcilable with the layered nature of the mantle or the seismic evidence of stress and strain differences in the mantle which flow should reduce. Where, in addition, was the evidence of tensional as well as compressional features of the earth's crust which would be predicted by convection current theory? Given these difficulties, combined with the absence of any direct evidence for convection currents, Wilson turned to the case for contraction. He argued that,

While agreement has not yet been reached, it seems quite safe to advance a theory of contraction. Recent data make cooling more probable than when Jeffreys advocated it. The version of the contraction hypothesis advocated here appears to fit the geological observations. It explains the pattern of seismicity. It is a sound and precise physical theory by means of which predictions can be made and checked. (1954, p. 170)

He then advanced a specific model of earth history based on the contraction hypothesis that he believed could act as a guide to further research. As the earth cooled, the outer surface solidified and then fractured, allowing gases, water, and lava to escape from the interior to form the first atmosphere, the first oceans, and the nuclei of our present continents. As a result of the processes of erosion, sedimentation, and further mountain building, the continents gradually expanded during geological time until they reached their present dimensions. Simultaneously, the addition of new water from the interior of the earth, combined with continental growth, caused the oceans to slowly deepen. This model, of which I have sketched only the barest outline, was readily admitted by Wilson to be tentative and speculative. But he stressed frequently that new geological and geophysical methods were becoming available which would allow a more thorough exploration of the earth's crust and that in these circumstances, it was essential that geologists have specific models available for test. In 1955, then, Wilson was prepared to consider two major programs of research dealing with the origin of continents and oceans, and to develop a specific version of one of them. Yet even the specific hypothesis he proposed can be considered a 'working' hypothesis in that he was by no means prepared to state that he believed it.

By 1959, the situation had changed somewhat, and Wilson (1959) reconsidered the rival hypotheses. For him, the major development was the announcement by Ewing and Heezen in 1956 that there was a *global* mid-oceanic ridge system. (*Ibid.*, p. 5) Such a system had not been predicted by any geological hypothesis, but Wilson still believed that it was more readily reconcileable with a contraction theory. "A vast amount remains to be done", he stressed, "but no other theory can explain so much" (p. 23). He dismissed the convection current theory even more firmly than he had four years previously

Continental drift is without a cause or a physical theory. It has never been applied to any but the last part of geological time. Convection currents may exist in the mantle but this is a pure hypothesis and it has never been shown how they form mountains with many of the properties of existing mountains. (*Ibid.*)

But despite the fact that the new evidence did not lead Wilson, unlike many other geologists, to consider convection currents, he did radically modify the contraction hypothesis. He played up the importance of the escape of lava from the interior of the earth to the point where it, rather than cooling, became the major cause of contraction in the earth. He suggested that the ocean basins were the original crust of the earth thinly covered with condensed lava. The Moho was given great significance as the original surface of the Earth, which had gradually shrunk as lava escaped from beneath it and built up the continents. In general, Wilson's attention was still directed primarily to the causes of formation of the continents and mountain building. He still considered the mid-ocean ridge system a less important feature, particularly as the earthquakes found along it were very shallow.

Thus the immediate result of the new geophysical discoveries was to convince Wilson that the contraction hypothesis, admittedly drastically altered, was still the better of the two available hypotheses. This situation was not to last for long, as by the following year Wilson was considering a third major hypothesis seriously enough to publish on it.

In a paper in *Nature*, entitled 'Some consequences of expansion of the earth', Wilson (1960, p. 880) considered the possibility put forward by Dicke and others that Dirac's suggestion that the gravitational constant was decreasing with time, causing the earth to expand, might provide an explanation of the mid-ocean ridge system. His estimate of the importance of this feature was now much greater, partly due to the discoveries of high heat flows over the ridge, and the presence of median rift valleys marked by earthquake disturbances and magnetic anomalies. Wilson points out that if the assumption is made that the globe expands only along the mid-ocean ridges, then the physical and the geological consequences of the theory are consistent. Once again, it is a working hypothesis that Wilson is considering. As he says, the point that has just been made

does not prove that the expansion of the Earth has occurred or if it has, that it is due to a change in G, but it is worth considering some of the consequences of accepting that hypothesis. (1960, p. 881)

He then lists a number of different kinds of evidence that bear on the question, and compares them with the three rival hypotheses. Once again, convection currents come off worst. He points out that one of the chief arguments in favor of them are large horizontal faults observed in certain parts of the world, but that these can also be explained on other models. Further, although rising convection currents could explain the high heat flow over mid-ocean ridges, he does not see how sinking currents could explain the form of continental coastlines or the mountain chains with which they are so frequently associated. For him, this means that he need consider the model no further in this paper. On certain issues he believes that the contraction and expansion models are equally successful. Both could explain, for example, why the earth's surface would buckle into mountain chains at some distance from the mid-ocean ridges. But there are other circumstances in which the expansion hypothesis is more successful. It removes a difficulty that he had not stressed in his previous papers, namely that "If continents had grown on a shrinking earth, the oceans would have flooded the continents, and this has not happened." (1960, p. 881)[4] Here is a nice example of the consideration of a rival hypothesis showing up previously unexamined, though not unnoticed, difficulties in other hypotheses. In the contraction model he had had to assume that the oceans slowly deepened, but not so much as to flood the continents. Thus in 1960, Wilson's conclusion was that the expansion hypothesis

has the merit of appearing to explain many features of the Earth's surface, though this does not constitute a proof. Even if true, expansion at the rate here postulated could conceivably be due to phase changes in the earth's interior or perhaps to differentiation of the core and mantle, but a decrease in G remains an inviting idea. (*Ibid.*, p. 882)

By the fall of the following year, 1961, Wilson was beginning to take convection currents more seriously as a possible explanation of the development of the crust. Replying to an exchange between Dietz and Bernal in *Nature*, Wilson (1961, p.126) does not discuss any of the new evidence for convection currents that Hess and Dietz had cited, but instead considers whether there is a possible global geometry of convection cells, and the extent to which, if they exist, they can offer alternative explanations of older phenomena. He concludes that, in spite of prima facie difficulties about descending limbs in the Antarctic region, a geometry of convection cells can be reconciled with the positions of the major mid-ocean ridges and mountain chains. Furthermore, he states without too much argument that in terms of relations to older orogenetic theories,

Besides admitting of some features of compression mountains, the new theory admits of continental growth, both by sweeping up sediments on the ocean floor and by the emission of andesitic lavas. (p. 127)

Thus here Wilson is suggesting that many, though not all, of the features that he had previously explained by a contraction model, could also be dealt with in terms of a convection model. One outstanding difficulty remains, however, and it is that of maintaining the continents above sea level. He rejects Dietz's and Hess's suggestion that this is due to thickening of the crust beneath continental plains, since he thinks that not one of their three proposed mechanisms for this would work, and instead once more tentatively suggests that an expanding earth would explain this puzzling fact. At this stage, then, Wilson is working with all three hypotheses simultaneously.

After another couple of years, Wilson (1963*a*) had turned his attention primarily to the question of the movement of the continents by convection currents. He conceded that there was still no direct evidence for such currents, but suggested two arguments that made their existence seem more likely. He cited Jeffreys's conclusion that an enormously high viscosity would be required to stop mantle convection and added that the currents explained the median position of the oceanic ridges. Having satisfied himself that mantle convection was theoretically possible, he also pointed out that there was now some evidence that continents had moved, particularly in the paleomagnetic record and the existence of large faults with horizontal displacements (which he had previously thought explicable on a contraction hypothesis). In his usual fashion, he then turned to suggesting some testable consequences of the theory which would enable a more thorough assessment to be made. First he suggested that the volcanic islands of the Atlantic increase in age with distance from the mid-Atlantic ridge. Second, he believed that non-seismic lateral ridges could be found at right angles to many of the major mid-ocean ridges and that these could be matched up, giving geologists a much better method of reconstructing old land masses than the rather hit-and-miss methods previously available. As it turned out, neither of these predictions proved very frutiful: determining the age of volcanic islands formed by periodic eruptions over long periods is very tricky, and despite the efforts that Wilson and his group devoted to it, the geological community was never very impressed; the lateral ridges that Wilson tried to trace turned out to be much more discontinuous features than he had hoped. Thus his search for testable consequences of the convection model continued, and in one of his best known papers, Wilson (1965), he finally made a prediction that, when confirmed several years later, seemed very compelling to the geological community. He suggested that the earth was divided into large rigid plates with three types of boundary; the mid-ocean ridges, the continental mountain chains, and what Wilson now described as 'transform faults'. He suggested that these were represented on the earth's surface by the large horizontal faults that had worried him for so long. He made specific predictions about their motion and claimed that

Transform faults cannot exist unless there is crustal displacement, and their existence would provide a powerful argument in favour of continental drift and a guide to the nature of the displacements involved. (1965, p. 345)

This was not the only prediction that led to the widespread acceptance of continental drift and its successor theory, plate tectonics, but it was one of the more important. What morals can we now draw from this story?

It seems to me that there is no doubt that Wilson did work in a number of programs of research for a number of years. Without accepting or believing any one of them to be true, he compared the explanatory power and predictive capacity of three major geological traditions. This constant comparison, I submit, enabled him to assess the relative strengths and weaknesses, and attempt to develop the consequences of each of the theories. Would progress have been made so rapidly if there had not been someone like Wilson pursuing the method of multiple working hypotheses? Historically, this is impossible to ascertain, but it can scarcely be accidental that he was one of the main proponents of theoretical innovations that are widely acknowledged to have made geology a more exciting discipline. But a critic of the method of multiple working hypotheses might claim that this was a strange and unusual period in the development of the earth sciences, and that they have now settled down once more, and rightly, to a period of normal science.[4] To this there are two replies. First, granted that this comment is true, the method of multiple working hypotheses still has much to say about the discovery of the paradigm or research programme that has been adopted, a topic neglected by the proponents of those approaches. Second, it seems to me that it would be unhealthy for the earth sciences to relapse too much into normal science and lose the critical and exploratory attitude that marked the 1960s. Furthermore, Wilson at any rate has not done so. Having accepted that continents moved as a consequence of accepting convection currents, he very quickly decided that the difficulties with convection currents as generally understood were so great that alternatives, such as the effects of possible 'hot spots' had to be explored. He, at any rate, has not abandoned the method of multiple working hypotheses.

*Center for Philosophy of Science*
*University of Pittsburgh*

## NOTES

[1] This example would bear out the case against 'Planck's Principle' made recently by David Hull (1978).

[2] For a discussion of the publication history of these articles, and a reprint of the most important, see Albritton (1975, pp. 126–131). All references are to this reprint. Chamberlin's methodology has been discussed recently by Pyne (1978). His conclusion that Chamberlin's methodology became less important with the advent of plate tectonics differs from mine.

[3] It might be argued that Chamberlin, like many other scientists who have written on methodology, did not practise what he preached, since he argued so strenuously, and

over such a long period for the Chamberlin-Moulton planetismal theory. (See Brush, 1978) Whether or not this is the case, it does not detract from either the value of the methodological insight, or from its influence.

[4] This is the closest approach I could find in Wilson's development to justifying Feyerabend's rationale for multiple hypotheses. However, the difficulty had been known for a considerable time, and just downplayed in the absence of an acceptable alternative theory.

[5] It is clear that even during the geological 'revolution', Wilson's method was not typical. As a fellow-participant in that revolution and an historian of the period, Uyeda (1978, p. 65) remarks of Wilson, "Scholars who change their opinion too often usually lose the respect of their colleagues, but Wilson's insight and originality appear to have made him an exception." But in examining the genesis of major new theories, we are not studying the methods of the average scientist under normal conditions.

## BIBLIOGRAPHY

Albritton, C. (ed.), 1975 *Philosophy of Geohistory*, Dowden, Hutchinson and Ross, Stroudsburg, Pennsylvania.

Bernal, J. D.: 1961, 'Continental and oceanic differentiation', *Nature* 192, 123–125.

Brush, S. G.: 1978, 'A Geologist Among Astronomers: The Rise and Fall of the Chamberlin-Moulton Cosmogony', *Journal for the History of Astronomy* 9, 1–41 and 77–104.

Chamberlin, T. C.: 1980, 'The method of multiple working hypotheses', *Science* 15, 92–96, reprinted in Albritton (1975), pp. 126–131. Page references are to this reprint.

Cox, A. (ed.): 1973, *Plate Tectonics and Geomagnetic Reversals*, Freeman, San Francisco.

Dietz, R. S.: 1961a, 'Continent and ocean basin evolution by spreading of the sea floor', *Nature* 190, 854–857.

Dietz, R. S.: 1961b, 'Continental and ocean differentiation', *Nature* 192, 124.

Feyerabend, P. K.: 1975, *Against Method*, New Left Books, London.

Griggs, D. T.: 1939, 'A theory of mountain building', *American Journal of Science* 237, 611–650.

Hallam, A.: 1973, *A Revolution in the Earth Sciences: From Continental Drift to Plate Tectonics*, Clarendon Press, Oxford.

Hess, H.: 1948, 'Major structural features of the western North Pacific: an interpretation of H. O. 5485, Bathymetric Chart, Korea to New Guinea', *Geological Society of America, Bulletin* 59, 417–446.

Hess, H.: 1962, 'History of ocean basins', in A. Engel *et al.* (eds.), *Petrologic Studies: A volume to honour A. F. Buddington*, Geological Society of America, pp. 599–620. Reprinted in Cox (1973).

Hull, D. L., P. D. Tessner, and A. M. Diamond: 1978. 'Planck's Principle', *Science* 202, 717–723.

Kuhn, T. S.: 1962, *The Structure of Scientific Revolutions*, Univ. of Chicago Press, Chicago.

Lakatos, I.: 1970, 'Falsification and the methodology of scientific research programmes', in I. Lakatos and A. Musgrave (eds.), *Criticism and the Growth of Knowledge*, Cambridge Univ. Press, Cambridge.

Marvin, U. B.: 1973, *Continental Drift: The Evolution of a Concept*, Smithsonian Institution Press, Washington, D. C.

Pyne, S.: 1978, 'Methodologies for geology: G. K. Gilbert and T. C. Chamberlin', *Isis* **69**, 413–424.

Runcorn, S. K. (ed.): 1962, *Continental Drift*, Academic Press, New York and London.

Schiller, F. S. C.: 1921, 'Hypotheses', in C. Singer (ed.), *Studies in History and Method of Science*, Oxford Univ. Press, Oxford, pp. 414–446.

Suppe, F. (ed.): 1977, *The Structure of Scientific Theories*, 2nd edition, Univ. of Illinois Press, Urbana.

Takeuchi, H. *et al.*: 1967, *Debate about the Earth: Approach to Geophysics through Analysis of Continental Drift*, Freeman, San Francisco.

Uyeda, S.: 1978, *The New View of the Earth*, Freeman, San Francisco.

Wilson, J. T.: 1954, 'The development and structure of the crust', in G. Kuiper (ed.), *The Earth as a Planet*, Univ. of Chicago Press, Chicago, pp. 138–207.

Wilson, J. T.: 1959, 'Geophysics and continental growth', *American Scientist* **47**, 1–24.

Wilson, J. T.: 1960, 'Some consequences of expansion of the earth', *Nature* **185**, 880–882.

Wilson, J. T.: 1961, 'Continental and oceanic differentiation', *Nature* **192**, 125.

Wilson, J. T.: 1963*a*, 'Hypothesis of earth's behavior', *Nature* **198**, 925–929.

Wilson, J. T.: 1963*b*, 'The movement of continents', in *Symposium on the Upper Mantle Project*, International Union of Geodesy, Berkeley.

Wilson J. T.: 1965, 'A new class of faults and their bearing on continental drift', *Nature* **207**, 343–347.

Wilson, J. T.: 1970, *Continents Adrift*, Freeman, San Francisco.

Wilson, J. T.: 1976, *Continents Adrift and Continents Aground*, Freeman, San Francisco.

HANK FRANKEL

# HESS'S DEVELOPMENT OF HIS SEAFLOOR SPREADING HYPOTHESIS

In 1960 Harry Hess, in the words of Robert Fisher, "put it all together."[1] What Hess put together was a mass of seemingly unrelated data when he proposed his seafloor spreading hypothesis.[2] This hypothesis had more to do with the eventual acceptance of continental drift theory in the form of plate tectonics by most researchers in the geosciences during the late sixties than any other conceptual innovation. The eventual acceptance of continental drift came with the confirmation of the Vine-Matthews hypothesis and J. T. Wilson's transform fault hypothesis. Both of these hypotheses were virtual corollaries of Hess's idea of seafloor spreading.[3] Hess grafted his hypothesis onto the existing continental drift tradition (hereafter DRIFT), and utilized his hypothesis as a plausible solution to the most serious problem faced by DRIFTers, namely, how on earth the continents could plow their way through the seafloor and remain intact. Consequently, Hess provided DRIFT with a new theory which solved the old empirical problems handled by DRIFT, offered a solution to their engineering problem by providing an adequate mechanism for the drift of the continents, and enabled DRIFT to take credit for solving new problems in oceanography and geophysics.

The subject matter of this paper, the development rather than the reception of Hess's seafloor spreading hypothesis, is particularly interesting because Hess was not a proponent of DRIFT until after the generation of his hypothesis. I shall argue the following: (1) Hess's seafloor spreading hypothesis arose out of his long time interests in solving a nest of problems in oceanography and geophysics, and many of these problems were ones for which he had offered and rejected previous solutions. (2) Hess's major epistemic aim in theorizing was to present hypotheses which effectively solved problems. On every occasion where he presented a solution to a problem, he defended it in terms of its problem-solving effectiveness and argued that it had greater effectiveness than competing solutions. (3) Hess's adoption of DRIFT with the grafting of his own seafloor spreading hypothesis was undertaken because it increased the problem-solving effectiveness of his own hypothesis and made DRIFT a much better problem-solver through elimination of its most serious deficiency. (4) Hess's major epistemic aim in theorizing, presenting hypotheses which were good problem-solvers, when coupled with his apparent lack of fear in

345

*T. Nickles (ed.), Scientific Discovery: Case Studies*, 345–366.
*Copyright © 1980 by D. Reidel Publishing Company.*

making mistakes, gave him the freedom for inventing, evaluating, rejecting and salvaging potential solutions.

## I. HISTORICAL ANALYSIS OF HESS'S WORK

There were two occasions when Hess's thought underwent fairly substantial change. The first was prompted by the influx of a massive amount of new data about the seafloor which began in the early fifties, and the second resulted from his seafloor spreading hypothesis. Thus Hess's career, like so many other good things, naturally is divisible into three parts: an early period when Hess developed his views on the basis of a relative scarcity of data, a middle period when Hess was able to take advantage of the data boom, and a final period when Hess presented and eventually expanded his seafloor spreading hypotheses.

### A. Hess's early career (1932–1950)

Although almost all of Hess's geological pursuits during these first two periods had a bearing on the construction of his seafloor spreading hypothesis, I shall discuss in detail only those aspects which were critical to its development.

*1. Hess's downbuckling or tectogene hypothesis: a solution to numerous problems connected with trenches, island arcs and Alpine mountain systems.* From 1932 through 1938, Hess published a number of papers wherein he offered solutions to several diverse problems: the problems concerned the presence and formation of magmatic serpentine belts in island arcs and Alpine-type mountain ranges, the presence of gravity anomalies by trenches in island arc regions, and the formation of island arcs and mountain ranges.[4] By 1937 he had carved out a unifying solution to all of these problems. He (along with Vening Meinesz and others) supposed downbuckling of the earth's crust which resulted in the formation of negative gravity anomalies surrounding trench regions, accompanying island arcs containing serpentine intrusions of magma, and eventual alteration of island arc structures into mountain ranges with the addition of sediments that are squeezed upward through continued action of the downbuckle. Hess supported this downbuckling hypothesis throughout his early career, and he extended its problem-solving effectiveness when in 1940 he offered an explanation for the generation of the Hawaiian Islands and accompanying swell which was derivative from the downbuckling hypothesis.[5] Then in 1950 he explicitly coupled the downbuckling hypothesis with convection currents in the earth's mantle so as to explain how the crust could remain downbuckled for an extended period of time, and he utilized the conjunction to account for the rather complicated pattern of deep foci earthquakes typically located on the continental side of an island arc (Hess [1951]).

2. *Origin and development of guyots.* During the second world war, Hess commanded the U.S.S. Cape Johnson, and he managed to take numerous soundings of the Pacific seafloor "on random traverses incidental to wartime cruising" (1946, p. 773). Through these soundings Hess discovered a number of submerged, reefless, flat-topped seamounts, which he named guyots (after the Swiss oceanographer Arnold Guyot). In 1946 he presented an ingenious hypothesis for their development and formation. The unique and puzzling characteristic of guyots is their absence of reefs, and consequently, Hess had to construct a solution for their origin and development which would prohibit reef development. He argued that guyots originally were Precambrian islands that already had submerged themselves sufficiently below sea level prior to the evolution of lime secreting organisms. Central to his analysis was the claim that sea level has risen with respect to oceanic structures since Precambrian because of the unending deposition of continental sediment upon the seafloor. The sediments would raise the ocean floor, and therefore, assuming a relative constancy of oceanic water since Precambrian, sea level would rise.

3. *Hess's position toward DRIFT circa 1950.* Hess opposed DRIFT. He never wrote favorably toward it — at least judging from what I have examined. The only reference I have seen by Hess directed toward DRIFT, albeit obliquely, was negative. When discussing his belief in a strong oceanic crust, a belief in conflict with most versions of DRIFT, Hess remarked:

The writer believes the oceanic crust is very strong though his opinion is at variance with existing textbooks and much of the literature. However, Jeffreys (1929), Daly (1940), and Longwell (1945) all favor a strong oceanic crust. The only bases for judging its strength are its behavior and the strength of the rocks of which it is thought to be composed. Both of these indicate strength. . . . Those favoring the hypothesis of continental drift assumed a very weak basaltic crust below the oceans without, so far as the writer is aware, presenting evidence other than the hypothesis of drift to substantiate the assumption. (1946, pp. 786–787)

This comment of Hess's is of special interest because it not only indicates his rejection of DRIFT but also that he recognized as a key objection to DRIFT the conceptual problem of finding a way for the continents to plow their way through the seafloor. DRIFTers needed to postulate a weak oceanic crust for movement to occur, but Hess thought it possessed strength. Added to this, Hess could not have been a proponent of DRIFT because of his belief in the permanency of the ocean basins.

## B. Hess's middle career (1952–1959)

In 1953 and 1955 Hess wrote several speculative papers (1954, 1955a and b) wherein he expanded his downbuckling theory, suggested a new solution to the problem of guyot formation and development arising from new concerns with serpentine formation, and devoted much of his attention to the nature, formation, and development of oceanic ridges. All of these pursuits were undertaken in light of Hess's new analysis of oceanic crust, which was based on the plethora of new information on the oceanic crust and layer of covering sediments unearthed by M. Ewing *et al*. at Lamont-Doherty and others at Scripps in the late forties and early fifties.

Except for the extension of his downbuckling hypothesis, all the other activities had a direct bearing on the later development of the seafloor spreading hypothesis. But even his extension of the downbuckling hypothesis was not irrelevant, for it continued to keep Hess a nonbeliever in DRIFT. In short, Hess combined with the old notion of growth of continents at their margins through accretion of island arcs during mountain building epochs, his downbuckling hypothesis, along with a novel suggestion that serpentine belts, which evolve into arcs and mountain ranges, develop along extensive, intercontinental strikes. Because his suggested strike pattern required the permanency of continents and oceans with respect to one another throughout geological time, Hess quite clearly was not in favor of drifting continents.

*1. The new data, and Hess's revised model of the oceanic crust and upper mantle.* Hess declared in 1953 that:

The most momentous discovery since the war is that the Mohorovicic discontinuity rises from its level of about 35 km under the continents to about 5 km below the seafloor . . . while . . . the second most important discovery is the recognition that we have at the earth's surface fragments of what are almost certainly the material from below the Mohorovicic discontinuity. (1954, pp. 341–342)

Because of this new data, Hess and others revised their view of the oceanic crust. His analysis, like many others at that time, was as follows: Oceanic crust was taken to be approximately only 5 km thick with an average of 0.7 km of unconsolidated sediments above it, while the mantle was considered to be made up of peridotite. Major differences between this new model and previous ones, including Hess's former model, were as follows: Elimination of any granitic material from oceanic crust, and a drastic thinning of the basalt layer along with the consequent raising of the Moho discontinuity. Moreover, the 0.7 km of unconsolidated sediments reduced by a factor of 3 to 5 times the size of former estimates, based upon extrapolation of established, present rates to Precambrian.

2. *The origin and development of mid-ocean ridges: 1953 and 1955 solutions*. Because "something [was] known of the character of oceanic crust," Hess thought it profitable to set up working hypotheses for the origin and development of oceanic ridges. (*ibid*., p. 344). Indeed, he thought the time ripe, even though he had only tentative suggestions at hand, because

Without hypotheses to test and prove or disprove, exploration tends to be haphazard and ill-directed. Even completely incorrect hypotheses may be very useful in directing investigation toward critical details. (*Ibid*.)

The relevant, new ridge data utilized by Hess in the development of his hypothesis was as follows: Almost all ridges are associated with basalt volcanism. Besides basaltic materials, the only other ridge samples have been peridotites, and no samples have been older than Cretaceous. There is little indication of folding in ridge structures as is most often exhibited by terrestrial mountain systems. And finally, in most cases it appears that ridge structures have subsided somewhat since initial formation.

Hess's hypothesis offered a solution for the origin and development of mid-oceanic ridges, fit together much of the above data, and utilized his new model of oceanic crust and mantle. Beginning with a thin basaltic crust and peridotite upper mantle, Hess envisioned the formation of a less dense layer of crust through magmatic intrusion of basalt mixed with peridotite. Since the basaltic intrusion would be less dense than the surrounding peridotite, the column would rise due to isostatic adjustment. The suggested process involved extension of the crust at right angles to the ridge without crustal foldings, and the resultant surface materials, basalt mixed with peridotite, matched the samples uncovered from the Mid-Atlantic ridge. Moreover, the subsequent cooling of the rising magma and cessation of the convection currents provided Hess with a solution to the problem of how the ridge decreased in height after its initial formation.[6]

Hess proposed an alternative model for the origin and development of the Mid-Atlantic Ridge along with other mid-oceanic ridges in 1955, and the genesis of this new model is found in his continued concern with serpentine.

In 1953 Hess advanced an hypothesis involving serpentine formation in oceanic mantle which could bring about reversible vertical movement of the seafloor. However, he did not apply this serpentinization model to problems concerning ridge formation and development until 1955, although in 1953 he employed it as an alternative solution to the problem of guyot formation and development since he had good reason for rejecting his 1946 hypothesis. The central reaction involving serpentine, isolated in 1949 by two petrologists, was as follows:

$$olivine + water = serpentine + heat$$

Both olivine and serpentine are forms of peridotite. At temperatures above approximately 500°C. the reaction proceeds to the left, while below the 500°C. mark it moves to the right. Because the reactive equivalents of serpentine are less dense than those of olivine, when the reaction proceeds to the right the resulting serpentine has a greater volume than the reactive olivine. Consequently, this reaction offered a mechanism for the reversible, vertical changes in the contour of the earth's surface. Serpentine production would bring about a rise, and olivine production would cause a decrease in elevation.

Hess, a lifelong student of serpentine, quickly realized that this reaction could take place on the seafloor. His new model of oceanic crust and upper mantle placing peridotite much closer to the surface, and the new data indicating surprisingly high heat flows in oceanic crust, both suggested to Hess the feasibility of this reaction occurring under the seafloor.

Hess fleshed out this serpentinization process in 1955, and the general production of serpentine with its ensuing deserpentinization involved the following steps: he placed the crucial 500°C. isotherm at a depth well below the Moho. Assuming the upward movement of water from the earth's interior, he proposed the transform of olivine peridotite to serpentine peridotite at the 500°C. isotherm. With the addition of more water, continued serpentinization would occur, and at the same time the whole mass of serpentinized peridotite would rise upward, if the 500°C. isotherm migrated upward. Hess suggested two possible causes for the isotherm rise, namely, either convective overturn in the mantle or intrusion of basalt. Once the rising isotherm reached the crust where there was no more olivine, then the deserpentinization below, brought about by the rising isotherm, would result in a net loss of serpentinized material. Rising of oceanic surface would occur whenever the serpentinization was greater than the deserpentinization, while subsequent lowering of the surface would result when deserpentinization outstretched serpentinization. Hess pledged his allegiance to this model in the following passage, where he applied it to the Mid-Atlantic Ridge:

Until recently most geologists would have supposed that the Mid-Atlantic Ridge was either a folded mountain system with consequent thickening of the curst above the Mohorovicic discontinuity or alternatively that it was a thick section of volcanic material lying on normal oceanic crust or intruded into it. Now it seems more likely that it represents a welt of serpentine. Serpentinized peridotite was dredged from large fault scarps on Ridge by Ewing *et al.* (Shand, 1949). The flanks of the Ridge once stood much higher than today as indicated by paired terraces along its east and west slopes. Presumably some deserpentinization has occurred since maximum serpentinization. The problem of why serpentinization was concentrated in the Atlantic along a median line can perhaps be explained in several ways. One hypothesis could be that convective circulation in the mantle occurs, and the ridge represents the trace of an upward limb of a cell. In this case water ejected from the top of the column might cause the later deserpentinization. Whatever hypothesis may be suggested to account for the localization of serpentinization along the Ridge is at present pure speculation. The idea that the topographic elevation of the Ridge may be due to serpentinization should be considered on its own merits apart from the above speculation. (1955a, pp. 404–405)

Apparently, Hess in 1955 took to heart the serpentinization of the peridotite found on the Mid-Atlantic Ridge in 1949, and consequently, he decided that his new hypothesis was more worthy of pursuit than his former one.

*3. Hess's new solution to the problem of guyot formation and development.* By 1953 Hess realized that his former solution to the development of guyots was untenable, and offered a new solution which incorporated the serpentine reaction. By 1955 Hess again presented his new hypothesis and succinctly summarized the problems with his 1946 solution:

Originally the writer (1946) postulated that guyots were truncated in Precambrian time before lime-secreting organisms were available to protect them, and that their submergence was due to relative rise of sea level largely by sedimentation in the oceans. Hamilton's (1953) discovery of Late Cretaceous shallow-water fossils on them cuts the time to one-fifth, and recent investigation of the total amount of sediment on the seafloor cuts the rate of sedimentation by about one-fifth so that submergence as called for in the original hypothesis is 25 times too slow. (1955*a*, p. 405)

In essence the problem of guyot development had become one of finding a mechanism for sinking guyots which was slow enough to allow truncation when guyots were at sea level, and fast enough to sink them sufficiently below sea level so as to prohibit reef formation. Hess's solution was to propose that guyots were formed above a well of serpentinized peridotite. Once the mass of serpentine began to deserpentinize, the resulting decrease in volume of the mass could cause the required downward movement.

*4. 1959 changes in Hess's thought.* In (1959*a* and *b*) Hess further developed his 1955 hypothesis for the origin and development of mid-ocean ridges, presented a new model for oceanic crust, and continued to recognize the importance of the fact that seismic profiles of accumulated oceanic sediment consistently yielded lower values than those established through extrapolation of present sedimentary rates.

In August of 1959 Hess reproposed his 1955 solution to the problem of mid-ocean ridge development, but he introduced two changes. He settled on convection currents as opposed to rising basalts as the driving force behind the serpentinization-deserpentinization transform because of two new discoveries, namely, higher than normal flows over several Pacific ridges and association of a central grabben along the Mid-Atlantic Ridge with shallow earthquakes.

Assuming that the process forming the ridges is serpentinization which involves about 100 cal/g in heat evolved, the possibility was investigated that this might account for the high heat flow. It fails to do so by about two orders of magnitude. Somewhat more heat could be obtained by supposing it resulted from basalt intrusions into the ridge but this too fails by more than an order of magnitude . . . . The hypothesis is advanced that the ridges represent the trace on the Earth's surface of upward flowing limbs of mantle convection cells. Water released at the top of the column produces the serpentinization, subcrustal drag of the horizontal flow produces extension and the grabben on the crust. Heat moving slowly upward by conduction accounts for the high flow and ultimately for deserpentinization and subsidence of the ridge. (1959*a*, p. 34)

In addition, Hess also suggested that mid-oceanic ridges were ephemeral in nature. Behind this claim was his identification of the region of guyots in the Pacific as a dying, but short-lived ridge. Its elevation and heat flow were lower than recorded on 'young' ridges like the Mid-Atlantic. Of course, Hess's suggestion that regions of high guyot concentration represent 'old' ridges made perfect sense and, perhaps, was partly suggested by his solution to the guyot problem in terms of serpentinization.

In December of 1959 Hess presented his new model of oceanic crust. There were two major differences between this version and his 1953/1955 versions. He replaced the 5 km basalt layer with a 5 km layer of serpentinized peridotite. Actually, because of the presence of serpentinized peridotite on the Mid-Atlantic Ridge, he had suggested this alternative in 1955 but opted for basalt. The overall reason for his 1959 switch was that seismic velocity data indicated either basalt or serpentinized peridotite. But the newly discovered fact that this layer of oceanic crust (layer 3) has a uniform thickness throughout indicated to Hess that it had to have been formed on location — it was just too even to have been formed, say, in the mantle and then transported piecemeal to its present location.[7] This requirement was enough to eliminate magma generation of basalt, and Hess opted for his favorite transformation, namely, serpentinization of peridotite because of peridotite's presence in the mantle. In short, Hess supposed that ascending water serpentinized the peridotite of the original mantle down to a level of 5 km where the critical 500°C. isotherm must have been present at the time. The other major difference between this model and his former ones was his more detailed account of the upper two layers. He argued that layer 1 was unconsolidated sediment, while layer 2 was "consolidated sedimentary rocks or volcanic rocks or both."

Hess also expressed his continued puzzlement over the fact that seismic profiles indicated a much thinner layer of accumulated sediment on the sea-floor than predicated through extrapolation of present rates, and he stated his three most obvious solutions to the dilemma without opting for any one of them in particular.

The most obvious alternatives are: (1) The oceans are relatively young. At 1cm/100 yrs the sediment could be accounted for if sedimentation only started in the Cretaceous. (2) The pre-Cretaceous sediments have in some manner been removed; for example, by incorporation into the continents by continental drift. (3) Nondeposition of any sediment over much of the ocean floor was a common attribute of the past. In any case those who expect a complete record far back to billions of years ago are doomed to dissappointment. (1959b, p. 343)

All that he was sure of was that the "sedimentary-rock sections of layers 1 and 2 will be very incomplete," although he was "rooting against" such a prediction! (15, p. 343). In the year to follow, Hess would have chosen an option that was not extremely obvious to him (or anyone else) at this time, namely, that the ocean floors but not oceans are very young since they are continually being created and destroyed through seafloor spreading. However, in 1959 Hess had not developed his seafloor spreading hypothesis.

*C. Hess's later career – seafloor spreading and continental drift.*

In December of 1960, Hess in a preprint (1960) proposed his seafloor spreading hypothesis. The overall hypothesis provided new solutions for several of Hess's favorite problems; and its various aspects reflected his continual concerns over developing an adequate model of oceanic crust, figuring out how serpentine can be generated from peridotite and is present in island arcs, trenches and Alpine mountain ranges, and making sense out of the role of convection currents.

The central aspect of the seafloor spreading hypothesis was its new solution to the problem of the origin and development of mid-ocean ridges. Hess realized that his 1955/1959 solution was inadequate, but that if he were to propose a slightly different solution incorporating many of the same elements, he could both avoid the problems with his former solution and simultaneously solve the problem of how layer 3 of the oceanic crust had been formed. Moreover, Hess was quick to realize that his hypothesis worked equally well as a solution to the problems of guyot formation and development and as an answer to why sedimentary deposits were much less than predicted by present accumulation rates. Hess at this time also realized that he had solutions, or at least the essential ingredients, for problems concerned with trench, island arc, and mountain formation. Added to all of the above, Hess was fully aware that his seafloor spreading hypothesis offered a solution to the number one nemesis faced by DRIFTers, namely, how to move continents through the seafloor without calling on God. Hess simply proposed that the continents do not plow their way through the seafloor, but are carried passively atop the spreading seafloor. Therefore, Hess grafted his seafloor spreading hypothesis onto DRIFT, and thereby greatly enhanced the overall problem-solving effectiveness, not only of his seafloor spreading hypothesis but of DRIFT. His hypothesis offered solutions to all the old problems solved by DRIFT and the new ones connected with paleomagnetism, while DRIFT now had a solution to its 'engineering problem' and much better solutions to problems concerned with island arc and mountain formation.

*1. Hess's seafloor spreading hypothesis.* Young mid-ocean ridges are located on upward moving convection currents and are the site for the generation of new seafloor. That is, they are where layer 3 of oceanic crust, which is composed of serpentinized peridotite, is created; the place where the peridotite of the mantle undergoes 70% serpentinization. The layer of generated serpentinized peridotite has a thickness of 5 km because 5 km is the position of the critical 500°C. isotherm. The rising convection currents elevate the 500°C. isotherm from its normal position up to the 5 km level but are incapable of pushing it any higher. Once the serpentinized peridotite is generated, it is forced outward from the ridge axis through the action of the parting convection currents. Eventually, the convection cell subsides, and the ephemeral ridge disappears. Meanwhile, the outward moving new seafloor ultimately sinks into the mantle on the backs of the descending convection currents, and thereby forms oceanic trenches. Continents, pushed along by the convection currents, cease moving once the currents stop their horizontal movement and descend into the mantle; and their leading edges "are strongly deformed when they impinge upon the downward moving limbs of convecting mantle" (1960, p. 33). In addition, since the spreading seafloor moves away from the ridge axis at equal rates, the ridges naturally have a median position with respect to the drifting continents. The continents do not sink into the mantle because of their relatively low density, and sometimes seafloor sediments are added onto them when the sediments ride down "into the jaw-crusher of the descending limbs" of a convection cell and are metamorphosed in the process. The oceanic crust is deserpentinized when it descends back into the mantle below the 500°C. isotherm. As a result of this continual destruction of seafloor, the seafloor is always young. Consequently, relatively little sediment accumulates on its surface despite high rates of deposition, and hardly any sediment is around ridge sites during the generation of new seafloor. Just as the crustal material moves outward from the ridge site, so do the formed guyots. When on the ridge axis, truncation occurs through erosion; and once the guyots have moved off the ridge, they drown themselves well below sea level.

*2. Development and presentation of the seafloor spreading hypothesis.* Hess initiated his study, 'an essay in geopoetry', by supposing the formation of a relatively solid earth, in accordance with the prevailing cosmogony; and then postulated a great catastrophe involving a two-cell convective overturn resulting in the separation of core from mantle and formation of a single primordial continent. Then he introduced and defended his 1959 model of oceanic crust, and he stressed its importance for the remainder of the study.

Hess initiated his discussion of the hypothesis itself by turning his attention to the origin and development of mid-oceanic ridges. Here he summarized the important data, uncovered errors in his 1955/1959 solution to their origin and development, and proposed part of his new solution to ridge origin and development along with most of his solution to the origin of layer 3 of oceanic crust.

Hess in his summary of the relevant data on mid-ocean ridges, listed the 1959 data, added the newly discovered datum that sediments are almost completely absent from ridge sites, and stressed the seismic velocity data which indicated that the "M discontinuity is not found or is represented by a transition from layer 3 to velocities near 7.4 km/sec."

Immediately after citing this data, Hess presented his devastating criticism of his 1955/1959 solution to the problem of ridge origin and development. Because the refutation is so important for understanding Hess's development of his seafloor spreading hypothesis, it is cited in full:

Formerly the writer (1955, 1955a) attributed the lower velocities (ca. 7.4 km/sec) in what should be mantle material to serpentinization, olivine reacting with water released from below. The elevation of the ridge itself was thought to result from the change in density (olivine 3.3 g/cc to serpentine 2.6 g/cc). A 2 km rise of the ridge would require 8 km of complete serpentinization below, however a velocity of 7.4 km/sec is equivalent to only 40% of the rock serpentinized. Thus serpentinization would have to extend to 20 km depth to produce the required elevation of the ridge. But this reaction cannot take place at a temperature much above 500°C. which considering the heat flow probably lies at the bottom of layer 3, about 5 km below the seafloor, and cannot reasonably be 20 km deep. Layer 3 is thought to be peridotite 70% serpentinized. It would appear that the highest elevation that the 500°C. isotherm can reach is approximately 5 km below the seafloor and this supplies the reason for the very uniform thickness of layer 3. The cause of the actual elevation of the ridge will be considered in the succeeding section. (1960, pp. 10–11)

Hess left no stone unturned in this refutation of his 1955/1959 solution. Given the high heat flow data at ridge sites, there was good reason to believe that the critical isotherm must be at about 5 km below the seafloor. But in order for the phase change of peridotite into 70% serpentinized peridotite to bring about sufficient rising of the upper mantle and crust and produce a ridge, the isotherm must be 20 km deep. Consequently, Hess gave up his 1955/1959 solution for the origin of mid-ocean ridges, along with his former belief that serpentinization of peridotite could occur in the earth's mantle below ridges.

This critique of his earlier solution was of utmost importance to Hess in the construction of his 1960 seafloor spreading hypothesis. Hess used it as a guide for what he had to delete in his new hypothesis, and it demonstrated to Hess that the problems of ridge origin and origin of layer 3 of oceanic crust were intimately related. The keystone to Hess's refutation was his realization that the critical 500°C. isotherm is at a depth of 5 km. Because he fully understood the significance of this fact, he realized that serpentinization of peridotite could not account for the uplift of a ridge and that serpentinization at ridge sites offered a solution to the problem of the origin of layer 3 of oceanic crust. The volume change brought about by serpentinization was inadequate. Hess knew he had to postulate some other mechanism for ridge formation. "The cause of the actual elevation of the ridge" he said "will be considered in the succeeding section" (1960, p. 11). But at the same time he fully realized that he should not dismiss the occurrence of serpentinization at ridge sites. There was an awfully good reason for not giving it up, namely, it provided a mechanism for creation of layer 3 of oceanic crust. The very fact which destroyed the utility of the serpentinization of peridotite as a solution to the problem of mid-ocean ridge origin, that serpentinization occurs at 5 km under ridge sites, demonstrated to Hess that serpentinization at ridge sites offered a solution to the problem of the formation of layer 3 of oceanic crust, and that he had to find some other mechanisms to account for the origin of mid-oceanic ridges.

At his juncture in his presentation, Hess still had to present his new solution to the origin and development of mid-oceanic ridges and complete his solution to the *formation* of layer 3. Although he had solved the problem of the *origin* of the third layer, he still needed to propose some mechanism allowing for its removal from ridge sites. Moreover, whatever mechanism he proposed for removal of newly created crust had to be such that the new material flowed outward from its point of origin without alteration of its 5 km thickness. Otherwise a new seafloor would be no wider than a ridge, or if it managed to cover a greater area through some violent method of removal, it would not possess uniform thickness. Of course, Hess's solution to both these problems was the postulation of slowly moving convection currents. In short, Hess argued as follows:

As mentioned above, this hypothesis [my former one] is no longer tenable because the high heat flow requires that the 500°C. isotherm be a very shallow depth. The topographic rise of the ridge must be attributed to the fact that a rising column of a mantle convection cell is warmer and hence less dense than normal or descending columns. (*Ibid.*, p. 21)

Once the rising convection currents reached the surface, they split apart along the horizontal, taking the newly formed layer 3 along with them.

What, however, is perhaps even more interesting than the use of convection currents to solve both of these problems is the fact that Hess spent considerable time arguing for the reasonableness of the postulation of convection currents and his overall seafloor spreading hypothesis, through listing all the other problems which his seafloor spreading hypothesis could solve before presenting the solutions to these two problems. He had good reason to employ this circuitous technique; for convection currents, let alone continental drift, were not held in the highest esteem by most geologists and geophysicists. Apparently, Hess thought that the best way to present the overall hypothesis was by demonstrating its problem-solving effectiveness. In all, Hess, before fleshing out solutions to these two problems so central to his hypothesis, argued that his overall hypothesis could solve eight additional problems, which, along with their solutions were as follows: The problem, faced by those espousing the permanentist thesis that the continents and ocean basins had not changed their positions since their formation, of making sense out of the recent paleomagnetic data.

Paleomagnetic data presented by Runcorn (1959), Irving (1959) and others strongly suggest that the continents have moved by large amounts in geologically comparatively recent times. One may quibble over the details but the general picture on paleomagnetism is sufficiently compelling that it is much more reasonable to accept it than to disregard it. . . . This strongly indicates independent movement in direction and amount of large portions of the Earth's surface with respect to the rotational axis. This could be most easily accomplished by a convecting mantle system which involves actual movement of the Earth's surface passively riding on the upper part of the convecting cell. (*Ibid.*, pp. 14–15)

The problems of why the mid-ocean ridges are median in location, and for-mation of trench, island-arc systems in the Pacific.

Menard's theorem that mid-ocean ridge crests correspond to median lines, now takes on new meaning. The mid-ocean ridges could represent the traces of the rising limbs of con-vection cells while the circum-Pacific belt of deformation and volcanism represents descending limbs. The Mid-Atlantic Ridge is median because the continental areas on each side of it have moved away from it at the same rate – a centimeter a year. (p. 16)

The problem faced by previous DRIFT hypotheses in explaining how on earth the continents plow through the seafloor: Hess simply pointed out that his view didn't require their plowing through the seafloor.

This [my view] is not exactly the same as continental drift. The continents do not plow through oceanic crust impelled by unknown forces, rather they ride passively on mantle material as it comes to the surface at the crest of the ridge and then moves laterally away from it. (p. 16)

The problem of why the amount of accumulated ocean sediment is much less than predicted through extrapolation of present deposition rates, as well as why the ridges are almost completely free of sediments: Here Hess simply turned the reasoning inside-out. The difficulty lay in assuming the permanency of the seafloor. Once this assumption was eliminated and it was supposed that the seafloor was young, the discrepancies dissolved. Hess, with his seafloor spreading hypothesis, now could argue for ancient oceans but young seafloors. The next problem Hess solved, one which he had not considered before and one closely related to the sedimentation problem, was why the number of existing oceanic volcanoes was much less than was estimated through extra-polation of existing formation rates. He again solved the problem by arguing for youthful seafloors, which could be expected given his hypothesis. Next, Hess offered a solution to the problem of why it appears that mid-ocean ridges are ephemeral in nature. His answer was that they are ephemeral, and he argued that this would be expected given his hypothesis since the life span of a convection cell would be equally as short. Hess also offered a new solution to the problem of guyot formation which, like his 1953 hypothesis, was simply that the area of guyots represented an old ridge. But unlike his earlier hypothesis, this solution obviously employed his new analysis of ridge formation and suggested that the absence of reefs upon their tops could now be accounted for through continental drift and polar wandering, in that they would have formed in a region too cold to support lime secreting organisms. Finally, Hess suggested that oceanic trenches were formed when seafloor descended back into the mantle upon the backs of descending convection currents.[8]

Of all these additional problem solutions, Hess devoted more space to those which had occupied his attention throughout his middle career, namely, the thinness of the sedimentary deposits, the ephemeral nature of the ridge and guyot formation. Surely this was no accident; Hess must have considered these problems significant. Once Hess had his seafloor spreading concept, he had a different point of view from which to solve them. Guyots were recent but had no reefs upon them because, due to continental drift and polar wandering, they were formed in a region too cold for the flourishing of lime secreting organisms.

Hess (1946) had difficulty in explaining why the guyots of the mid-Pacific mountain area did not become atolls as they subsided. He postulated a PreCambrian age for their upper flat surfaces moving the time back to an era before lime secreting organisms appeared in the oceans. This became untenable after Hamilton found shallow water Cretaceous fossils on them. Looking at the same problem today and considering that the North Pole in early Mesozoic time as determined from paleomagnetic data from North America and Europe, was situated in southeastern Siberia, it seems likely that the Mid-Pacific mountain area was too far north for reef growth when it was subsiding. (1960, pp. 20–21)

The sedimentary layer was so thin because the existing seafloors had only been around since Cretaceous, not because it only appeared to be thin or had been deposited at extremely uneven rates throughout the earth's history since Cambrian. And the ridges were ephemeral, like the seafloor itself, because convection cells, geologically speaking, exist for only a short time. With his new point of view, Hess saw a common solution to these three long-time concerns.

## II. SOME METHODOLOGICAL IMPLICATIONS OF HESS'S WORK

At the outset, I proposed to defend four claims; it is time to present these arguments in explicit form.

*Claim 1:* Hess's seafloor spreading hypothesis arose out of his longtime interests in solving a nest of problems in oceanography and geophysics, and many of these were ones for which he had offered and rejected previous solutions.

Hess brought to bear on his seafloor spreading hypothesis his long-time concerns with the origin of serpentine from peridotite, the structure of the oceanic crust and mantle, the formation of guyots, the absence of large amounts of oceanic deposits, the formation of ocean ridges, trenches, island arcs and mountain ranges, and the use of convection in this formation. He had been working on problems of serpentine formation since the beginning of his career. He discovered guyots in the mid-forties and by 1946 proposed his first of three solutions to their origin and development. With the uncovering of much of the new data about oceanic features, he revised his model of the curst, suggested a mechanism for the production of serpentine from peridotite in the earth's mantle under the seafloor, and used the mechanism in his second solution to the problem of guyot formation and development. At the same time he offered his first of three solutions to the problem of mid-ocean ridge development, suggesting a solution which involved convection currents, magma generation of basalt, and reliance on his 1953/1955 model of the earth's crust. By 1955 he had worked out his mechanism for serpentine production in sufficient detail to offer it as a solution to mid-ocean ridge development, and such a ploy allowed him to account for the 1949 dredging of serpentinized peridotite from the Mid-Atlantic Ridge. In 1959 he revised his model of the earth's oceanic crust, by substituting serpentinized peridotite for basalt. With this revision he began to speculate about the formation of layer 3, and he thought that serpentinization of peridotite played an essential role in its production. He also refined his second solution to the problem of mid-ocean ridge development by settling upon convection currents as the cause of the rising of the 500°C. isotherm, because it offered the best available account of the higher than normal heat flows on ridge sites. He also became more concerned with the ephemeral nature of mid-ocean ridges. Meanwhile, Hess continued to express puzzlement over the thinness of the sedimentary deposits upon the seafloor and in 1959 suggested that they indeed were incomplete, although he was unsure as to why they were so incomplete.

With the development of his seafloor spreading hypothesis in 1960, Hess brought every one of these concerns together. He argued that his 1955/1959 solution to the problem of ridge origin and development was inadequate — no serpentinization of peridotite could occur in the mantle below ridge sites. The heat flow data required that the critical isotherm be only 5 km below the seafloor. But serpentinization could occur at and above 5 km, precisely where layer three should be created. Convection currents still were important – even more so than before – because they accounted for the formation of the ridge, the passive horizontal movement of layer 3 away from ridge sites, and the eventual subsidence of the ridge and formation of trenches with the descent of layer 3 into the earth's mantle. Guyots formed at ridge sites and drowned themselves as they moved off the ridge along with layer 3. Moreover, once Hess had the idea that the seafloor was shortlived, he had an answer to the incompleteness of the sedimentary layers; it was no longer puzzling because seafloor itself is youthful.

> *Claim 2:* Hess's major epistemic aim in theorizing was to present hypotheses which effectively solved problems. It seems that on every occasion where he presented a solution to a problem, he defended it in terms of its problem-solving effectiveness, and argued that it had greater effectiveness than competing solutions.

In his public writings – at least the ones I have examined – Hess never explicitly stated that his major aim in theorizing was to solve problems; but his method of presenting his various hypotheses as well as his defense of them certainly suggests that this was his major aim. Whenever Hess presented an hypothesis, he isolated the problems it was supposed to solve, and defended the hypothesis on the basis of its problem-solving effectiveness. This is evidenced in his presentation, defense, and extension of his downbuckling hypothesis. When he first presented it in the thirties, he argued that it offered a unifying solution to a nest of problems which he carefully isolated, and he claimed that it was the only hypothesis that could effectively solve so many problems. Then in the forties and fifties, Hess expanded the downbuckling hypothesis in terms of its ability to solve additional problems. He employed it as a solution to the formation of the Hawaiian Islands and accompanying swell (1940); explicitly used it (1950) in conjunction with convection currents to account for the observed pattern of deep focus earthquakes on the continental side of island arcs and the needed longevity of a downbuckle, a problem he maintained could not be solved by the contractionist theory of mountain building; and extended it (1955) so as to encompass the old idea of continental accretion. Moreover, the claim that Hess's major concern was to present hypotheses which would solve problems, is evidenced in his work on guyot and ridge formation. He considered them to be intriguing problems and presented successive solutions, while arguing that the most recent possessed a better problem solving effectiveness than its predecessor. The discovery of

Cretaceous fossils atop guyots (1953) destroyed his previous (1946) hypothesis. Hess proposed a new solution (1953 and 1955) which employed serpentinization of peridotite, which avoided the problem faced by his former solution, and was suggested by his new analysis of oceanic crust and mantle, as well as by the unexpectedly high oceanic heat. Likewise, his 1960 solution to the problem, which was one part of his seafloor spreading hypothesis, had the advantage of being part of a unifying solution to a number of problems. The same pattern is illustrated in his treatment of his various solutions to the origin and development of mid-ocean ridges. When he presented a new solution, he always argued that it was a more effective problem solver than his former one, and eventually he was able to incorporate it into his unifying seafloor spreading hypothesis. Hess's aim in solving problems is suggested equally by his recognition that the discrepancy in predicted and observed amounts of seafloor sedimentation was itself problematic. He pointed out the discrepancy as early as 1953, admitting that he could not solve it in 1959 (though he did offer three suggestions), and in 1960 he was quick to realize that his seafloor spreading offered a straightforward solution to the problem. Of course, the most obvious case wherein Hess displayed his overall goal to propose hypotheses which are effective problem solvers, was his presentation and defense of seafloor spreading. He explicitly refuted his 1955/1959 solution to the problem of ridge formation, used his refutation as a guide to the development of his new solution, and explicitly defended his overall hypothesis on the grounds that it solved so many seemingly unrelated problems. Moreover, Hess even argued that his seafloor spreading hypothesis was a more viable solution to several problems which were also solved by the expanding earth hypothesis suggested by Carey, Heezen, and others.

> *Claim 3:* Hess's adoption of DRIFT with the grafting of his seafloor spreading hypothesis was undertaken because it increased the problem-solving effectiveness of his own hypothesis and made DRIFT a much better problem-solver through elimination of its most serious deficiency.

This action on Hess's part is really just another instance of his supporting an hypothesis on the basis of its problem solving effectiveness, which coincides with his general aim in presenting hypotheses. However, because of its historical importance to the eventual acceptance of continental drift in the form of plate tectonics, it deserves special attention. I have argued that Hess did not endorse DRIFT in 1959. But I suggest that Hess surely realized that DRIFT was on the upswing because of its ability to make sense out of new paleomagnetic studies which had been going on for several years. In addition, Hess obviously realized in 1959 that DRIFT could solve the problem of the scarcity of seafloor sediments, if it was supposed that the "pre-Cretaceous sediments have in some manner been removed, for example, by continental drift." But, at this time Hess knew of no adequate solution to the engineering problem faced by DRIFTers. With the invention of his seafloor spreading hypothesis, Hess typically realized that he had a solution to the engineering problem and explicitly argued that it was a solution. With this problem solved, he became a proponent of DRIFT, albeit his own version of DRIFT; and he argued for DRIFT along with his seafloor spreading hypothesis on their mutual basis to solve numerous problems in a unified manner. His seafloor spreading hypothesis, when conjoined with DRIFT, offered solutions to the traditional problems solved by DRIFT and made sense out of the paleomagnetic data; while DRIFT when conjoined with his own hypothesis could solve every problem solved by his hypothesis which was connected with the seafloor, and no longer was faced with the question of how on earth the continents drifted.

> *Claim 4:* Hess's major epistemic aim in theorizing, solving problems, when coupled with his apparent lack of fear in making mistakes, gave him the freedom for inventing, evaluating, rejecting and salvaging potential solutions.

This claim is clearly speculative and is offered in the spirit of this conference. From his public writings it certainly appears that Hess found no psychological difficulties in either proposing solutions to problems, or, at the same time, dismissing one of his former solutions and replacing it with an alternative. Both of these abilities may have a common psychological explanation, namely, that Hess was not afraid to make mistakes. He was always one to propose solutions, and often introduced his theoretical discussions with comments to the effect that it is much better to present solutions rather than record data in a supposedly objective manner — even suggesting on one occasion that the mere recording of data led to blindness rather than objectivity. Moreover, Hess introduced hypotheses fully aware that he probably was mistaken or at least that there was a good chance that he was wrong. On almost every occasion, he introduced his prospective solution with some remark indicating that he was proposing it in the spirit of pursuit rather than outright acceptance. Indeed, Hess was not afraid to take chances and make mistakes. Furthermore, I think that his lack of fear of being wrong provided him with the freedom to examine critically his former solutions. As has been demonstrated, Hess constantly recognized and often discovered difficulties with his former solutions, and he responded either by altering them or by proposing new ones.

Needless to say, sentiments like the above are as hackneyed and common-place as when first expressed. Nevertheless, I should like to push them one step further. An immediate problem with the above is as follows: How can the lack of fear of being wrong ever serve as a guide for dismissing a solution faced with difficulties, and for proposing an alternative as opposed to rework-ing the troubled solution? Quite obviously it cannot. It is not a guide for making epistemic or any other type of decisions. However, if the proclivity for taking chances is coupled with a working criterion for decision making, the combination can be quite fruitful. As I have already argued, Hess had a deci-sion procedure, namely, maintenance of that constructed solution which possesses the greater problem solving effectiveness. In every case, Hess switched positions when he thought the new solution solved more problems than the former without producing ensuing difficulties. His overall aim in theorizing was to present solutions with the greatest problem-solving effectiveness. Indeed, perhaps it was because he had this overall aim that he wasn't afraid to make mistakes, that he desired to ferret out his own mistakes and remedy the situation either through invention of a new hypothesis or reworking a former one. Because of this overall aim, he would not have found difficulty in either reworking a former solution or pursuing a new one. With such an aim he would have felt no essential tension between either strategy, and he probably did pursue them in concert.[9]

*Department of Philosophy*
*University of Missouri, Kansas City*

## NOTES

[1] R. L. Fisher, then and presently a geologist at Scripps, unearthed much of the data concerning oceanic trenches, and in 1961 co-authored an article with Hess, Fisher (1963), wherein they further expanded Hess's concept of seafloor spreading to consider in some detail the development of oceanic trenches and island arcs.

[2] The expression 'seafloor spreading' was coined by R. S. Dietz rather than by Hess. Actually, Hess did not publish his thesis until 1962, while Dietz published his version of seafloor spreading in 1961. However, Hess deserves credit, for he distributed his un-published version of his thesis in 1960. Thanks to Professor Sheldon Judson, Chairman, Department of Geological and Geophysical Sciences, Princeton University, I have had the opportunity to examine the 1960 version. There are no substantial changes between the two versions, although I shall cite only the unpublished version. Dietz has freely

admitted that Hess first developed the notion of 'seafloor spreading': "As regards sea-floor spreading, Hess deserves full credit for the concept. . . I have done little more than introduce the term seafloor spreading. . . ." (From R. S. Dietz, 'Reply', *Journal of Geophysical Research* 73, p. 6567.) Dietz's 1961 article on DRIFT appeared as 'Continent and Ocean Basin Evolution by Spreading of the Sea Floor', *Nature* 190, 854–857; and Hess's 1962 version appeared as 'History of Ocean Basins', in *Petrologic Studies: A Volume to Honor A. F. Buddington*, pp. 599–620.

[3] I have considered the reception of Hess's seafloor spreading and DRIFT in several other papers: cf. Frankel (1979a and 1979b). In *a* I apply Lakatos's account of scientific growth and change to the development, reception and acceptance of DRIFT, while in *b*, Laudan's account is applied. Of the two, Laudan's account provides a better 'fit' – in fact, a rather good one. Indeed, the accent in this paper on Hess's aim to solve problems in the construction of his seafloor spreading hypothesis is an outgrowth of *b* and Laudan's overall account.

[4] *Cf.* Hess's papers (1932, 1937a, 1937b, and 1938a and 1938b).

[5] Betz and Hess (1942). They first presented their thesis in 1940.

[6] Hess was not the only person suggesting that ridges are created by a process involving tensional forces, for both S. K. Carey and Bruce Heezen were suggesting their own alternatives – both of which involved the hypothesis of an expanding earth.

[7] Hess at this juncture failed to consider a slightly different alternative, namely, passive transport of newly created seafloor by a passive process that would not bring about an uneven layer. Indeed, movement of newly created seafloor from ridge sites by a passive process is central to his seafloor spreading concept. However, in 1959 he had not developed his concept of seafloor spreading.

[8] Actually, Hess only suggested a solution to the problem of trench formation in his 1960 paper. However, Hess expanded his suggestion shortly after, when he co-authored with R. Fisher 'Trenches.' (The manuscript for the paper was completed in May of 1961, which was six months after Hess finished his original preprint on seafloor spreading.) In the closing section of the paper there is an account of the origin and development of trenches in accordance with the seafloor spreading hypothesis. According to Fisher, the section was written by Hess.

[9] Research for this paper has been supported by NSF. I should also like to thank Bob Arnold and Bill Glen for commenting on earlier drafts of this paper.

## BIBLIOGRAPHY

Betz, F., and H. H. Hess: 1942, 'The Floor of the North Pacific Ocean', *Geographical Review* 32, 99–116.

Fisher, R. L., and H. H. Hess: 1963, 'Trenches', in M. N. Hill (ed.), *The Sea*, Vol. 3, John Wiley & Sons, New York, pp. 411–436.

Frankel, H.: 1979a, 'The Career of Continental Drift Theory: An Application of Imre Lakatos' analysis of scientific growth to the rise of drift theory', *Studies in History and Philosophy of Science* 10, 21–66.

Frankel, H.: 1979b, 'The Reception and Acceptance of Continental Drift Theory as a Rational Episode in the History of Science', in S. H. Mauskopf (ed.), *The Reception of Unconventional Science*, American Assn. for the Advancement of Science, Washington, D. C., pp. 51–89.

Hess, H. H.: 1932, 'Interpretation of Gravity Anomalies and Sounding-profiles obtained in the West Indies by the International Expedition to the West Indies in 1932', *Transactions of the American Geophysical Union*, 13th Annual Meeting, pp. 26–32.

Hess, H. H.: 1937a, 'Island Arcs, Gravity Anomalies, and Serpentine Intrusions: A contribution to the Ophiolite Problem', *17th International Geological Congress, Moscow Report*, Vol. 2, pp. 263–283.

Hess, H. H.: 1937b, 'Geological interpretation of data collected on cruise of *U.S.S. Barracuda* in the West Indies – preliminary report', *Transactions of the American Geophysical Union*, 18th Annual Meeting, pp. 69–77.

Hess, H. H.: 1938a, 'A Primary Peridotite Magma', *American Journal of Science* 35, 322–344.

Hess, H. H.: 1938b, 'Gravity Anomalies and Island Arc Structure with Particular Reference to the West Indies', *Proceedings of the American Philosophical Society* 79, 71–96.

Hess, H. H.: 1946, 'Drowned Ancient Islands of the Pacific Basin', *American Journal of Science* 244, 772–791.

Hess, H. H.: 1951, 'Comment on Mountain Building', in 1950 Colloquium on Plastic Flow and Deformation within the Earth, *Transactions of the American Geophysical Union* 32, 528–531.

Hess, H. H.: 1954, 'Geological hypotheses and the earth's crust under the oceans', *Royal Society of London Proceedings* 222 A, 341–348.

Hess, H. H.: 1955a, 'Serpentines, Orogeny, and Epeirogeny', *Geological Society of America, Special Paper* 62, pp. 391–406.

Hess, H. H.: 1955b, 'The Oceanic Crust', *Journal of Marine Reserach* 14, 423–439.

Hess, H. H.: 1959a, 'Nature of the Great Oceanic Ridges', *International Ocean Congress preprints*, American Assn. for the Advancement of Science, Washington, D. C., pp. 33–34.

Hess, H. H.: 1959b, 'The AMSOC Hole to the Earth's Mantle', *Transactions of the American Geophysical Union* 40, 340–345.

Hess, H. H.: 1960, 'Evolution of Ocean Basins', preprint, 37pp.

# INDEX OF NAMES

# INDEX OF SUBJECTS

# BOSTON STUDIES IN THE PHILOSOPHY OF SCIENCE

*Editors:*
ROBERT S. COHEN and MARX W. WARTOFSKY
(Boston University)

24. Don Ihde, *Technics and Praxis. A Philosophy of Technology.* 1978.
25. Jaakko Hintikka and Unto Remes, *The Method of Analysis. Its Geometrical Origin and Its General Significance.* 1974.
26. John Emery Murdoch and Edith Dudley Sylla, *The Cultural Context of Medieval Learning.* 1975.
27. Marjorie Grene and Everett Mendelsohn (eds.), *Topics in the Philosophy of Biology.* 1976.
28. Joseph Agassi, *Science in Flux.* 1975.
29. Jerzy J. Wiatr (ed.), *Polish Essays in the Methodology of the Social Sciences.* 1979.
32. R. S. Cohen, C. A. Hooker, A. C. Michalos, and J. W. van Evra (eds.), *PSA 1974: Proceedings of the 1974 Biennial Meeting of the Philosophy of Science Association.* 1976.
33. Gerald Holton and William Blanpied (eds.), *Science and Its Public: The Changing Relationship.* 1976.
34. Mirko D. Grmek (ed.), *On Scientific Discovery.* 1980.
35. Stefan Amsterdamski, *Between Experience and Metaphysics. Philosophical Problems of the Evolution of Science.* 1975.
36. Mihailo Marković and Gajo Petrović (eds.), *Praxis. Yugoslav Essays in the Philosophy and Methodology of the Social Sciences.* 1979.
37. Hermann von Helmholtz: *Epistemological Writings. The Paul Hertz/Moritz Schlick Centenary Edition of 1921 with Notes and Commentary by the Editors.* (Newly translated by Malcolm F. Lowe. Edited, with an Introduction and Bibliography, by Robert S. Cohen and Yehuda Elkana.) 1977.
38. R. M. Martin, *Pragmatics, Truth, and Language.* 1979.
39. R. S. Cohen, P. K. Feyerabend, and M. W. Wartofsky (eds.), *Essays in Memory of Imre Lakatos.* 1976.
42. Humberto R. Maturana and Francisco J. Varela, *Autopoiesis and Cognition. The Realization of the Living.* 1980.
43. A. Kasher (ed.), *Language in Focus: Foundations, Methods and Systems. Essays Dedicated to Yehoshua Bar-Hillel.* 1976.
46. Peter L. Kapitza, *Experiment, Theory, Practice.* 1980.
47. Maria L. Dalla Chiara (ed.), *Italian Studies in the Philosophy of Science.* 1980.
48. Marx W. Wartofsky, *Models: Representation and the Scientific Understanding.* 1979.
50. Yehuda Fried and Joseph Agassi, *Paranoia: A Study in Diagnosis.* 1976.
51. Kurt H. Wolff, *Surrender and Catch: Experience and Inquiry Today.* 1976.
52. Karel Kosík, *Dialectics of the Concrete.* 1976.
53. Nelson Goodman, *The Structure of Appearance.* (Third edition.) 1977.
54. Herbert A. Simon, *Models of Discovery and Other Topics in the Methods of Science.* 1977.
55. Morris Lazerowitz, *The Language of Philosophy. Freud and Wittgenstein.* 1977.
56. Thomas Nickles (ed.), *Scientific Discovery, Logic, and Rationality.* 1980.
57. Joseph Margolis, *Persons and Minds. The Prospects of Nonreductive Materialism.* 1977.
58. Gerard Radnitzky and Gunnar Andersson (eds.), *Progress and Rationality in Science.* 1978.

59. Gerard Radnitzky and Gunnar Andersson (eds.), *The Structure and Development of Science*. 1979.
60. Thomas Nickles (ed.), *Scientific Discovery: Case Studies*. 1980.
61. Maurice A. Finocchiaro, *Galileo and the Art of Reasoning*. 1980.